T0206958

# Communications in Computer and Information Science 1765

## Rationale

The CCIS series is devoted to the publication of proceedings of computer science conferences. Its aim is to efficiently disseminate original research results in informatics in printed and electronic form. While the focus is on publication of peer-reviewed full papers presenting mature work, inclusion of reviewed short papers reporting on work in progress is welcome, too. Besides globally relevant meetings with internationally representative program committees guaranteeing a strict peer-reviewing and paper selection process, conferences run by societies or of high regional or national relevance are also considered for publication.

## Topics

The topical scope of CCIS spans the entire spectrum of informatics ranging from foundational topics in the theory of computing to information and communications science and technology and a broad variety of interdisciplinary application fields.

## Information for Volume Editors and Authors

Publication in CCIS is free of charge. No royalties are paid, however, we offer registered conference participants temporary free access to the online version of the conference proceedings on SpringerLink (http://link.springer.com) by means of an http referrer from the conference website and/or a number of complimentary printed copies, as specified in the official acceptance email of the event.

CCIS proceedings can be published in time for distribution at conferences or as post-proceedings, and delivered in the form of printed books and/or electronically as USBs and/or e-content licenses for accessing proceedings at SpringerLink. Furthermore, CCIS proceedings are included in the CCIS electronic book series hosted in the SpringerLink digital library at http://link.springer.com/bookseries/7899. Conferences publishing in CCIS are allowed to use Online Conference Service (OCS) for managing the whole proceedings lifecycle (from submission and reviewing to preparing for publication) free of charge.

## Publication process

The language of publication is exclusively English. Authors publishing in CCIS have to sign the Springer CCIS copyright transfer form, however, they are free to use their material published in CCIS for substantially changed, more elaborate subsequent publications elsewhere. For the preparation of the camera-ready papers/files, authors have to strictly adhere to the Springer CCIS Authors' Instructions and are strongly encouraged to use the CCIS LaTeX style files or templates.

## Abstracting/Indexing

CCIS is abstracted/indexed in DBLP, Google Scholar, EI-Compendex, Mathematical Reviews, SCImago, Scopus. CCIS volumes are also submitted for the inclusion in ISI Proceedings.

## How to start

To start the evaluation of your proposal for inclusion in the CCIS series, please send an e-mail to ccis@springer.com.

Ling Zhenhua · Gao Jianqing · Yu Kai · Jia Jia
Editors

# Man-Machine Speech Communication

17th National Conference, NCMMSC 2022
Hefei, China, December 15–18, 2022
Proceedings

 Springer

*Editors*
Ling Zhenhua
University of Science and Technology
of China
Anhui, China

Yu Kai
Shanghai Jiaotong University
Shanghai, China

Gao Jianqing
Hefei University
Anhui, China

Jia Jia
Tsinghua University
Beijing, China

ISSN 1865-0929     ISSN 1865-0937 (electronic)
Communications in Computer and Information Science
ISBN 978-981-99-2400-4     ISBN 978-981-99-2401-1 (eBook)
https://doi.org/10.1007/978-981-99-2401-1

This Springer imprint is published by the registered company Springer Nature Singapore Pte Ltd.
The registered company address is: 152 Beach Road, #21-01/04 Gateway East, Singapore 189721, Singapore

# Preface

This volume contains the papers from the 17th National Conference on Man–Machine Speech Communication (NCMMSC), the largest and most influential event on speech signal processing in China, which was hosted in Hefei, China, December 15–18, 2022 by the Chinese Information Processing Society of China and China Computer Federation, jointly co-organized by iFLYTEK Co., Ltd., the University of Science and Technology of China and the National Engineering Research Center for Speech and Language Information Processing.

NCMMSC is also the academic annual meeting of the technical committee of Speech Dialogue and Auditory Processing of China Computer Federation (CCF TFSDAP). As an important stage for experts, scholars, and researchers in this field to share their ideas, research results and experiences, NCMMSC strongly promotes continuous progress in this field and development work.

Papers published in the Special Issue on "National Conference on Man–Machine Speech Communication (NCMMSC 2022)" are focused on the topics of speech recognition, synthesis, enhancement and coding, as well as experimental phonetics, speech prosody analysis, pathological speech analysis, speech analysis, acoustic scene classification and human–computer dialogue understanding. Each paper was assigned 3 or more reviewers for peer review. And each reviewer was assigned 6–10 papers. The total number of submissions was 91, the number of full papers accepted was 21, and the number of short papers accepted was 7.

The proceedings editors wish to thank the dedicated Scientific Committee members and all the other reviewers for their contributions. We also thank Springer for their trust and for publishing the proceedings of NCMMSC 2022.

December 2022

Zhenhua Ling
Jianqing Gao
Kai Yu
Jia Jia

# Organization

## Chairs

| | |
|---|---|
| Changchun Bao | Beijing University of Technology, China |
| Jianwu Dang | Tianjin University, China |
| Fang Zheng | Tsinghua University, China |
| Lirong Dai | University of Science and Technology of China, China |

## Program Chairs

| | |
|---|---|
| Zhenhua Ling | University of Science and Technology of China, China |
| Jianqing Gao | iFLYTEK Co., Ltd., China |
| Kai Yu | Shanghai Jiao Tong University, China |
| Jia Jia | Tsinghua University, China |

## Program Committee

| | |
|---|---|
| Longbiao Wang | Tianjin University, China |
| Rui Liu | Inner Mongolia University, China |
| Heng Lu | Himalaya Technology Co., Ltd., China |
| Shengchen Li | Xi'an Jiaotong-Liverpool University, China |
| Bin Dong | iFLYTEK Co., Ltd., China |
| Yan Song | University of Science and Technology of China, China |
| Ya Li | Beijing University of Post and Telecommunications, China |
| Jie Zhang | University of Science and Technology of China, China |
| Qing Wang | University of Science and Technology of China, China |
| Jiajia Yu | iFLYTEK Co., Ltd., China |
| Xiaolei Zhang | Northwestern Polytechnical University, China |
| Zhiyong Wu | Tsinghua Shenzhen International Graduate School, China |

| | |
|---|---|
| Jun Du | University of Science and Technology of China, China |
| Min Li | Duke Kunshan University, China |
| JunFeng Li | Institute of Acoustics, Chinese Academy of Sciences, China |
| Yanmin Qian | Shanghai Jiao Tong University, China |
| Xuan Zhu | Samsung Electronics China Research Institute, China |
| Wu Guo | University of Science and Technology of China, China |
| Lei Xie | Northwestern Polytechnical University, China |
| Jiangyan Yi | Institute of Automation, Chinese Academy of Sciences, China |
| Yanhua Long | Shanghai Normal University, China |
| Yang Ai | University of Science and Technology of China, China |
| Li Kong | iFLYTEK Co., Ltd., China |

# Contents

# MCPN: A Multiple Cross-Perception Network for Real-Time Emotion Recognition in Conversation

Weifeng Liu[1,3] and Xiao Sun[2,3(✉)]

[1] AHU-IAI AI Joint Laboratory, Anhui University, Hefei, China
[2] Hefei University of Technology, Hefei, China
[3] Institute of Artificial Intelligence, Hefei Comprehensive National Science Center, Hefei, China
sunx@iai.ustc.edu.cn

**Abstract.** Emotion recognition in conversation (ERC) is crucial for developing empathetic machines. Most of the recent related works generally model the speaker interaction and context information as a static process but ignore the temporal dynamics of the interaction and the semantics in the dialogue. At the same time, the misclassification of similar emotions is also a challenge to be solved. To solve the above problems, we propose a Multiple Cross-Perception Network, MCPN, for multimodal real-time conversation scenarios. We dynamically select speaker interaction intervals for each time step, so that the model can effectively capture the dynamics of interaction. Meanwhile, we introduce the multiple cross-perception process to perceive the context and speaker state information captured by the model alternately, so that the model can capture the semantics and interaction information specific to each time step more accurately. Furthermore, we propose an emotion triple recognition process to improve the model's ability to recognize similar emotions. Experiments on multiple datasets demonstrate the effectiveness of the proposed method.

**Keywords:** Real-time Emotion Recognition in Conversation · Multimodal · Natural Language Processing

## 1 Introduction

With the development of network technology and social networks, communication through the Internet has become part of the daily life of people in today's society. In the face of this booming social network, the need to develop an empathic dialogue system is also becoming more and more obvious. And emotion recognition in conversation (ERC) is an important foundation.

The task of ERC is to recognize the emotion of each utterance in a conversation [14]. Because the dialogue takes place in real-time, the historical utterances may have different effects on the target utterances at different moments of dialogue [15]. In addition, there is a complex interaction process between the speakers in the dialogue, which increases the difficulty of the ERC task. Furthermore,

L. Zhenhua et al. (Eds.): NCMMSC 2022, CCIS 1765, pp. 1–15, 2023.
https://doi.org/10.1007/978-981-99-2401-1_1

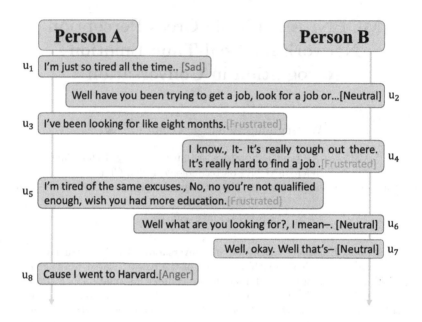

**Fig. 1.** An instance of a conversation.

the existing ERC methods still can not effectively solve the problem of misclassification of similar emotions, such as *happiness* and *excited*, *anger* and *frustrated*. Like the examples shown in Fig. 1, the utterance $u_3$ is easily misclassified as an *anger* emotion.

Recent related works [3,5,6,9,17,18,20] usually utilize transformers [19] to encode each utterance in the dialogue from a global perspective or gather information from the dialog graph with graph neural networks (GNN). However, these methods model the interaction between speakers through static methods, thus ignoring the dynamics of interaction in the conversation. Moreover, these methods model the context information in the dialogue as a single static semantic flow, while ignoring the fact that context has different understandings at different times, i.e., the semantic dynamics.

In this paper, we propose a Multiple Cross-Perception Network (MCPN) to solve the above problems.

Firstly, inspired by the dynamical influence model [13], we select the state interaction intervals for each time step dynamically, so that the model can model the various information in the conversation in a dynamic manner.

Secondly, we introduce the multiple cross-perception process to alternately perceive semantic and interactive information in the dialogue, so that the model can capture the semantic and interactive information in each time step more accurately.

Thirdly, to alleviate the problem of misclassification of similar emotions, we propose an Emotion Triple-Recognition (ETR) process, which introduces additional fuzzy classification probability through multiple recognition processes, thus enhancing the discriminative ability of the classifier for similar emotions.

We experiment our MCPN on three publicly available ERC datasets, and our model has achieved considerable performance on them. The contributions of this work are summarized as follows:

- We propose to dynamically select the state interaction interval of each time step in the dialogue, thus capturing the dynamics of interaction between speakers from it.
- We utilize the multiple cross-perception process to re-perceive the context and interaction at each time step of the dialog, thus capturing the semantic and interactive information specific to each time step more accurately.
- We introduce an Emotion Triple-Recognition process to alleviate the problem of misclassification of similar emotions by the multi-recognition process.
- Experiments on three datasets demonstrate the effectiveness of our approach.

## 2   Related Work

### 2.1   Emotion Recognition in Conversation

A large number of methods have been proposed for ERC. In the following paragraphs, we present the relevant methods in two categories according to the different input objects of the model.

**Sequence-Based Methods.** Such methods usually regard dialogue as an utterance sequence and encode each utterance with RNNs or transformers. For example, DialogueRNN [11] uses GRUs to capture the interactions in the dialog and introduces the attention mechanism [19] to summarize the context information, thus generating the emotional states of each speaker. Based on DialogueRNN, COSMIC [2] introduces commonsense knowledge into the model. AGHMN [7] proposes a speaker-free model for real-time ERC through GRUs and an improved attention mechanism. CESTa [20] utilizes a network containing LSTM [4] and transformers to model the context and introduces a Conditional Random Field to learn emotional consistency in dialogs. DialogXL [17] utilizes the improved Xlnet [21] to model information from the global perspective. Inspired by multi-task learning, CoG-BART [9] introduces the response generation task as an auxiliary task to improve the model's ability to handle context information.

**Graph-Based Models.** Graph-based methods describe the conversation as a graph network and treat the task of ERC as a node classification task. For example, DialogueGCN [3] constructs a relation graph and gathers node information within a fixed window through the GNN. RGAT [6] proposes a novel relative position encoding to enhance the position-sensing ability of graph networks. I-GCN [12] proposes an incremental graph network to represent the temporal information in the dialog. DAG-ERC [18] regards the dialogue as a directed

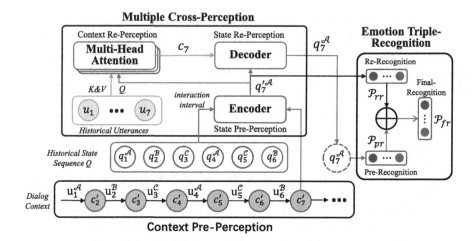

**Fig. 2.** The framework of MCPN, which is a snapshot at time step 7, where $u_i^p$ is the $i$-th utterance in the dialog, and $p \in \{\mathcal{A}, \mathcal{B}, \mathcal{C}\}$ denotes the corresponding speaker. At each time step, the model will eventually generate a speaker state $q_t^p$ corresponding to this time step, which will be stored in the historical state sequence $Q$.

acyclic graph, which is more appropriate to describe the flow of data in a conversation. MM-DFN [5] proposes a graph-based multimodal fusion network to effectively utilize the multi-modal information of the dialogue.

## 2.2   Dynamical Influence Model

The dynamical influence model [13] is a promotion of the Hidden Markov model. It gives an abstract definition of influence: the state of an entity in the system will be affected by the recent historical state of all entities and changes accordingly. Suppose there is an entity set $E = \{e_1, ... e_n\}$ in the system, in which each entity $e$ is linked to a finite set of possible states $\mathcal{S} = \{1, ..., s\}$. At each time step $t$, each entity is in one of these states, which is denoted as $q_t^e \in \mathcal{S}$. The influence between entities is regarded as a conditional dependence of the state of each entity at the current time step $t$ and the state of all entities at the previous time step $t - 1$. This process can be described by the following formula:

$$P(q_t^e | q_{t-1}^1, q_{t-1}^2, ... q_{t-1}^n) \tag{1}$$

where $q_t^e$ represents the state of an entity $e$ in the system at time step $t$ and $q_{t-1}^1, q_{t-1}^2, ... q_{t-1}^C$ denote the state of all entities $e_1, ..., e_n$ in the system at time step $t - 1$.

## 3   Methodology

Our MCPN consists of four components: Multimodal Utterance Feature Extraction, Context Pre-Perception (CPP) module, Multiple Cross-Perception (MCP) module, and Emotion Triple-Recognition (ETR) process. Figure 2 illustrates the overall structure.

## 3.1  Problem Definition

First, we define the ERC task. Given a dialogue $D = \{u_1, u_2 \ldots, u_T\}$ with $T$ utterances, which is participated by $M$ speakers who come from the speaker set $S = \{s_1, s_2, \ldots, s_M\}$. For the target utterance $u_t$ of time step $t$, the corresponding target speaker is $s_{p(u_t)}$ who is in a specific state $q_t^{p(u_t)}$, where $p(\cdot)$ is the mapping function between each utterance and the corresponding speaker index. The task is to identify the emotion of each utterance $u_t$ in dialog $D$.

## 3.2  Multimodal Utterance Feature Extraction

For each sentence in the dialogue, we extract two types of features.

**Textual Features:** Following [2, 18], we utilize a pre-trained language model, RoBERTa-Large [10], to extract textual feature $v^\mathcal{T}$ for each utterance.

**Acoustic Features:** Following [11], we utilize the *OpenSmile* toolkit with IS10 configuration [16] to extract acoustic features $v^a$ for each utterance.

Finally, the representation of each utterance is obtained by combining two types of features.

$$u_t = [v^\mathcal{T}; v^a] \tag{2}$$

## 3.3  CPP: Context Pre-perception Module

The task of the Context Pre-Perception Module is to obtain the pre-perceived context for each time step of the dialog, which is regarded as the historical global semantic information in each time step. For time step $t$, we use the LSTM cell to obtain the pre-perceived context representation $c_t'$ as follows:

$$c_t' = LstmCell(c_{t-1}', u_{t-1}) \tag{3}$$

where $c_{t-1}'$ is the pre-perceived context representation of time step $t - 1$ and it should be noted that $c_t'$ only contains semantic information of historical utterances.

## 3.4  MCP: Multiple Cross-Perception Module

The task of the MCP is to accurately capture the semantics and interaction information specific to each time step through multiple perceptions, and then obtain the corresponding state representation of the target speaker. It consists of three processes.

**State Pre-perception Process.** This process obtains the preliminary speaker state $q'_t$ with the pre-perceived context representation $c'_t$ and the interaction information. To capture the dynamics of interaction between speakers, we modify the dynamical influence model [13] to apply it to conversation scenarios. Since our task is to identify the emotion of each utterance in the dialogue, we assume that only when a speaker speaks, he/she will have the corresponding speaker state. Under this premise, a more reasonable assumption is: in time step $t$, the speakers who interact with the target speaker are the other speakers who spoke after the last time the target speaker spoke. This process is shown as follows:

$$P\left(q_t^{p(u_t)}|q_h^{p(u_h)}, q_{h+1}^{p(u_{h+1})}, \ldots, q_{t-1}^{p(u_{t-1})}\right) \qquad (4)$$

where $p(u_h) = p(u_t)$ and $\nexists j \in (h, t), p(u_j) = p(u_t)$, i.e., the state $q_h^{p(u_h)}$ is the latest historical state of the target speaker. The state sequence $\{q_h^{p(u_h)}, \ldots, q_{t-1}^{p(u_{t-1})}\}$ is the state interaction interval of utterance $u_t$. We capture this interval through $SII_t = f(t-1)$ from the historical state sequence $Q$, which is shown in the following procedure:

$$f(k) = \begin{cases} \emptyset, & \text{if } t = 1 \\ \{q_k\}, & \text{if } p(u_k) = p(u_t) \\ \{q_k\} \cup f(k-1), & \text{otherwise} \end{cases} \qquad (5)$$

Then we combine the pre-perceived context information $c'_t$ to capture the influence of other speakers on the target speaker from the dynamic state interaction interval $SII_t$, and thus obtaining the preliminary state $q'_t$ of the target speaker. We use the structure of transformer encoder [19] to achieve this purpose, and the specific process is as follows:

$$q'_t = LN(Mhatt_{SPP}(c'_t, SII_t, SII_t) + c'_t) \qquad (6)$$

$$q'_t = LN(W'q'_t + b' + q'_t) \qquad (7)$$

where $Mhatt(\cdot, \cdot, \cdot)$ is the Multi-head attention mechanism in the transformer encoder, and its calculation process is displayed in Eq. (8)–(9). $LN(\cdot)$ denotes the Layer Normalization, and $W'$ and $b'$ are trainable weight matrix and bias of feedforward neural network in encoder.

$$h_i = \text{Softmax}\left(\frac{f_i^q(q)f_i^k(k)}{\sqrt{d_h}}\right)f_i^v(v) \qquad (8)$$

$$MHatt(q, k, v) = \text{Concat}\{h_1, h_2, \ldots h_m\} \qquad (9)$$

where $f_i^\varphi(\varphi) = W_i^\varphi \varphi + b_i^\varphi, \varphi \in \{q, k, v\}$ represents the transformation function of query, key and value in the attention mechanism.

**Context Re-perception Process.** Due to the complicated interaction between speakers, each speaker may have a different understanding of historical utterances at different times. Therefore, we utilize the Context Re-Perception Process to re-perceive the historical utterance under the condition of the target speaker's preliminary state $q'_t$, so as to obtain an accurate context representation $c_t$ specific to the current time step.

In practice, We utilize the multi-head attention mechanism to achieve this goal, and the detailed process is as follows:

$$c_t = Mhatt_{CRP}(q'_t, H, H) \tag{10}$$

where $q'_t$ is the preliminary state representation of target speaker and $H = \{u_1, ..., u_t\}$, which is the global history utterance sequence up to time step $t$.

It should be noted that in this stage, we also include the target utterance $u_t$ in the global historical utterance sequence $H$. The purpose is to capture the context representation specific to the current time step and provide more information about the target speaker in the current time step for the subsequent state re-perception process.

**State Re-perception Process.** The purpose of the state re-perception process is to refine the pre-perceived state representation $\hat{q}'_t$ by the re-perceived context obtained by the second process, so as to obtain a more comprehensive speaker state representation. We introduce the GRU as the state decoder to refine $q'_t$ with the context representation $c_t$. The specific process is as follows:

$$q_t = GRU(c_t, q'_t) \tag{11}$$

where $q_t$ is the speaker state representation of the target speaker at time step $t$. We will store $q_t$ in the historical state sequence $Q$ shared by each time step for emotional recognition of the target utterance and the use of subsequent time steps.

### 3.5   Emotion Triple-Recognition Process

Since the existing methods tend to misclassify similar emotions, to alleviate this problem, we propose the Emotion Triple-Recognition process. We not only utilize the speaker state $q_t$ for emotion recognition but also introduce the preliminary speaker state $q'_t$. As an intermediate vector obtained from the model, $q'_t$ has a certain probability to point to other emotional categories, especially when misclassification occurs, it can play the role of bias to balance the classification probability.

The Triple-Recognition process consists of three steps. The first step, which we call the *emotion pre-recognition step*, we use the re-perceived speaker state representation $q_t$ for emotion recognition.

$$\mathcal{P}_{pr} = Softmax(MLP_{pr}(q_t)) \tag{12}$$

where $MLP$ refers to the multilayer perceptron, $\mathcal{P}_{pr}$ is the probability of all the emotion labels.

The second step is called the *emotion re-recognition step*. In this step, we use the preliminary speaker state $q'_t$ obtained in the State Pre-Perception Process to carry out emotion recognition. The specific process is as follows:

$$\mathcal{P}_{rr} = Softmax(MLP_{rr}(q'_t)) \tag{13}$$

In the last step, the *final recognition step*, we introduced a gating mechanism to synthesize the probability distribution obtained in the first two steps to obtain the final classification probability $\mathcal{P}_{fr}$.

$$\mathcal{P}_{fr} = Softmax(z \cdot \mathcal{P}_{pr} + (1-z) \cdot \mathcal{P}_{rr}) \tag{14}$$

$$z = \sigma(W_{fr}[\mathcal{P}_{pr}; \mathcal{P}_{rr}; \mathcal{P}_{pr} - \mathcal{P}_{rr}; \mathcal{P}_{pr} \odot \mathcal{P}_{rr}]) \tag{15}$$

where $W_{fr}$ is a trainable weight matrix, $\odot$ denotes the element-wise multiplication.

### 3.6   Loss Function

Because our emotion recognition process is divided into three steps, the loss function we use is needed to be adjusted accordingly. For each recognition step, we introduce the cross-entropy loss as the objective function:

$$\mathcal{L}^{step} = -\sum_{d=1}^{N} \sum_{t=1}^{L(d)} \log \mathcal{P}_{step}^{d,t}[y_{d,t}] \tag{16}$$

where $step \in \{pr, rr, fr\}$ represents different recognition step, $N$ is the number of conversations, $L(d)$ is the number of utterances in conversation $d$.

Finally, our loss function is:

$$\mathcal{L} = \alpha\mathcal{L}^{fr} + \beta\mathcal{L}^{pr} + (1 - \alpha - \beta)\mathcal{L}^{rr} \tag{17}$$

where $\alpha$ and $\beta$ are hyperparameters that control the weight of each recognition step.

## 4   Experimental Settings

### 4.1   Datasets

We evaluate our model on three ERC datasets, and their statistical information is shown in Table 1.

**IEMOCAP** [1]: A multimodal ERC dataset in which each conversation is attended by two speakers. Each utterance in the dialogue is labeled as one of 6 emotions, namely *neutral, happiness, sadness, anger, frustrated,* and *excited*. Since there is no validation set in it, follow [18], we choose the last 20 dialogues of the training set for validation.

**Table 1.** The overall statistics of three datasets.

| Dataset | Dialogues | | | Utterances | | | utt/dia | classes |
|---------|-------|-----|------|-------|------|------|---------|---------|
|         | Train | Val | Test | Train | Val  | Test |         |         |
| IEMOCAP | 100   | 20  | 31   | 4835  | 975  | 1623 | 48.4    | 6       |
| MELD    | 1038  | 114 | 280  | 9989  | 1109 | 2610 | 9.6     | 7       |
| EmoryNLP| 713   | 99  | 85   | 9934  | 1344 | 1328 | 14.0    | 7       |

**MELD** [14]: A multimodal ERC dataset collected from the TV show *Friends*, in which each conversation consists of two or more speakers participating. It contains 7 emotion labels including *neutral, happiness, surprise, sadness, anger, disgust,* and *fear*.

**EmoryNLP** [22]: It is an ERC dataset only contains textual data. The labels of each utterance in this dataset include *neutral, sad, mad, scared, powerful, peaceful,* and *joyful*.

For the evaluation metrics, we choose the weighted F1 (w-F1) score for each dataset.

### 4.2 Implementation Details

we adopt Adam [8] as the optimizer with learning rate of $\{0.0002, 0.0002, 0.0003\}$ and L2 weight decay of $\{0.0003, 0.0003, 0.0001\}$ for IEMOCAP, MELD and EmoryNLP, respectively. The number of attention heads in each module are set to 6, and the dropout rates for all datasets are set to 0.25 as default. The hyperparameters in our loss function are set to $\alpha = 0.5, \beta = 0.3$ for all datasets. The training and testing process are running on a single GTX 1650 GPU. Each training process contains 60 epochs at most and we report the average score of 10 random runs on test set.

### 4.3 Baseline Methods

To verify the validity of our model, we compare our approach with the following baseline approaches:

*Sequence-based models:* DialogueRNN [11], COSMIC [2], CESTa [20], DialogXL [17], and CoG-BART [9].

*Graph-based methods:* DialogueGCN [3], RGAT [6], I-GCN [12], DAG-ERC [18] and MM-DFN [5].

## 5  Results and Analysis

### 5.1  Overall Performance

The overall performance is presented in Table 2 and the results are statistically significant under the paired t-test ($p < 0.05$). According to the results in Table 2, our model has achieved considerable performance on the three datasets.

**Table 2.** Overall performance. The symbol * indicates that the model is not for real-time ERC. The results shown in italics are the results of our re-implementation of the corresponding model, and the corresponding original results are shown in parentheses. 'A', 'V' and 'T' are used to indicate the modality used by each model, corresponding to the acoustic, visual, and textual data.

| Model | Modality | IEMOCAP | | | | | | | MELD | EmoryNLP |
|---|---|---|---|---|---|---|---|---|---|---|
| | | Happy | Sad | Neutral | Angry | Excited | Frustrated | w-F1 | w-F1 | w-F1 |
| DialogueRNN* | A+V+T | 33.18 | 78.80 | 59.21 | 65.28 | 71.86 | 58.91 | 62.75 | 57.03 | 31.70 |
| COSMIC | T | *46.83* | *81.40* | *64.11* | *60.98* | *68.88* | *64.43* | *65.90* (65.28) | 65.21 | 38.11 |
| CESTa* | T | 47.70 | 80.82 | 64.74 | 63.41 | 75.95 | 62.65 | 67.10 | 58.36 | – |
| DialogXL* | T | *44.06* | *77.10* | *64.67* | *61.59* | *69.73* | *66.98* | *65.88* (65.94) | 62.41 | 34.73 |
| CoG-BART* | T | – | – | – | – | – | – | 66.18 | 64.81 | 39.04 |
| DialogueGCN* | T | *41.95* | *80.04* | *63.18* | *64.07* | *62.74* | *65.63* | *64.43* (64.18) | 58.10 | – |
| RGAT* | T | **51.62** | 77.32 | 65.42 | 63.01 | 67.95 | 61.23 | 65.22 | 60.91 | 34.42 |
| I-GCN | T | 50.00 | **83.80** | 59.30 | 64.60 | 74.30 | 59.00 | 65.40 | 60.80 | – |
| DAG-ERC | T | *46.69* | *80.12* | *66.92* | *68.35* | *70.53* | *67.28* | *68.02* (68.03) | 63.65 | 39.02 |
| MM-DFN* | A+V+T | 42.22 | 78.98 | 66.42 | 69.77 | 75.56 | 66.33 | 68.18 | 59.46 | – |
| Ours-A | A | 35.32 | 63.47 | 54.75 | 55.89 | 61.34 | 53.13 | 55.31 | 42.88 | – |
| Ours-T | T | 47.32 | 78.17 | 66.25 | 69.41 | 75.68 | 66.74 | 68.56 | 65.97 | **39.95** |
| Ours | A+T | 49.40 | 79.06 | **67.84** | **70.27** | **76.54** | **67.56** | **69.69** | **66.52** | – |

For the IEMOCAP dataset, the weighted-F1 score of our model is 69.69%, which is 1.51% higher than the best baseline MM-DFN. According to Table 1, the conversation length of the IEMOCAP dataset is the longest among the three datasets, with each conversation containing 48.4 utterances on average. Therefore, for conversations in this dataset, it is important to correctly understand the context information in the conversation. Thanks to the re-perception of context information, our MCPN can capture the context representation specific to each time step more accurately, which is the reason why it can achieve better performance on IEMOCAP. At the same time, our model outperforms the baseline model that used bidirectional context information, such as DialogXL and CoG-BART, while we used only one-way historical context information, which indicates the efficient utilization ability of our proposed model for context information. Furthermore, the classification performance of our model on most emotional categories in the IEMOCAP dataset is better than that of the baseline model compared, such as *Neutral, Angry, Excited,* and *Frustrated.*

For the MELD and EmoryNLP datasets, our model's weighted-F1 scores are 66.52% and 39.95%, respectively, which achieves 1.31% and 0.91% improvements to the corresponding optimal baseline models. As the dialogue length of these two datasets is relatively short, the average number of utterances in each dialogue is 9.6 and 14, respectively. Furthermore, these two datasets are multi-person conversation datasets, and the interaction between speakers is more complex than IEMOCAP. Therefore, the interaction information between speakers in the conversation is very important for the recognition of utterance emotion. According to Table 2, The performance of our model on these two datasets is better than that of other baseline models, which indicates that by capturing the dynamics of interactions between speakers, our MCPN can model interactions in conversations more effectively.

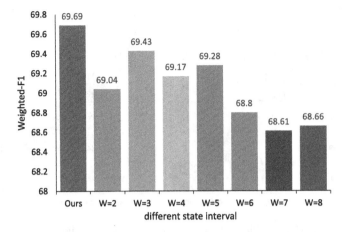

**Fig. 3.** The performance on IEMOCAP when the model uses our state interaction interval (Ours) and different sliding windows, $W = n$ denotes different size of windows.

## 5.2   Variants of Various Modalities

Table 2 also reports the performance of our model when only acoustic or textual features are used. It can be seen that the performance when using only unimodal features is worse than that when using two modalities at the same time, which indicates that the information of multiple modalities can complement each other, thus further improving the identification ability of the model. Meanwhile, the performance of textual modality is better than that of acoustic modality, which indicates that for a dialogue, the semantic content of utterance contains more emotional information.

## 5.3   Effectiveness of State Interaction Interval

Since we are inspired by the dynamical influence model [13] to obtain the dynamic state interaction interval of each time step from the state sequence $Q$, we will verify its effectiveness in this section. We report the performance when using our state interaction interval and when using a fixed-size sliding window. The specific results are shown in Fig. 3.

According to the results in Fig. 3, the performance of using the dynamic state interaction interval is better than of using the fixed sliding windows, which illustrates the effectiveness of our dynamic state interaction interval. By dynamically selecting interaction intervals for each time step, our model can capture the interaction between speakers more specifically. In addition, we also find that when sliding windows of different sizes are used, the performance shows a trend of increasing first and then decreasing. We infer that the state interval can not provide enough interaction information when the window size is small, so the performance of the model increases with the increase of the window size. However, when the window size increases to a certain extent, the selected state interval

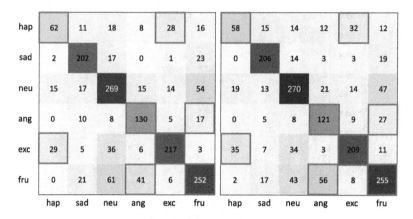

**Fig. 4.** The Confusion matrices of the model on IEMOCAP dataset when using our Emotion Triple-Recognition process (left one) and only using $q_t$ for emotion recognition (right one). The green squares mark similar emotions that are easily misclassified. (Color figure online)

contains too many obsolete states of early times, which prevents the model from capturing the interaction information for a specific time step.

### 5.4 Performance on Similar Emotion Classification

In this section, we verify the effectiveness of our proposed Emotion Triple-Recognition process. We make statistics on the classification performance of the model on the IEMOCAP dataset when the fully Triple-Recognition process is used. At the same time, we also calculate the classification performance when only the pre-recognition step is used. The results are shown in Fig. 4.

As can be seen from the two confusion matrices in Fig. 4, when the proposed Triple-Recognition process is used, the classification performance of emotions that are easily misclassified, such as *happiness* and *excited, anger* and *frustrated*, are better than that when only the pre-recognition step is used. Although the model is still easy to misclassify similar emotions after using the Triple-Recognition process proposed by us, compared with the commonly used single classification method, our method can distinguish similar emotions more effectively.

### 5.5 Ablation Study

To investigate the role of each component of our model, we perform ablation experiments on three datasets. Because some components are the precondition for executing other components, we made corresponding modifications to the model when removing these key components to ensure that subsequent components can work properly. These results are shown in Table 3.

**Table 3.** The performance (weighted-F1) of ablation study on three datasets. The 'SPP', 'CRP' and 'SRP' denote the corresponding process in MCP module.

| Component | IEMOCAP | MELD | EmoryNLP |
|---|---|---|---|
| Origin | 69.69 | 66.52 | 39.95 |
| w/o SPP | 68.93 ($\downarrow$0.76) | 65.66 ($\downarrow$**0.86**) | 38.60 ($\downarrow$**1.35**) |
| w/o CRP | 68.64 ($\downarrow$**1.05**) | 66.15 ($\downarrow$0.37) | 39.28 ($\downarrow$0.67) |
| w/o SRP | 68.71 ($\downarrow$0.98) | 65.87 ($\downarrow$0.65) | 38.82 ($\downarrow$1.13) |
| w/o ETR | 68.82 ($\downarrow$0.87) | 66.08 ($\downarrow$0.44) | 38.96 ($\downarrow$0.99) |

**Table 4.** Test accuracy of MCPN on samples with emotional shift and without it.

| Dataset | Emotional shift | | w/o Emotional shift | |
|---|---|---|---|---|
| | Samples | Accuracy | Samples | Accuracy |
| IEMOCAP | 576 | 60.41% | 1002 | 75.10% |
| MELD | 1003 | 62.03% | 861 | 71.35% |
| EmoryNLP | 673 | 38.74% | 361 | 43.56% |

It can be seen from the results in Table 3 that when a certain component is removed, the performance of the model decreased, which shows that every component in our model is essential. In particular, when we remove the CPR process, the performance of the model decreases most on the IEMOCAP dataset, indicating that context information has a greater impact on the utterance emotion in long conversations. On the contrary, when we remove the SPP or SRP processes in the MCP module, the model performance decreases the most in the other two datasets, indicating that the speaker interaction in the dialogue also cannot be ignored.

## 5.6  Error Study

In this section, we analyze the problem of emotional shift, which refers to the sudden change of the emotion of the speaker in the dialogue, that is, two consecutive utterances of the same speaker express different emotions [11]. Existing methods usually do not work well in the emotional shift. As shown in Table 4, the classification accuracy of our model on samples without emotional shift is much higher than that on samples with emotional shift, which indicates that our model also cannot effectively deal with this problem. However, the classification accuracy of our model on samples with emotional shift is still better than that of previous models. For example, the accuracy of MCPN in the case of emotional shift is 60.76% on IEMOCAP, which is higher than 47.5% achieved by DialogueRNN [11] and 57.98% achieved by DAG-ERC [18]. To solve this problem effectively, further research is needed.

# 6    Conclusion

In this paper, we propose a Multiple Cross-Perception Network, MCPN, for real-time emotion recognition in conversation. We dynamically select state interaction intervals for each time step so that the model can capture the dynamics of interactions between speakers. On this basis, with the help of multiple cross-perception processes, our MCPN can capture the semantic and interactive information in each time step of the conversation more accurately, so as to obtain the speaker state representation rich in emotional information. Furthermore, we propose an Emotion Triple-Recognition process, which effectively alleviates the misclassification problem of similar emotions in the model by introducing fuzzy classification probability and multiple recognition processes.

# References

1. Busso, C., et al.: IEMOCAP: interactive emotional dyadic motion capture database. Lang. Resour. Eval. **42**(4), 335–359 (2008). https://doi.org/10.1007/s10579-008-9076-6

2. Ghosal, D., Majumder, N., Gelbukh, A.F., Mihalcea, R., Poria, S.: COSMIC: commonsense knowledge for emotion identification in conversations. In: EMNLP, pp. 2470–2481. Findings of ACL (2020). https://doi.org/10.18653/v1/2020.findings-emnlp.224

3. Ghosal, D., Majumder, N., Poria, S., Chhaya, N., Gelbukh, A.F.: DialogueGCN: a graph convolutional neural network for emotion recognition in conversation. In: EMNLP-IJCNLP, pp. 154–164 (2019). https://doi.org/10.18653/v1/D19-1015

4. Hochreiter, S., Schmidhuber, J.: Long short-term memory. Neural Comput. **9**(8), 1735–1780 (1997). https://doi.org/10.1162/neco.1997.9.8.1735

5. Hu, D., Hou, X., Wei, L., Jiang, L., Mo, Y.: MM-DFN: multimodal dynamic fusion network for emotion recognition in conversations. In: 2022 IEEE International Conference on Acoustics, Speech and Signal Processing (ICASSP), pp. 7037–7041 (2022). https://doi.org/10.1109/ICASSP43922.2022.9747397

6. Ishiwatari, T., Yasuda, Y., Miyazaki, T., Goto, J.: Relation-aware graph attention networks with relational position encodings for emotion recognition in conversations. In: EMNLP, pp. 7360–7370 (2020). https://doi.org/10.18653/v1/2020.emnlp-main.597

7. Jiao, W., Lyu, M.R., King, I.: Real-time emotion recognition via attention gated hierarchical memory network. In: AAAI, pp. 8002–8009 (2020). https://aaai.org/ojs/index.php/AAAI/article/view/6309

8. Kingma, D.P., Ba, J.: Adam: A method for stochastic optimization. In: 3rd International Conference for Learning Representations (ICLR) (2015). http://arxiv.org/abs/1412.6980

9. Li, S., Yan, H., Qiu, X.: Contrast and generation make BART a good dialogue emotion recognizer. In: AAAI, pp. 11002–11010 (2022). https://ojs.aaai.org/index.php/AAAI/article/view/21348

10. Liu, Y., et al.: RoBERTa: a robustly optimized BERT pretraining approach. arXiv preprint arXiv:1907.11692 (2019)

11. Majumder, N., Poria, S., Hazarika, D., Mihalcea, R., Gelbukh, A.F., Cambria, E.: DialogueRNN: an attentive RNN for emotion detection in conversations. In: AAAI, pp. 6818–6825 (2019). https://doi.org/10.1609/aaai.v33i01.33016818

12. Nie, W., Chang, R., Ren, M., Su, Y., Liu, A.: I-GCN: incremental graph convolution network for conversation emotion detection. IEEE Trans. Multimedia (2021). https://doi.org/10.1109/TMM.2021.3118881
13. Pan, W., Dong, W., Cebrian, M., Kim, T., Pentland, A.: Modeling Dynamical Influence in Human Interaction. Tech. rep, Technical Report, Media Lab, MIT, Cambridge (2011)
14. Poria, S., Hazarika, D., Majumder, N., Naik, G., Cambria, E., Mihalcea, R.: MELD: a multimodal multi-party dataset for emotion recognition in conversations. In: Proceedings of the 57th Annual Meeting of the Association for Computational Linguistics (ACL), pp. 527–536 (2019). https://doi.org/10.18653/v1/p19-1050
15. Poria, S., Majumder, N., Mihalcea, R., Hovy, E.H.: Emotion recognition in conversation: research challenges, datasets, and recent advances. IEEE Access **7**, 100943–100953 (2019). https://doi.org/10.1109/ACCESS.2019.2929050
16. Schuller, B.W., Batliner, A., Steidl, S., Seppi, D.: Recognising realistic emotions and affect in speech: state of the art and lessons learnt from the first challenge. Speech Commun. **53**(9–10), 1062–1087 (2011). https://doi.org/10.1016/j.specom.2011.01.011
17. Shen, W., Chen, J., Quan, X., Xie, Z.: DialogXL: all-in-one XLNet for multi-party conversation emotion recognition. In: AAAI, pp. 13789–13797 (2021). https://ojs.aaai.org/index.php/AAAI/article/view/17625
18. Shen, W., Wu, S., Yang, Y., Quan, X.: Directed acyclic graph network for conversational emotion recognition. In: ACL/IJCNLP, pp. 1551–1560 (2021). https://doi.org/10.18653/v1/2021.acl-long.123
19. Vaswani, A., et al.: Attention is all you need. In: NeurIPS, pp. 5998–6008 (2017)
20. Wang, Y., Zhang, J., Ma, J., Wang, S., Xiao, J.: Contextualized emotion recognition in conversation as sequence tagging. In: Proceedings of the 21th Annual Meeting of the Special Interest Group on Discourse and Dialogue (SIGdial), pp. 186–195 (2020). https://aclanthology.org/2020.sigdial-1.23/
21. Yang, Z., Dai, Z., Yang, Y., Carbonell, J.G., Salakhutdinov, R., Le, Q.V.: XLNet: generalized autoregressive pretraining for language understanding. In: Neural Information Processing Systems (NeurIPS), pp. 5754–5764 (2019)
22. Zahiri, S.M., Choi, J.D.: Emotion detection on TV show transcripts with sequence-based convolutional neural networks. In: AAAI. AAAI Workshops. vol. WS-18, pp. 44–52. AAAI Press (2018). https://aaai.org/ocs/index.php/WS/AAAIW18/paper/view/16434

# Baby Cry Recognition Based on Acoustic Segment Model

Shuxian Wang, Jun Du$^{(\boxtimes)}$, and Yajian Wang

University of Science and Technology of China, Hefei, Anhui, China
{sxwang21,yajian}@mail.ustc.edu.cn, jundu@ustc.edu.cn

**Abstract.** Since babies cannot speak, they can only communicate with the outside world and express their emotions and needs through crying. Considering the variety of reasons why babies cry, it is a challenging task to accurately understand the meaning of baby crying. In this paper, we propose a baby cry recognition method based on acoustic segment model (ASM). Firstly, based on Gaussian mixtures models - hidden Markov models (GMM-HMMs), baby cry recordings are transcribed into ASM sequences composed of ASM units. In this way, different baby cry recordings are segmented in more detail, which can better capture the similarities and differences between acoustic segments. Then, by using latent semantic analysis (LSA), these ASM sequences are converted into feature vectors, and the term-document matrix is obtained. Finally, a simple classifier is adopted to distinguish different types of baby crying. The effectiveness of the proposed method is evaluated on two infant crying databases. The ASM-based approach can achieve higher accuracy compared with the approach based on residual network (ResNet). And through experiments, we analyze the reasons for the better performance of the ASM-based method.

**Keywords:** baby cry recognition · acoustic segment model · latent semantic analysis · deep neural network

## 1 Introduction

Baby cry recognition (BCR) is a task to identify the needs contained in a baby's cry [1]. Since babies do not yet have the ability to speak, crying has become the most important way for them to convey their physical and psychological needs to the outside world [2–7]. However, novice parents usually have little parenting experience. When babies cry, they are often at a loss. What's more serious is that when a baby cries because of pathological pain, if the novice parent cannot quickly and accurately understand the meaning of the baby's cry and make a wrong judgment, it is likely to miss the best time for treatment. Therefore, how to quickly understand the meaning of a baby's cry and make timely and accurate judgments is an urgent problem for every novice parent. It can be seen that the task of baby cry recognition has important research significance.

© The Author(s), under exclusive license to Springer Nature Singapore Pte Ltd. 2023
L. Zhenhua et al. (Eds.): NCMMSC 2022, CCIS 1765, pp. 16–29, 2023.
https://doi.org/10.1007/978-981-99-2401-1_2

Recently, many technical methods have been applied to the research of BCR, including traditional classifier methods such as Gaussian Mixture Models - Universal Background Models (GMM-UBM), i-vectors methods [8,9] and some methods based on deep learning: feed-forward neural networks (FNN) [2,10], time-delay neural networks (TDNN) [11], and convolutional neural networks (CNN) [12,13]. Although these methods mentioned above have achieved certain results in the recognition of baby crying, there are still some problems worthy of discussion. On the one hand, traditional methods such as GMM cannot learn deep non-linear feature transformation. On the other hand, deep learning-based methods such as CNN require a sufficient amount of data, and the difficulty of network training will increase as the number of network layers increases. In addition, it classifies infant crying by learning the feature information corresponding to the entire audio. As a result, it tends to be disturbed by longer but indistinguishable segments of the audio, while ignoring shorter but critical segments, so it cannot accurately locate the key segments that distinguish different types of infant crying.

Therefore, this paper proposes an infant cry recognition method based on the acoustic segment model (ASM), which combines the advantages of traditional methods and deep learning methods well. It can accurately mine acoustic information and segment the entire audio into more detailed segments according to whether the acoustic features have changed, so as to locate the key segments that can distinguish different categories of infant crying. ASM has been successfully applied to many tasks, such as automatic speech recognition (ASR) [14], speech emotion recognition [15,16], speaker recognition [17], music genre classification [18] and acoustic scene classification (ASC) [19]. Just as the basic building blocks of language are phonemes and grammars, baby crying signals that contain different needs of babies are also composed of fundamental units, and these fundamental units are related to each other. The proposed ASM method aims to find a universal set of acoustic units from baby cries to distinguish different types of baby cries.

The ASM framework generally consists of two steps, namely initial segmentation and iterative modeling. In the initial segmentation step, there are many different segmentation methods to obtain the basic acoustic units, such as maximum likelihood segmentation [14,20], even segmentation [15], K-means clustering algorithm [21], etc. The segmentation method used in this paper is GMM-HMMs, that is, each type of baby crying is modeled by GMM-HMMs [22–24]. Specifically, according to the similarities and differences of acoustic characteristics, the segments with similar acoustic characteristics are grouped together and marked with the same hidden state. Each hidden state corresponds to an ASM unit. In this way, through the initial segmentation, each baby cry recording is divided into variable length segments, so that we get the initial ASM sequences. Then, for iterative modeling, each ASM unit is modeled by a GMM-HMM and then baby cries are decoded into a new sequence of ASM units. After transcribing a baby cry into an ASM sequence, each baby cry is composed of ASM units, which is similar to a text document composed of terms. Therefore, we can use latent semantic analysis (LSA) to generate the term-document matrix. Each column of the matrix is a feature vector of a baby cry recording, and then these feature vectors are sent to the backend classifier.

The remainder of the paper is organized as follows. In Sect. 2, we introduce the proposed model and method used in baby cry recognition. In Sect. 3, experimental results and analysis are presented. Finally, we make the conclusions of this study and summarize our work in Sect. 4.

## 2    Method

For BCR, this paper proposes an ASM-based analysis method. The framework of the method is shown in Fig. 1. For the training data, through the two steps of initial segmentation and iterative modeling, they are all transcribed into ASM sequences. At the same time, the acoustic segment model generated in the iteration can be used to transcribe the test data into ASM sequences. In this way, each acoustic recording is transcribed into a sequence composed of ASM units, which is similar to a text document composed of terms. Therefore, text classification methods widely used in the field of information retrieval, such as LSA, can be used to analyze this problem. Through LSA and singular value decomposition (SVD) [25], an ASM sequence can be converted into a vector, so that the ASM sequences transcribed from all training data can be mapped to a term-document matrix. Each sample in the test set is processed in the same way. After the above processing, we can get the feature vector corresponding to each sample in the training and test sets, and then send these vectors to the backend DNN for classification.

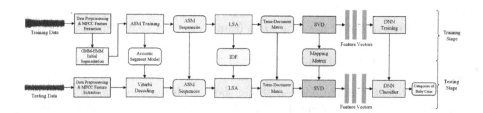

**Fig. 1.** ASM-based system framework.

### 2.1    Acoustic Segment Model

The main function of the acoustic segment model is to convert a baby cry recording into a sequence composed of basic acoustic units, just like a sentence is composed of words. The ASM method usually consists of two stages, namely initial segmentation and model training.

**Initial Segmentation.** The initial segmentation affects the result of the ASM method, so it is a very critical step. Many methods have been proposed to do the segmentation. Considering that GMM-HMMs achieve outstanding performance in ASR and can well explore the boundaries of acoustic feature changes, we use GMM-HMMs to perform initial segmentation.

First of all, each baby cry is modeled by GMM-HMMs, and the HMM is a left-to-right topology. However, considering that similar baby cries may occur at different time periods, we add a swivel structure to the topology, so that similar frames can be represented by the same hidden state. Assuming that there are $B$ kinds of baby crying in the database, each kind of baby crying is modeled by a GMM-HMM with $M$ hidden states. And then, through the Baum-Welch algorithm [26], we can update the parameters of GMM-HMMs. Through decoding, each baby cry recording is transcribed into a sequence composed of hidden states, and each hidden state is corresponding to a segment of the baby cry recording. Thus, the $C = B \times M$ hidden states are used as the corpus to initialize each baby cry recording as an ASM sequence.

**Model Training.** After completing the initial segmentation, each baby cry recording is converted into a sequence composed of ASM units. In the model training step, first, we use the GMM-HMM with a left-to-right HMM topology to model each ASM unit. Then, the Baum-Welch algorithm is adopted to update the parameters of the GMM-HMM model. Next, we use the Viterbi decoding algorithm to transcribe the training set data into new ASM sequences. These new ASM sequences are used as new labels for the training recordings in the next iteration of model training. The above process is repeated until the ASM sequences corresponding to the training data converge stably.

## 2.2 Latent Semantic Analysis

After the processing of the above steps, each baby cry recording is transcribed into a sequence composed of ASM units, which is like a text document composed of terms. Therefore, in view of the outstanding performance of LSA in the field of text processing, we use LSA for analysis, that is, through the LSA method, the correspondence between ASM units and baby cry recordings can be described by a term-text document matrix. Each column of the matrix corresponds to each baby cry recording that has been transcribed, and each row corresponds to one ASM unit or two adjacent ASM units in the ASM sequence. Therefore, if there are $C$ elements in the baby cry corpus, then the dimension of the vector in each column of the matrix is $D = C \times (C + 1)$.

Similar to the processing method in the field of information retrieval, the value of each element in the matrix is determined by term frequency (TF) and inverse document frequency (IDF) [27]. TF reflects the frequency of a word in the current text, and IDF reflects the frequency of the word in all texts. If a word has a high probability of appearing in all texts, even if it appears many times in a certain text, its importance to the text is also very low. Therefore, only by integrating TF and IDF, can it more accurately reflect the importance of a word to a text. The formula for calculating the TF of the $i$-th ASM term in the $j$-th baby cry recording is as follows:

$$TF_{i,j} = \frac{x_{i,j}}{\sum_{d=1}^{D} x_{d,j}}, \tag{1}$$

where $x_{i,j}$ is the number of times the $i$-th term appears in the ASM sequence corresponding to the $j$-th baby cry recording. The IDF calculation formula is as follows:

$$IDF_i = \log\frac{Q+1}{Q(i)+1}, \tag{2}$$

where $Q$ is the number of baby cry recordings in the training set, and $Q(i)$ is the number of texts in which the $i$-th term has appeared. In this way, each element of the matrix $W$ is defined as follows:

$$w_{i,j} = TF_{i,j} \times IDF_i. \tag{3}$$

Due to the use of bigrams with sparsity problems, the term-document matrix $W$ with dimension $D \times Q$ is sparse. We can use the SVD to reduce the dimension of the matrix $W$, as follows:

$$W = U\Sigma V^T. \tag{4}$$

The matrix $W$ is decomposed into the product of three matrices: the left-singular $D \times D$ matrix $U$, the diagonal $D \times Q$ matrix $\Sigma$ and the right-singular $Q \times Q$ matrix $V$. Among them, the diagonal elements of matrix $\Sigma$ are the singular values of matrix $W$, and these singular values are arranged from large to small. We take the first $t$ singular values and the first $t$ rows of the matrix $U$ to form a mapping space $U_t$, and then multiply this matrix with the original matrix $W$ to get a new matrix $W_t$ after dimensionality reduction. And $W_t$ is used as the training data of the backend classifier. The value of $t$ is determined based on the percentage of the sum of the squares of the singular values.

For the test data, the processing method is similar to the above steps of the training data, but what we need to pay attention to is that the test data needs to use the IDF and matrix $U_t$ obtained in the training phase to obtain the matrix $W_t^{test}$.

## 2.3   DNN Classifier

The classifier used in this paper is DNN, which is used to distinguish different types of baby crying. Since the feature vectors of baby crying extracted by the ASM method are easy to distinguish, a simple DNN structure is adopted.

**Table 1.** The details of two baby crying databases.

| Database | A | B |
|---|---|---|
| Recording Scene | Home | Hospital |
| Number of Categories | 6 | 2 |
| Training Set | 8.03 h | 8.54 h, 2020.02-2020.12 |
| Validation Set | 3.44 h, seen babies | N/A |
| Test Set | 2.41 h, unseen babies | 1.34 h, 2021.01 |

## 3   Experiments and Analysis

### 3.1   Database and Data Preprocessing

There are few high-quality infant crying databases that have been published, which brings certain challenges to the research of baby cry recognition. In this study, we adopt two infant crying databases, which were recorded at home and in the hospital, respectively, to evaluate our method. It is worth mentioning that the baby crying data recorded at home was annotated by parents based on their own experience or subsequent processing response, while the data recorded in the hospital was annotated by hospital pediatric experts. In addition, in the data set recorded in the hospital, some of the cries were recorded when the baby was given injections due to illness, so these cries were labeled "pain". Therefore, for the crying data recorded in the hospital, we conduct a two-class analysis of "pain" and "non-pain". The baby crying data recorded at home has six types of labels: "hungry", "sleepy", "uncomfortable", "pain", "hug", and "diaper", which are analyzed in six categories.

Before analyzing these baby crying data, we need to preprocess the data. First, down-sampling processing is performed to unify the sampling rate of all baby cry recordings 16000 Hz. Then, we apply voice activity detection (VAD) to detect the start and end of the silent interval in a baby cry recording according to the difference between the energy of the silent interval and the non-silent interval. The VAD method is used to remove silent redundancy and expand the number of effective samples. Finally, all the non-silent intervals of a baby cry recording are combined and then divided into 10-second segments.

After the above data preprocessing, the detailed information of the two infant crying databases is shown in Table 1. For database A, it was recorded at home and consists of crying recordings of 65 babies aged 0–6 months, and a certain amount of crying data is collected for each baby. We randomly select the crying data of 10 babies as the test set, and the rest of the crying data are mixed together and divided into training set and validation set. Therefore, the babies in the validation set have been seen in the training set, and the babies in the test set have not been seen in the training set. For database B, it was recorded in the hospital. The recording time is from February 2020 to January 2021. We select the crying data recorded in January 2021 as the test set, and the rest of the crying data as the training set. Since there are many babies in the delivery room of the hospital, and considering the protection of the privacy of the newborn, this part of the data cannot record the ID of the baby. Therefore, the database B is not divided according to the baby ID but the recording month, and because the neonatal hospitalization time is limited and the infants hospitalized in the ward are highly mobile, only one or two cry data are recorded for each baby, so it can be considered that the crying in the test set comes from the infants that have not been seen in the training set.

### 3.2   Ablation Experiments

**Baseline System.** Residual network (ResNet) [28] has been successfully used in the research of infant cry recognition [29], and has performed well on audio classification tasks in recent years, so ResNet is adopted as the baseline system.

In our previous work, ResNet excelled in the acoustic scene classification (ASC) task [30]. Considering that these two tasks are similar, we adopt the ResNet structure used in the ASC task [30], and then the ResNet is trained on these two baby cry databases. The log-Mel spectrogram is used as the input to the baseline system. To generate log-Mel spectrogram, a short-time Fourier transform (STFT) with 2048 FFT points is applied, using a window size of 2048 samples and a frameshift of 1024 samples, and log-Mel deltas and delta-deltas without padding are also computed.

**ASM-DNN Baby Cry Recognition Model.** The structure and principle of the ASM-DNN baby cry recognition model have been introduced in Sect. 2, which uses MFCC as the input feature. 60-dimensional Mel-frequency cepstral coefficients (MFCC) features are extracted by using a 40-ms window length with a 20-ms window shift to train GMM-HMMs. The DNN we used has 3 hidden layers, each with 512 neurons, and a fixed dropout rate of 0.1. The parameters of the DNN are learned using the SGD [31] algorithm, and the initial learning rate is set to 0.1. We perform ablation experiments on database A to explore the impact of two important parameters in the ASM-DNN model (the number of ASM units and the percentage of dimensionality reduction after singular value decomposition (SVD)) on the recognition accuracy.

- **Number of ASM units**

Obviously, if the number of ASM units is too small, the differences between different baby cry clips cannot be captured well, but if the number of ASM units is too large, it may cause overfitting problems, so we need to find a suitable value through experiments.

The number of ASM units is equal to the total number of hidden states, that is, assuming that there are $B$ kinds of baby crying in the database, and each kind of baby crying is modeled by a GMM-HMM with $M$ hidden states, then the number of ASM units is $C = B \times M$. As mentioned earlier, database

A contains six types of crying. We adjust the number of ASM units by adjusting $M$. Table 2 shows the experimental results with different number of ASM units. It is worth noting that in these experiments, the first 70% of the singular values are retained after singular value decomposition. It can be observed from Table 2 that the best result achieved when each kind of baby crying is modeled with 6 hidden states. Meanwhile, an accuracy of 29.99% is achieved on the test set.

**Table 2.** Performance comparisons with different ASM units.

| Hidden States | ASM Units | The Accuracy of Validation Set |
|:---:|:---:|:---:|
| 4 | 24 | 45.45% |
| 5 | 30 | 46.33% |
| 6 | 36 | **47.21%** |
| 7 | 42 | 46.57% |

• **Dimensionality Reduction in SVD**

As mentioned earlier, the term-document matrix obtained by LSA is sparse, so we can reduce the dimensionality of the matrix by retaining the largest singular values after SVD. The dimension of the new matrix obtained after dimensionality reduction is determined by the percentage of the sum of squares of singular values. Keeping 36 ASM units, we adjust the percentage of dimensionality reduction and the results are presented in Table 3. We can observe from Table 3 that when the percentage is set to 70%, the model achieves the highest accuracy on the validation set. At this time, the model's recognition accuracy of the test set is 29.99%.

**Table 3.** Performance comparisons with different reduced dimensions in SVD.

| Percentage | The Accuracy of Validation Set |
|:---:|:---:|
| 60% | 47.05% |
| 70% | **47.21%** |
| 80% | 46.81% |

In summary, the optimal number of hidden states for modeling each type of baby crying is 6, so for databases A and B, the optimal number of ASM units is 36 and 12, respectively. Meanwhile, the best SVD dimension reduction dimension is 70%.

### 3.3   Overall Comparison

**Overall Comparison on Baby Crying Database A.** The results of these two approaches on database A are shown in Table 4. By comparing ResNet and ASM-DNN, we can observe that better results are obtained for ASM-based approach. Specifically, the recognition accuracy of crying for both seen babies and unseen babies is improved by adopting the ASM-based approach. In addition, it can be seen that the recognition accuracy of these two approaches on the test set is 28.49% and 29.99%, respectively, and neither exceeds 30%. Considering that the individual differences of babies will cause different babies' crying to be different, this may cause the model to have lower accuracy when recognizing the crying of babies that have not been seen in the training set.

**Table 4.** The accuracy comparisons between ResNet and ASM-DNN on database A.

| System | Validation Set | Test Set |
|---|---|---|
| ResNet | 39.58% | 28.49% |
| ASM-DNN | **47.21%** | **29.99%** |

**Overall Comparison on Baby Crying Database B.** The results of these two approaches on database B are shown in Table 5. Compared with the ResNet-based approach, although the proposed ASM-based approach has a slightly lower recognition accuracy of pain crying, however, for non-pain crying, the performance of the ASM-based approach is significantly improved. Specifically, the recognition accuracy of non-pain crying is improved from 60.73% to 69.09% by adopting the ASM-based approach, that is, the performance can be improved by 8.36%. Therefore, for the entire test set, compared with ResNet, the ASM-based approach improves the recognition accuracy from 66.46% to 71.01%, which is an increase of about 5% points. Obviously, similar to the experimental results of database A, the ASM-based approach also performs better on database B, which further demonstrates the effectiveness of the ASM-based approach.

**Table 5.** The accuracy comparisons between ResNet and ASM-DNN on database B.

| System | Pain Crying | Non-pain Crying | Overall Test Set |
|---|---|---|---|
| ResNet | **74.04%** | 60.73% | 66.46% |
| ASM-DNN | 73.56% | **69.09%** | **71.01%** |

### 3.4   Results Analysis

**Spectrogram Analysis.** Figures 2 and 3 show two examples of pain crying and non-pain crying from data set recorded in the hospital. The two recordings

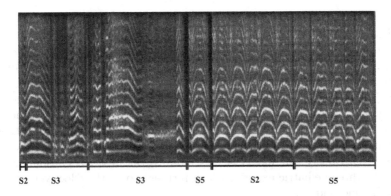

S2    S3         S3         S5        S2            S5

**Fig. 2.** The spectrogram and ASM sequence of an example recording of pain baby crying. This example was misclassified by ResNet as the non-pain baby crying but correctly classified by our ASM-DNN approach.

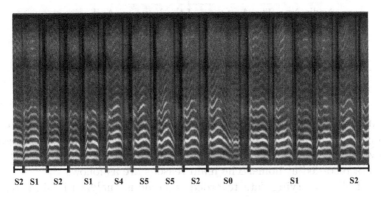

S2 S1  S2    S1    S4   S5   S5   S2   S0          S1          S2

**Fig. 3.** The spectrogram and ASM sequence of an example recording of non-pain baby crying. This example was misclassified by ResNet as the pain baby crying but correctly classified by our ASM-DNN approach.

are confused by the ResNet, but the ASM-based model can correctly distinguish these two kinds of crying. For a more intuitive analysis, we show the results of transcribing these two cry recordings into ASM sequences through the ASM method. The ASM units are named from S0 to S11. It can be observed that similar parts in the spectrograms are represented by the same ASM units, such as S2. At the same time, the different parts can be captured by the ASM units such as S3 for pain crying and S0 for non-pain crying. It can be seen that by using the ASM-based method, we can capture the differences between acoustic segments in more detail, and then distinguish different types of baby crying more accurately. Therefore, the result of the ASM-based method is better than that of the ResNet.

**Comparison of Model Judgment Results and Expert Audiometry Results.** In the database of baby crying recorded by the hospital (that is, database B), we randomly select 300 pieces of data, and invite three experienced pediatric experts to conduct audiometry, and each expert conduct audiometry on 100 cries. The results of the experts' audiometry can be regarded as the "ceiling" of the recognition accuracy of the model. The audiometric results are shown in Table 6. It can be seen from Table 6 that the accuracy rates of the three experts' audiometry are 66%, 72%, and 62%, respectively. At the same time, it can be found from Table 5 that the recognition accuracy of the ASM-DNN model is 71.01%. Obviously, our model is very close to the best recognition accuracy among the three pediatric experts, which further proves the effectiveness of the ASM-DNN method.

**Table 6.** The results of the experts' audiometry.

| Audio No. | The Accuracy of the Experts' Audiometry |
|-----------|------------------------------------------|
| 1–100     | 66%                                      |
| 101–200   | 72%                                      |
| 201–300   | 62%                                      |

**Comparative Analysis of Experimental Results on Two Databases.** By observing the previous experimental results, it can be found that the recognition accuracy of the crying data recorded in the hospital is higher than that of the crying data recorded at home. The main reasons are as follows:

First of all, the crying data recorded at home is analyzed in six categories, while the crying data recorded in the hospital is analyzed in two categories. It should be noted that the more categories, the greater the uncertainty. Hence the recognition accuracy of baby crying will be correspondingly improved when only two classes are needed to be predicted.

Secondly, the category labels of baby crying data collected in the hospital are marked by experienced pediatric experts. The experts mark the crying based on their years of experience, combined with the baby's facial expressions, movements, breathing state, and the intensity of crying. However, the category labels of baby crying data recorded at home are marked by parents. Pediatric experts have more experience, so the category labels of the crying data marked by them are more accurate and reliable, which makes the recognition accuracy for data collected in the hospital is also higher.

## 4    Conclusions

In this study, we propose an ASM-based analysis method for baby cry recognition. We first transcribe all baby cry recordings into ASM sequences composed of ASM units through the two steps of initial segmentation and iterative modeling, so that the similarities and differences between the segments of baby cry recordings can be well captured. Then, using LSA and SVD, a dimensionality-reduced term-document matrix is obtained. Finally, a classifier with a relatively simple structure can be used in the backend to achieve the purpose of identifying baby crying. Experiments conducted on two databases show that the ASM combined with a simple DNN classifier achieves better results than ResNet for baby cry recognition, which demonstrates the effectiveness of the ASM-based model.

# References

1. Drummond, J.E., McBride, M.L., Wiebe, C.F.: The development of mothers' understanding of infant crying. Clin. Nurs. Res. **2**(4), 396–410 (1993)
2. Garcia, J.O., Garcia, C.R.: Mel-frequency cepstrum coefficients extraction from infant cry for classification of normal and pathological cry with feed-forward neural networks. In: Proceedings of the International Joint Conference on Neural Networks, pp. 3140–3145 (2003)
3. Rusu, M.S., Diaconescu, Ş.S., Sardescu, G., Brătilă, E.: Database and system design for data collection of crying related to infant's needs and diseases. In: 2015 International Conference on Speech Technology and Human-Computer Dialogue (SpeD), pp. 1–6 (2015)
4. Wasz-Höckert, O., Partanen, T.J., Vuorenkoski, V., Michelsson, K., Valanne, E.: The identification of some specific meanings in infant vocalization. Experientia **20**(3), 154–154 (1964)
5. Orlandi, S., et al.: Study of cry patterns in infants at high risk for autism. In: Seventh International Workshop on Models and Analysis of Vocal Emissions for Biomedical Applications (2011)
6. Farsaie Alaie, H., Tadj, C.: Cry-based classification of healthy and sick infants using adapted boosting mixture learning method for gaussian mixture models. Model. Simul. Eng. **2012**(9), 55 (2012)
7. Chittora, A., Patil, H.A.: Classification of pathological infant cries using modulation spectrogram features. In: The 9th International Symposium on Chinese Spoken Language Processing, pp. 541–545 (2014)
8. Bǎnicǎ, I.A., Cucu, H., Buzo, A., Burileanu, D., Burileanu, C.: Baby cry recognition in real-world conditions. In: 2016 39th International Conference on Telecommunications and Signal Processing (TSP), pp. 315–318 (2016)
9. Bǎnicǎ, I.A., Cucu, H., Buzo, A., Burileanu, D., Burileanu, C.: Automatic methods for infant cry classification. In: 2016 International Conference on Communications (COMM), pp. 51–54 (2016)
10. Abdulaziz, Y., Ahmad, S.M.S.: Infant cry recognition system: a comparison of system performance based on mel frequency and linear prediction cepstral coefficients. In: 2010 International Conference on Information Retrieval & Knowledge Management (CAMP), pp. 260–263 (2010)
11. Reyes-Galaviz, O.F., Reyes-Garcia, C.A.: A system for the processing of infant cry to recognize pathologies in recently born babies with neural networks. In: 9th Conference Speech and Computer, pp. 552–557 (2004)

12. Chang, C.Y., Li, J.J.: Application of deep learning for recognizing infant cries. In: 2016 IEEE International Conference on Consumer Electronics-Taiwan (ICCE-TW), pp. 1–2 (2016)
13. Yong, B.F., Ting, H.N., Ng, K.H.: Baby cry recognition using deep neural networks. In: World Congress on Medical Physics and Biomedical Engineering 2018, pp. 809–813 (2019)
14. Lee, C.H., Soong, F.K., Juang, B.H.: A segment model based approach to speech recognition. In: International Conference on Acoustics, Speech, and Signal Processing (ICASSP), pp. 501–502 (1988)
15. Lee, H.Y., et al.: Ensemble of machine learning and acoustic segment model techniques for speech emotion and autism spectrum disorders recognition. In: INTER-SPEECH, pp. 215–219 (2013)
16. Zheng, S., Du, J., Zhou, H., Bai, X., Lee, C.H., Li, S.: Speech emotion recognition based on acoustic segment model. In: 2021 12th International Symposium on Chinese Spoken Language Processing (ISCSLP), pp. 1–5 (2021)
17. Tsao, Y., Sun, H., Li, H., Lee, C.H.: An acoustic segment model approach to incorporating temporal information into speaker modeling for text-independent speaker recognition. In: 2010 IEEE International Conference on Acoustics, Speech and Signal Processing, pp. 4422–4425 (2010)
18. Riley, M., Heinen, E., Ghosh, J.: A text retrieval approach to content-based audio retrieval. In: International Society for Music Information Retrieval (ISMIR), pp. 295–300 (2008)
19. Bai, X., Du, J., Wang, Z.R., Lee, C.H.: A hybrid approach to acoustic scene classification based on universal acoustic models. In: Interspeech, pp. 3619–3623 (2019)
20. Svendsen, T., Soong, F.: On the automatic segmentation of speech signals. In: International Conference on Acoustics, Speech, and Signal Processing (ICASSP), pp. 77–80 (1987)
21. Hu, H., Siniscalchi, S.M., Wang, Y., Bai, X., Du, J., Lee, C.H.: An acoustic segment model based segment unit selection approach to acoustic scene classification with partial utterances. In: INTERSPEECH, pp. 1201–1205 (2020)
22. Rabiner, L.R.: A tutorial on hidden Markov models and selected applications in speech recognition. Proc. IEEE **77**(2), 257–286 (1989)
23. Su, D., Wu, X., Xu, L.: GMM-HMM acoustic model training by a two level procedure with Gaussian components determined by automatic model selection. In: 2010 IEEE International Conference on Acoustics, Speech and Signal Processing, pp. 4890–4893 (2010)
24. Karpagavalli, S., Chandra, E.: Phoneme and word based model for tamil speech recognition using GMM-HMM. In: 2015 International Conference on Advanced Computing and Communication Systems, pp. 1–5 (2015)
25. Wall, M.E., Rechtsteiner, A., Rocha, L.M.: Singular value decomposition and principal component analysis. In: A Practical Approach to Microarray Data Analysis, pp. 91–109 (2003)
26. Elworthy, D.: Does Baum-Welch re-estimation help taggers?. arXiv preprint cmp-lg/9410012 (1994)
27. Hull, D.: Improving text retrieval for the routing problem using latent semantic indexing. In: SIGIR1994, pp. 282–291 (1994)
28. He, K., Zhang, X., Ren, S., Sun, J.: Deep residual learning for image recognition. In: Proceedings of the IEEE Conference on Computer Vision and Pattern Recognition, pp. 770–778 (2016)
29. Xie, X., Zhang, L., Wang, J.: Application of residual network to infant crying recognition. J. Electron. Inf. Technol. **41**(1), 233–239 (2019)

30. Hu, H., Yang, C.H.H., Xia, X., et al.: A two-stage approach to device-robust acoustic scene classification. In: 2021 IEEE International Conference on Acoustics, Speech and Signal Processing, pp. 845–849 (2021)
31. Bottou, L.: Large-scale machine learning with stochastic gradient descent. In: Proceedings of COMPSTAT 2010, pp. 177–186 (2010)

# A Multi-feature Sets Fusion Strategy with Similar Samples Removal for Snore Sound Classification

Zhonghao Zhao[1,2], Yang Tan[1,2], Mengkai Sun[1,2], Yi Chang[3], Kun Qian[1,2(✉)], Bin Hu[1,2(✉)], Björn W. Schuller[3,4], and Yoshiharu Yamamoto[5]

[1] Key Laboratory of Brain Health Intelligent Evaluation and Intervention, Ministry of Education (Beijing Institute of Technology), Beijing 100081, China
{zhonghao.zhao,yang_tan,smk,qian,bh}@bit.edu.cn
[2] School of Medical Technology, Beijing Institute of Technology, Beijing 100081, China
[3] GLAM – Group on Language, Audio, Music, Imperial College London, London SW7 2AZ, UK
y.chang20@imperial.ac.uk, schuller@ieee.org
[4] Chair for Embedded Intelligence for Health Care and Wellbeing, University of Augsburg, 86159 Augsburg, Germany
[5] Graduate School of Education, The University of Tokyo, Tokyo 113-0033, Japan
yamamoto@p.u-tokyo.ac.jp

**Abstract.** Obstructive sleep apnoe (OSA) is a common clinical sleep-related breathing disorder. Classifying the excitation location of snore sound can help doctors provide more accurate diagnosis and complete treatment plans. In this study, we propose a strategy to classify snore sound leveraging 'classic' features sets. At training stage, we eliminate selected samples to improve discrimination between different classes. As to unweighted average recall, a field's major measure for imbalanced data, our method achieves 65.6 %, which significantly ($p < 0.05$, one-tailed z-test) outperforms the baseline of the INTERSPEECH 2017 COMPARE Snoring Sub-challenge. Moreover, the proposed method can also improve the performance of other models based on the original classification results.

**Keywords:** Computer Audition · Feature Fusion · Digital Health · Snore Sound Classification · Obstructive Sleep Apnea

## 1 Introduction

Snoring is a very common condition among both adults and children. Sometimes, it is even regarded as a sign of patients being sleeping well. In the interview population of Chang et al. [1], the prevalence of snoring in individuals was 51.9 %. Among them, the prevalence rate of males is higher than that of females, with prevalence rates of 60 % and 50 %, respectively. But in fact, snoring is a disease, which is the sound that occurs during sleep due to the narrowing of the upper airway causing the uvula (palatine) to vibrate. Snoring, so-called "high-profile sleep killer", not only causes poor sleep at night [2] and daytime sleepiness [3], but also seriously reduces the sleep quality of the bed partner

© The Author(s), under exclusive license to Springer Nature Singapore Pte Ltd. 2023
L. Zhenhua et al. (Eds.): NCMMSC 2022, CCIS 1765, pp. 30–43, 2023.
https://doi.org/10.1007/978-981-99-2401-1_3

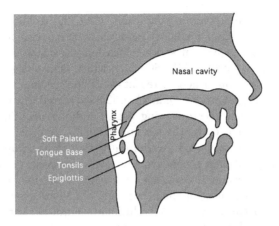

**Fig. 1.** The upper airways anatomy.

[4–6]. More seriously, loud and frequent snorers are thought more likely to have Obstructive Sleep Apnoea (OSA) [7]. It is reported that snoring intensity correlates with OSA severity [8,9]. OSA is generally manifested as a cessation of airflow through the mouth and nose for 10 seconds or longer, accompanied with a decrease in blood oxygen saturation, when the patients have each attack. During an adult's seven hours sleep each night, the number of seizures is often more than 30. In other words, the patients experience repeated episodes of respiratory airflow interruption, involuntary breath-holding and intermittent hypoxia during sleep. As a result, a series of hidden risks of diseases such as cardiovascular system, respiratory system, mental system, endocrine, and sexual dysfunction may occur [10–14]. In particular, because OSA increases the vascular burden of patients, they have three times the risk of having hypertension [15]. To this end, we should pay more attention to this common and neglected phenomenon of snoring and treat OSA in time.

Snoring originates from vibrations at different locations in the upper airway, i.e., palatal snoring (V), oropharyngeal snoring (O), tongue base snoring (T), and epiglottal snoring (E) [16,17] and thus, surgery is accordingly performed at the different point of relative OSA [18]. Fig 1 shows the anatomy of upper airways. In order to select the OSA surgical method in a targeted manner, the doctor first needs an accurate localisation of the snoring [16,19,20]. This is critical to the success of the surgery. Drug Induced Sleep Endoscopy (DISE) has been increasingly used to help determine the presence and location of airway obstruction in patients in recent years, but it has some limitations. On the one hand, the patients are induced to sleep by the drug during this process, which is not enough to fully simulate the reality. On the other hand, it is time consuming and costly for patients [21,22]. Therefore, snoring sound, as an inexpensive, easy-to-collect, universal and non-invasive audio signal, has been widely used in acoustic analysis to determine the vibration position in the upper airway during snoring. Thereby, good results have been achieved [23–28].

The rapidly developing computer audition (CA) [29] technology is becoming a popular topic of digital medicine research in the search for new digital phenotypes [30]. Several works on snoring location classification based on the Munich Passau Snoring Corpus (MPSSC) [31], which is also the database used in this paper, have been published in recent years for snoring as a subclass of an audio signal. In our past works, Qian *et al.* [23] extracted wavelet energy features and quantised feature vectors using bag-of-audio-words. The unweighted average recall (UAR) achieved by the method is 69.4 %. In order to balance the dataset, Zhang *et al.* [24] proposed a novel data augmentation approach based on semi-supervised conditional generative adversarial networks (scGANs). Compared with work on the same dataset, the best-achieved results are competitive with, or even superior to, most of the other state-of-the-art systems. Similarly, Ding *et al.* [25] demonstrated that a Convolutional Neural Network (CNN) with six-layer convolution and complement-cross-entropy loss function could well solve the problem of imbalance distribution of the dataset to yield the highest UAR with the value of 78.85 % on the development set and 77.13 % on the test set under all test conditions. In addition, Sun *et al.* [28] fused two feature types, the zero-crossing rate and Mel Frequency Cepstral Coefficients (MFCCs), and used Principal Component Analysis (PCA) and Support Vector Machines (SVM) for feature dimension reduction and classification, respectively. Most of the above works have shown good classification performance. However, they focus more on the effect and ignore the interpretability of the features, which can build trust between the model and doctors and patients [32–34]. Furthermore, Zhang *et al.* [24] balanced the datasets through data augmentation and Ding *et al.* [25] applied a complement-cross-entropy loss function from the model perspective. In a different direction, we compute the similarity between samples to alleviate the problem of dataset imbalance.

In this work, we propose a strategy to utilise distinct features to classify the snore sound excitation location. We extract different features and calculate the similarity between samples. Then, we eliminate similar samples in the data set at different feature scales. Finally, a machine learning model is used to classify the remaining audios.

The remainder of this article is organised as follows: Firstly, we present the materials and methods of this work in Sect. 2. Then, experimental details are shown in Sect. 3. Subsequently, Sect. 4 describes the results and discussion. Finally, a conclusion is made in Sect. 5.

## 2     Materials and Methods

### 2.1   MPSSC Database

MPSSC is an audio database dedicated to OSA research. It was released in the INTERSPEECH 2017 Computational Paralinguistics ChallengE (ComParE) [35]. The audio recordings from MPSSC were taken from three medical centres in Germany, i. e., Klinikum rechts der Isar, Technische Universitat Mu nchen, Munich, Alfried Krupp Hospital, Essen, and University Hospital, Halle. All of MPSSC were saved as wav files (16 bit, 16 000 Hz).

The snore sounds events were labelled based on VOTE classification. VOTE classification is a widely used scheme distinguishing four structures that can be involved in the upper airway: the level of the velum (V), the oropharyngeal area including the palatine tonsils (O), the tongue base (T), and the epiglottis (E). The details of the data distribution information can be found in Table 1. From the table, it can be seen that there is a data imbalance problem in MPSSC. MPSSC contains 828 snore sound events collected from 219 independent subjects. Moreover, the total time duration and average time duration are 1 250.11 s and 1.51 s (ranging from 0.73 to 2.75 s). In this study, we utilise all of the snore sound events in the MPSSC database.

**Table 1.** Data distribution of MPSSC. Train: train set, Dev: development set, Test: test set.

|   | Train | Dev | Test | $\sum$ |
|---|---|---|---|---|
| V | 168 | 161 | 155 | 484 |
| O | 76 | 75 | 65 | 216 |
| T | 8 | 15 | 16 | 39 |
| E | 30 | 32 | 27 | 89 |
| $\sum$ | 282 | 283 | 263 | 828 |

### 2.2 Feature Extraction

OPENSMILE (open Speech and Music Interpretation by Large Space Extraction) is a highly modular and flexible acoustic feature extraction toolkit which is widely applied in signal processing and machine learning [36]. Due to its modular operation mode, users can extract multiple features by various configuration files. In order to extract feature sets conveniently, the whole feature sets are extracted by OPENSMILE.

**COMPARE.** COMPARE was first released in the INTERSPEECH 2013 COMPARE [37]. The overall set contains 6 373 features, including energy, spectral, MFCCs, and voicing related low-level descriptors (LLDs). Moreover, a few LLDs includes logarithmic harmonic-to-noise ratio (HNR), spectral harmonicity, and psychoacoustic spectral sharpness. Table 2 shows the LLDs for the COMPARE. Table 3 describes the functionals applied to LLDs in the COMPARE feature set.

**eGeMAPS.** eGeMAPS [38] is an extension parameter set of GeMAPS (Geneva Minimalistic Standard Parameter Set). GeMAPS contains 18 LLDs, i.e., frequency related parameters, energy/amplitude related parameters, and spectral (balance) parameters. Combined with GeMAPS, eGeMAPS includes 88 features totally.

**Table 2.** The LLDs for COMPARE feature set.

| 55 Spectral LLDs | Group |
| --- | --- |
| MFCCs 1–14 | Cepstral |
| Spectral roll-off point 0.25, 0.5, 0.75, 0.9 | Spectral |
| Spectral energy 250–650 Hz, 1 k–4 kHz | Spectral |
| Spectral variance, skewness, kurtosis | Spectral |
| Spectral flux, centroid, entropy, slope | Spectral |
| Psychoacoustic sharpness, harmonicity | Spectral |
| RASTA-filtered auditory spectral bands 1–26 (0–8 kHz) | Spectral |
| **6 Voicing related LLDs** | **Group** |
| $F_0$ (SHS and Viterbi smoothing) | Prosodic |
| Probability of voicing | Voice Quality |
| log HNR, jitter (local and $\delta$), shimmer (local) | Voice Quality |
| **4 Energy related LLDs** | **Group** |
| RMSE, zero-crossing rate | Prosodic |
| Sum of auditory spectrum (loudness) | Prosodic |
| Sum of RASTA-filtered auditory spectrum | Prosodic |

**emo_large** emo_large (the large openSMILE emotion feature set) includes 6 552 features [39], which is larger than the above feature sets. More features potentially support more detailed information about the audios. 39 statistical functionals are applied to the LLDs (e. g., the fundamental frequency (F0), formants (F1-F3), energy) to obtain the features.

## 2.3 Classification Model

SVM, as an established, stable, and robust classifier, is applied in many tasks because of its excellent performance. It is a common method using kernel learning and can achieve non-linear classification. To render our work reproducible, an SVM classifier with a linear kernel is leveraged in this work. In our experiments, we carry out feature normalisation for each feature set before training. Then, the SVM models are trained with the complexity parameter in the range of $5\times10^{-6}$, $5\times10^{-5}, ..., 5\times10^{-1}, 5\times10^{0}$. The optimum complexity is selected based on the classification result on the development set. Further, we retrain the SVM model with the optimum complexity on the train and development set to predict on the test data set. In addition, we change the utilisation sequence of the feature sets to obtain a better classification result.

**Table 3.** Part of the functionals below would be applied to LLDs in the COMPARE feature set.

| Functionals |
| --- |
| Arithmetic or positive arithmetic mean |
| Inter-quartile ranges 1–2, 2–3, 1–3, |
| Linear regression slope, offset |
| Linear prediction gain and coefficients 1–5 |
| Linear regression quadratic error |
| Mean and std. dev. of peak to peak distances |
| Peak and valley range (absolute and relative) |
| Peak-valley-peak slopes mean and std. dev. |
| Peak mean value and distance to arithmetic mean |
| Quadratic regression quadratic error |
| Quadratic regression coefficients |
| Root-quadratic mean, flatness |
| Rise time, left curvature time |
| Range (difference between max. and min. values) |
| Relative position of max. and min. value |
| Standard deviation, skewness, kurtosis, quartiles 1–3 |
| Segment length mean, min., max., std. dev. |
| Temporal centroid |
| Up-level time 25 %, 50 %, 75 %, 90 % |
| 99-th and 1-st percentile, range of these |

## 3  Experimental Setups

After we extract the three feature sets shown previously, we train SVM models for each feature set, respectively. From the predictions, we find that, at distinct feature scales, correctly predicted audios are not similar. In other words, it is effortless to comprehend that audios from diverse classes may have similar features. As shown in Fig 2, we get a various data distribution after feature extraction. Fig 2a shows that a certain kind of class can be different from other data at a certain feature set scale. Moreover, as indicated in Fig 2b, when the distribution of data samples is imbalanced, categories classified wrongly with a small amount of data can have a great impact. Fig 2c demonstrates that the similar samples between two categories will affect the classification results. Therefore, we try to eliminate the similar samples and use the remaining samples to train the model. We hope that the fusion of diverse feature sets could make an improvement in the whole classification result.

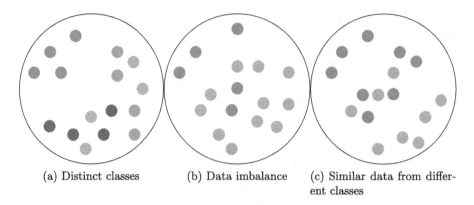

(a) Distinct classes        (b) Data imbalance        (c) Similar data from differ-
                                                      ent classes

**Fig. 2.** Three distinct data distributions.

Firstly, the three acoustic feature sets (i. e., COMPARE, eGeMAPS, and
emo_large) are extracted by the OPENSMILE v3.0 toolkit. Then, we train an
SVM model for each feature set and select the optimum complexity as the basic
hyper-parameter, respectively. As mentioned above, similar samples from different classes make the model pay more attention to fit the samples, which leads to
the distinguishing of the data becoming harder. Besides, imbalanced data distribution causes the model to more likely predict the data as belonging to the
more frequent classes. In order to improve the performance of the SVM model,
we eliminate selected mixed samples for each feature set during the training
process. To be specific, we calculate the Euclidean distance between samples in
the train set. Afterwards, a certain number of samples is omitted according to
the similarity result, which may help represent the train(ing) set sample space
more clearly. Moreover, the number of eliminated data for each class is different. A mixed set is composed of those eliminated samples. For the development
set, we calculate the Euclidean distance between it and the original train set.
We remove samples from the development set if one of the most similar samples
with a certain number is from the mixed set. The numbers of eliminated samples
for the train and development sets are regarded as optimised parameters. Furthermore, the criteria by which to eliminate samples should be different between
different feature sets. We select the optimum parameter for each feature set that
performed best on the development set, respectively. Due to the imbalance of
the data distribution, UAR is used as evaluation metric.

At the prediction stage, we also calculate the Euclidean distance between the
test set and the train set. For one sample of the test set, it is removed if one
of the most similar samples with a certain number is from the mixed set, which
is calculated from the train set and the development set. The 'certain number'
for the test set is the same as that for the development set. The specific number
of eliminated data in the test set is the same as that of the development set.
After that, we predict the classes in different orders. The unpredicted audios
would be predicted by the next feature set, and the last feature set would be

used to predict all of the remaining audios. Fig 3 shows the overall process of this method.

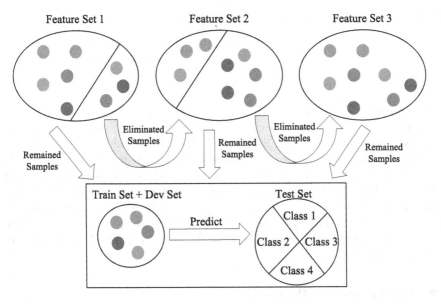

**Fig. 3.** The overall flow of this study includes feature extraction and snore sound classification.

## 4    Results and Discussion

### 4.1    Classification Results

In our experiments, we analyse variable combination modes of the three feature sets. For comparison, Table 4 presents the classification results of each feature set. In order to show more details of all experiments, Fig 4 shows specific confusion matrices for the test set with different orders.

Table 4 shows the results without deleting any sample. From Fig. 4, we can see that the best classification result resembles a UAR of 65.6 %. Compared with the specific feature set, the combination of different feature sets performs better. Because of the distinct samples eliminated from each feature set, we conduct experiments with different orders of feature extraction to explore whether this method can be used to improve the classification performance. When the first feature sets are the same, Fig 4 presents that we can obtain similar classification results for each class. The reason is that the majority of the samples is classified by the first feature set, which determines the final result. But in spite all of

**Table 4.** Specific classification results by unweighted average recall (UAR). (Unit: %). Num.: number of features, $C$: complexity.

|             | Num  | Dev  | Test | $C$               |
|-------------|------|------|------|-------------------|
| COMPARE     | 6373 | 37.5 | 54.4 | $5\times10^{-3}$  |
| eGeMAPSv01a | 88   | 41.3 | 60.1 | $5\times10^{-2}$  |
| emo_large   | 6552 | 35.6 | 64.4 | $5\times10^{-5}$  |

this, comparing the results in Fig 4 and Table 4, it can be found that using the combination of the first feature set and other feature sets is better than using only the first feature set. Specifically, the three feature sets have different representation capabilities for a certain class. The recalls of type 'V' and type 'E' are generally better than those for the other two classes. Considering the first feature set, COMPARE and emo_large represent type 'V' and type 'E' better, respectively. Moreover, as shown in Table 4e and Table 4f, the recall of the type 'E' classified by emo_large is up to 85.2 % UAR. It is obvious that type 'O' and type 'T' are wrongly distinguished as type 'V', which is probably caused by the data imbalance problem.

### 4.2    Limitations and Perspectives

In Table 5, the experimental results of different methods in the MPSSC are listed. Compared with other models, it is regrettable that our method is not state of the art. Up to now, the best result is up to 77.13 % UAR, which used a prototypical network to classify the location [25]. Demir *et al.* [40] considered local binary patterns and a histogram of oriented gradients extracted from colour spectrograms, which achieved a UAR of 72.6 %. Qian *et al.* [23] utilised a bag-of-audio-words approach and selected a Naïve Bayes model as the classifier, which reached a UAR of 69.4 %. Schmitt *et al.* [41] presented the performance of an end-to-end learning method with a UAR of 67.0 %. Although the result of our method is not the best, this method also bears noticeable improvements and could be combined with others for further gain.

Firstly, feature boundaries for similar samples from different classes are established by many researchers. Unlike the above methods, we focus on eliminating the similar samples, which could help distinguish different classes. Moreover, abundant features provide deep information of audios. Utilising diverse features reasonably could construct a complete data space. On the other hand, all of our experiments are reproducible, which is especially important for snoring sound analysis or health tasks in more general. Furthermore, we extract low-level descriptors and corresponding functionals features for each feature subset, which are explainable.

It is remarkable that we use different feature sets, each of which represents a distinct representation spaces of audio. In our experiments, we find that some classes could be easily distinguished by a certain feature. Therefore, if each class

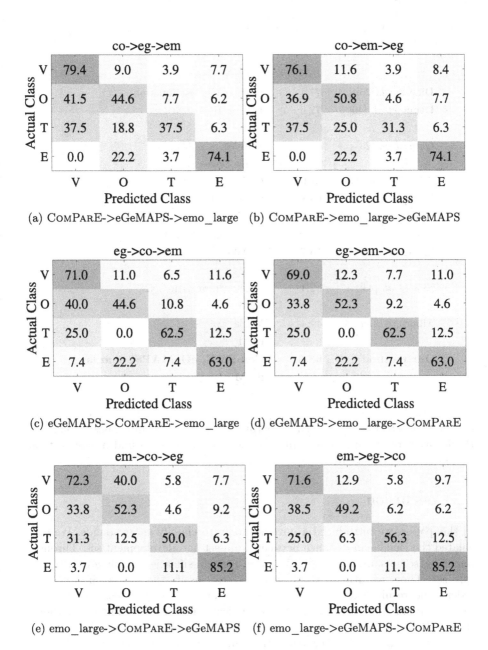

(a) COMPARE->eGeMAPS->emo_large   (b) COMPARE->emo_large->eGeMAPS

(c) eGeMAPS->COMPARE->emo_large   (d) eGeMAPS->emo_large->COMPARE

(e) emo_large->COMPARE->eGeMAPS   (f) emo_large->eGeMAPS->COMPARE

**Fig. 4.** Confusion matrices for test set with different orders. co: COMPARE, e.g.: eGeMAPS, em: emo_large.

**Table 5.** Classification results of other models. CCE: Complement-Cross-Entropy loss function. LBP: Local Binary Patterns. HOG: Histogram of Oriented Gradients. BoAWs: Bag-of-Audio-Words. GRU: Gated Recurrent Unit. MLSa: Mean of the Log-spectrum. ELM: Extreme Learning Machines.

|  | UAR (%) | Main Methods |
|---|---|---|
| Ding *et al.* [25] | 77.1 | CNN (CCE loss function) |
| Demir *et al.* [40] | 72.6 | LBP + HOG features |
|  |  | SVM |
| Qian *et al.* [23] | 69.4 | BoAWs |
|  |  | Naïve Bayes |
| Schmitt *et al.* [41] | 67.0 | CNN |
| Wang *et al.* [42] | 63.8 | CNN+GRU |
| Official baseline [35] | 58.5 | ComParE features |
|  |  | SVM (linear kernel) |
| Zhang *et al.* [24] | 56.7 | BoAWs |
|  |  | SVMs |
| New *et al.* [43] | 52.4 | Spectrogram with CNN |
|  |  | SVM |
| Albornoz *et al.* [44] | 49.4 | MLSa |
|  |  | ELM |
| **Our method** | **65.6** | **ComParE, eGeMAPS, emo_large** |
|  |  | **SVM** |

can be characterised by a certain feature, we can easily and quickly diagnose the location of snoring sound. In other words, we need to find a way to take advantage of each feature.

## 5 Conclusion

In this paper, we introduced a novel strategy to classify snore sounds. We calculated the Euclidean distance between train set and development set. Through removing some samples, we reached the best classification result for each feature set. Then, we utilised the feature sets to classify the test set in a certain order. The results show that this method can promote the result of classification compared with the baseline. Statistical dimensional reduction methods such as Principal Component Analysis and Random Forests can reduce redundant features and avoid overfitting. In future work, we will use such statistical dimensional reduction methods to improve the model performance.

**Acknowledgements.** This work was partially supported by the Ministry of Science and Technology of the People's Republic of China (2021ZD0201900), the National Natural Science Foundation of China (62272044), the BIT Teli Young Fellow Program from the Beijing Institute of Technology, China, and the Grants-in-Aid for Scientific Research (No. 20H00569) from the Ministry of Education, Culture, Sports, Science and Technology (MEXT), Japan.

# References

1. Chuang, L., Hsu, S., Lin, S., Ko, W., Chen, N., Tsai, Y., et al.: Prevalence of snoring and witnessed apnea in taiwanese adults. Chang Gung Med. J. **31**(2), 175 (2008)
2. Kang, J.M., Kim, S.T., Mariani, S., Cho, S.E., Winkelman, J.W., Park, K.H., Kang, S.G.: Difference in spectral power density of sleep eeg between patients with simple snoring and those with obstructive sleep apnoea. Sci. Rep. **10**(1), 1–8 (2020)
3. Yang, K.I., et al.: Prevalence of self-perceived snoring and apnea and their association with daytime sleepiness in korean high school students. J. Clin. Neurol. (Seoul, Korea) **13**(3), 265 (2017)
4. Sharief, I., Silva, G.E., Goodwin, J.L., Quan, S.F.: Effect of sleep disordered breathing on the sleep of bed partners in the sleep heart health study. Sleep **31**(10), 1449–1456 (2008)
5. Zarhin, D.: Sleep as a gendered family affair: Snoring and the "dark side" of relationships. Qualitative Health Research **26**(14), 1888–1901 (2016)
6. Blumen, M.B.: Is snoring intensity responsible for the sleep partner's poor quality of sleep? Sleep Breath **16**(3), 903–907 (2012)
7. Arnold, J., Sunilkumar, M., Krishna, V., Yoganand, S., Kumar, M.S., Shanmugapriyan, D.: Obstructive sleep apnea. J. Pharm. Bioallied Sci. **9**(Suppl 1), S26 (2017)
8. Maimon, N., Hanly, P.J.: Does snoring intensity correlate with the severity of obstructive sleep apnea? J. Clin. Sleep Med. **6**(5), 475–478 (2010)
9. Nakano, H., Furukawa, T., Nishima, S.: Relationship between snoring sound intensity and sleepiness in patients with obstructive sleep apnea. J. Clin. Sleep Med. **4**(6), 551–556 (2008)
10. Jehan, S., et al.: Obstructive sleep apnea and stroke. Sleep Med. Disorders: Int. J. **2**(5), 120 (2018)
11. Kuvat, N., Tanriverdi, H., Armutcu, F.: The relationship between obstructive sleep apnea syndrome and obesity: A new perspective on the pathogenesis in terms of organ crosstalk. Clin. Respir. J. **14**(7), 595–604 (2020)
12. Kaufmann, C.N., Susukida, R., Depp, C.A.: Sleep apnea, psychopathology, and mental health care. Sleep Health **3**(4), 244–249 (2017)
13. Lavrentaki, A., Ali, A., Cooper, B.G., Tahrani, A.A.: Mechanisms of endocrinology: Mechanisms of disease: the endocrinology of obstructive sleep apnoea. Eur. J. Endocrinol. **180**(3), R91–R125 (2019)
14. Skoczyński, S., et al.: Sexual disorders and dyspnoea among women with obstructive sleep apnea. Adv. Med. Sci. **65**(1), 189–196 (2020)
15. Baguet, J., Barone-Rochette, G., Pépin, J.: Hypertension and obstructive sleep apnoea syndrome: Current perspectives. J. Hum. Hypertens. **23**(7), 431–443 (2009)
16. Kezirian, E.J., Hohenhorst, W., de Vries, N.: Drug-induced sleep endoscopy: the vote classification. Eur. Arch. Otorhinolaryngol. **268**(8), 1233–1236 (2011)
17. Qian, K., et al.: Can machine learning assist locating the excitation of snore sound? A review. IEEE J. Biomed. Health Inform. **25**(4), 1233–1246 (2021)

18. Giralt-Hernando, M., Valls-Ontañón, A., Guijarro-Martínez, R., Masià-Gridilla, J., Hernández-Alfaro, F.: Impact of surgical maxillomandibular advancement upon pharyngeal airway volume and the apnoea-hypopnoea index in the treatment of obstructive sleep apnoea: Systematic review and meta-analysis. BMJ Open Respir. Res. **6**(1), e000402 (2019)

19. Ephros, H.D., Madani, M., Yalamanchili, S.C., et al.: Surgical treatment of snoring and obstructive sleep apnoea. Indian J. Med. Res. **131**(2), 267 (2010)

20. Ephros, H.D., Madani, M., Geller, B.M., DeFalco, R.J.: Developing a protocol for the surgical management of snoring and obstructive sleep apnea. Atlas Oral Maxillofac. Surg. Clin. North Am. **15**(2), 89–100 (2007)

21. Bergeron, M., et al.: Safety and cost of drug-induced sleep endoscopy outside the operating room. Laryngoscope **130**(8), 2076–2080 (2020)

22. Pang, K.P., et al.: Does drug-induced sleep endoscopy affect surgical outcome? A multicenter study of 326 obstructive sleep apnea patients. Laryngoscope **130**(2), 551–555 (2020)

23. Qian, K., et al.: A bag of wavelet features for snore sound classification. Ann. Biomed. Eng. **47**(4), 1000–1011 (2019)

24. Zhang, Z., Han, J., Qian, K., Janott, C., Guo, Y., Schuller, B.W.: Snore-gans: Improving automatic snore sound classification with synthesized data. IEEE J. Biomed. Health Inform. **24**(1), 300–310 (2019)

25. Ding, L., Peng, J.: Automatic classification of snoring sounds from excitation locations based on prototypical network. Appl. Acoust. **195**, 108799 (2022)

26. Dogan, S., Akbal, E., Tuncer, T., Acharya, U.R.: Application of substitution box of present cipher for automated detection of snoring sounds. Artif. Intell. Med. **117**, 102085 (2021)

27. Tuncer, T., Akbal, E., Dogan, S.: An automated snoring sound classification method based on local dual octal pattern and iterative hybrid feature selector. Biomed. Signal Process. Control **63**, 102173 (2021)

28. Sun, J., Hu, X., Peng, S., Peng, C.K., Ma, Y.: Automatic classification of excitation location of snoring sounds. J. Clin. Sleep Med. **17**(5), 1031–1038 (2021)

29. Qian, K., et al.: Computer audition for healthcare: Opportunities and challenges. Front. Digital Health **2**, 5 (2020)

30. Qian, K., Zhang, Z., Yamamoto, Y., Schuller, B.W.: Artificial intelligence internet of things for the elderly: From assisted living to health-care monitoring. IEEE Signal Process. Mag. **38**(4), 78–88 (2021)

31. Janott, C., et al.: Snoring classified: The Munich-Passau snore sound corpus. Comput. Biol. Med. **94**, 106–118 (2018)

32. Vellido, A.: The importance of interpretability and visualization in machine learning for applications in medicine and health care. Neural Comput. Appl. **32**(24), 18069–18083 (2020)

33. Stiglic, G., Kocbek, P., Fijacko, N., Zitnik, M., Verbert, K., Cilar, L.: Interpretability of machine learning-based prediction models in healthcare. Wiley Interdisc. Rev.: Data Mining Knowl. Disc. **10**(5), e1379 (2020)

34. Banegas-Luna, A., et al.: Towards the interpretability of machine learning predictions for medical applications targeting personalised therapies: A cancer case survey. Int. J. Mol. Sci. **22**(9), 4394 (2021)

35. Schuller, B.W., et al.: The interspeech 2017 computational paralinguistics challenge: Addressee, cold & snoring. In: Proceedings of the 18th Annual Conference of the International Speech Communication Association (INTERSPEECH), pp. 3442–3446. INTERSPEECH, Stockholm, Sweden (2017)

36. Eyben, F., Wöllmer, M., Schuller, B.W.: opensmile: The munich versatile and fast open-source audio feature extractor. In: Proceedings of the 18th ACM International Conference on Multimedia (ACM MM), pp. 1459–1462. ACM, Firenze, Italy (2010)

37. Schuller, B.W., et al.: The interspeech 2013 computational paralinguistics challenge: Social signals, conflict, emotion, autism. In: Proceedings of the 14th Annual Conference of the International Speech Communication Association (INTERSPEECH), pp. 148–152. INTERSPEECH, Lyon, France (2013)

38. Eyben, F., et al.: The geneva minimalistic acoustic parameter set (gemaps) for voice research and affective computing. IEEE Trans. Affective Comput. **7**(2), 190–202 (2015)

39. Eyben, F., Weninger, F., Gross, F., Schuller, B.W.: Recent developments in opensmile, the munich open-source multimedia feature extractor. In: Proceedings of the 21st ACM International Conference on Multimedia (ACM MM), pp. 835–838. ACM, Barcelona, Spain (2013)

40. Demir, F., Sengur, A., Cummins, N., Amiriparian, S., Schuller, B.W.: Low level texture features for snore sound discrimination. In: Proceedings of the 40th Annual International Conference of the IEEE Engineering in Medicine and Biology Society (EMBC), pp. 413–416. IEEE, Honolulu, HI, USA (2018)

41. Schmitt, M., Schuller, B.W.: End-to-end audio classification with small datasets - making it work. In: Proceedings of the 27th European Signal Processing Conference (EUSIPCO), pp. 1–5. IEEE, A Coruna, Spain (2019)

42. Wang, J., Strömfeli, H., Schuller, B.W.: A cnn-gru approach to capture time-frequency pattern interdependence for snore sound classification. In: Proceedings of the 26th European Signal Processing Conference (EUSIPCO), pp. 997–1001. IEEE, Rome, Italy (2018)

43. Nwe, T.L., Dat, T.H., Ng, W.Z.T., Ma, B.: An integrated solution for snoring sound classification using bhattacharyya distance based gmm supervectors with svm, feature selection with random forest and spectrogram with cnn. In: Proceedings of the 18th Annual Conference of the International Speech Communication Association (INTERSPEECH), pp. 3467–3471. INTERSPEECH, Stockholm, Sweden (2017)

44. Albornoz, E.M., Bugnon, L.A., Martínez, C.E.: Snore recognition using a reduced set of spectral features. In: 2017 XVII Workshop on Information Processing and Control (RPIC), pp. 1–5. IEEE, Mar del Plata, Argentina (2017)

# Multi-hypergraph Neural Networks for Emotion Recognition in Multi-party Conversations

Cheng Zheng[1] , Haojie Xu[1,3] , and Xiao Sun[2,3(✉)]

[1] AHU -IAI AI Joint Laboratory, Anhui University, Hefei, China
e20201139@stu.ahu.edu.cn
[2] Hefei University of Technology, Hefei, China
[3] Institute of Artificial Intelligence, Hefei Comprehensive National Science Center, Hefei, China
sunx@iai.ustc.edu.cn

**Abstract.** Emotion recognition in multi-party conversations (ERMC) is becoming increasingly popular as an emerging research topic in natural language processing. Although previous work exploited inter-dependency and self-dependency among participants, they paid more attention to the use of specific-speaker contexts. Specific-speaker context modeling can well consider the speaker's self-dependency, but inter-dependency has not been fully utilized. In this paper, two hypergraphs are designed to model specific-speaker context and non-specific-speaker context respectively, so as to deal with self-dependency and inter-dependency among participants. To this end, we design a multi-hypergraph neural network for ERMC, namely ERMC-MHGNN. In particular, we combine average aggregation and attention aggregation to generate hyperedge features, which can make better use of utterance information. Extensive experiments are conducted on two ERC benchmarks with state-of-the-art models employed as baselines for comparison. The empirical results demonstrate the superiority of this new model and confirm that further exploiting inter-dependency is of great value for ERMC. In addition, we also achieved good results on the emotional shift issue.

**Keywords:** Emotional shift · Emotion recognition in conversations · Emotion recognition in multi-party conversations

## 1 Introduction

Emotion recognition in conversations (ERC) has attracted more and more attention because of the prevalence of dialogue behaviour in various fields. The primary purpose of ERC is to recognize the emotion of each utterance in the dialogue. The recognized emotion can be used for opinion mining on social media such as Facebook and Instagram, building conversational assistants, and conducting medical psychoanalysis [1,15,18]. However, ERC, especially emotion recognition in multi-party conversations (ERMC), often exhibits more difficulties than traditional text sentiment analysis due to the emotional dynamics of the

L. Zhenhua et al. (Eds.): NCMMSC 2022, CCIS 1765, pp. 44–58, 2023.
https://doi.org/10.1007/978-981-99-2401-1_4

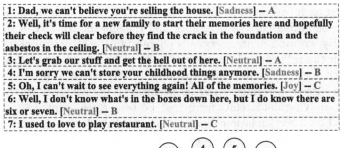

1: Dad, we can't believe you're selling the house. [Sadness] -- A

2: Well, it's time for a new family to start their memories here and hopefully their check will clear before they find the crack in the foundation and the asbestos in the ceiling. [Neutral] -- B

3: Let's grab our stuff and get the hell out of here. [Neutral] -- A

4: I'm sorry we can't store your childhood things anymore. [Sadness] -- B

5: Oh, I can't wait to see everything again! All of the memories. [Joy] -- C

6: Well, I don't know what's in the boxes down here, but I do know there are six or seven. [Neutral] -- B

7: I used to love to play restaurant. [Neutral] -- C

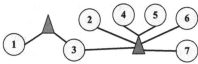

**Fig. 1.** Conversations as a hypergraph. Circles and triangles represent nodes and hyperedges, respectively.

dialogue [18]. There are two kinds of emotional dependencies among the participants in a dialogue, inter-dependency and self-dependency among participants. Self-dependency is the influence of what the speaker says on the current utterance. Inter-dependency is the influence of what others say on what the current speaker says. Therefore, identifying the emotion of an utterance in a multi-party dialogue depends not only on the itself and its context, but also on the speaker's self-dependence and the inter-dependency [5,21].

Some recent works [4,6,15] based on recurrent neural networks have begun to focus on conversational context modeling and speaker-specific modeling, and even some works [11] have carried out multi-task learning for speaker-specific modeling on this basis. They try to deal with speaker-dependent influences through speaker-specific modeling and conversational context modeling, but they cannot well use other speakers' utterances to influence the current utterance. While some works [5,8,10,20,21] based on graph neural networks use relational graph neural networks to distinguish different speaker dependencies, and some even use conversational discourse structure [21] or commonsense knowledge [10] to extend relationships between utterances. These models hope to establish more perfect utterance relations, and then aggregate according to the relations to form the influence of the surrounding utterances on the current utterance. However, the performance of such models will be affected by the type and quantity of inter-utterance relations. Moreover, the emotional change of a speaker may be caused by the joint influence of multiple utterances of multiple speakers. This influence may also be caused by the interaction of utterances under different relationships. So inter-dependency is more complex than self-dependency. We believe that it is necessary to build a graph network alone to model inter-dependency, especially for multi-dialogue, which can better identify the problem of emotional shift between consecutive utterances of the same speaker.

According to the hypergraph [3] structure, we know that a hyperedge may contain multiple utterances, and an utterance may belong to multiple hyperedges. We let each utterance generate a hyperedge. The nodes on the hyperedge are the corresponding current utterance and the specific context of the current utterance. Hypergraph neural networks [3] can use the structure of hypergraph to deal with the influence of multiple utterances from multiple speakers on an utterance, that is, to use multiple surrounding utterances to produce an influence on the current utterance. And by performing node-edge-node transformation, the underlying data relationship can be better represented [3], and more complex and high-level relationships can be established among nodes [25]. Previous work has shown that speaker-specific information is very important for ERMC [15, 21]. Therefore, how to use hypergraphs for speaker-specific modeling of ERMC is a very important issue. Second, the current utterance may be influenced by utterances from different speakers. Therefore, how to use hypergraphs for non-speaker-specific modeling of ERMC is also a very important issue.

In this paper, we construct two hypergraphs for speaker-specific and non-speaker-specific modeling, respectively. The hyperedges in the two hypergraphs are different. A hypergraph for speaker-specific modeling, where the nodes on the hyperedge are from the speaker of the current utterance, this hypergraph mainly deals with self-dependency. In a hypergraph for non-speaker-specific modeling, where nodes on a hyperedge contain the current utterance and utterances from other speakers, the hypergraph is primarily used to handle inter-dependency. In Fig. 1, we construct two kinds of hyperedges for the third utterance. The hyperedge of the green triangle indicates that the node of the hyperedge is from speaker B of the third utterance. The hyperedge of the blue triangle indicates that the nodes of the hyperedge are from speakers other than speaker B. Note that this hyperedge needs to contain the current utterance, so that the nodes within the hyperedge have an effect on the current utterance. We use the location information and node features to aggregate to generate hyperedge features. Here, we use the location information to obtain the weight of the average aggregation, use the node features to perform the attention aggregation to obtain the attention weight, and combine the two weights to obtain the hyperedge feature. Then, the hyperedge features are used to model the conversational context using a recurrent neural network. Finally, the hyperedge features are used to aggregate to obtain new node features. The hypergraph convolution of the two hypergraphs can be used to model specific speakers and non-specific speakers, so as to deal with inter-dependency and self-dependency among participants.

The main contributions of this work are summarized as follows:

- We construct hypergraphs for two different dependencies among participants and design a multi-hypergraph neural network for emotion recognition in multi-party conversations. To the best of our knowledge, this is the first attempt to build graphs for inter-dependency alone.
- We combine average aggregation and attention aggregation to generate hyperedge features, which can better utilize the information of utterances.
- We conduct experiments on two public benchmark datasets. The results consistently demonstrate the effectiveness and superiority of the proposed model. In addition, we achieved good results on the emotional shift issue.

# 2  Related Work

## 2.1  Emotion Recognition in Conversations

In the following paragraphs, we divide the related works into two categories according to their methods to model the conversation context. Note that here we consider some models that utilize Transformer [22] like DialogXL [19] without actually building a graph network as recurrent neural networks. DialogXL [19], BERT+MTL [11], and ERMC-DisGCN [21] have done some research on emotion recognition in multi-party conversations.

**Recurrence-Based Models.** DialogueRNN [15] uses three GRUs to model the speaker, the context given by the preceding utterances, and the emotion behind the preceding utterances, respectively. COSMIC [4] built on DialogueRNN using commonsense knowledge (CSK) to learn interactions between interlocutors participating. EmoCaps [13] introduce the concept of emotion vectors to multi-modal emotion recognition and propose a new emotion feature extraction structure Emoformer. BERT+MTL [11] exploit speaker identification as an auxiliary task to enhance the utterance representation in conversations. DialogueCRN [6] from a cognitive perspective to understand the conversation context and integrates emotional cues through a multi-turn reasoning module for classification. VAE-ERC [16] models the context-aware latent utterance role with a latent variable to overcome the lack of utterance role annotation in ERC datasets. TODKAT [28] proposes a new model in which the transformer model fuses the topical and CSK to predict the emotion label. DialogXL [19] improves XLNet [24] with enhanced memory and dialog-aware self-attention. CoG-BART [12] utilizes supervised contrastive learning in ERC and incorporates response generation as an auxiliary task when certain contextual information is involved.

**Graph-Based Models.** DialogueGCN [5] used a context window to connect the current utterance with surrounding utterances and treated each dialogue as a graph. RGAT [8] added positional encodings to DialogGCN. DAG-ERC [20] designed a directed acyclic graph neural network and provided a method to model the information flow between remote context and local context. SKAIG [10] utilizes CSK to enrich edges with knowledge representations and process the graph structure with a graph transformer. MMGCN [7] proposes a new model based on a multi-modal fused graph convolutional network. TUCORE-GCN [9] proposed a context-aware graph convolutional network model by focusing on how people understand conversations. ERMC-DisGCN [21] designed a relational convolution to lever the self-speaker dependency of interlocutors to propagate contextual information, and proposed an utterance-aware graph neural network.

## 2.2  Hypergraph Neural Network

HGNN [3] propose a hypergraph neural network framework and demonstrate its ability to model complex high-order data dependencies through hypergraph structures. HyperGAT [2] used subject words to construct hypergraphs for text classification. HGC-RNN [25] adopted a recurrent neural network structure to learn temporal dependencies from data sequences and performed hypergraph convolution operations to extract hidden representations of data. SHARE [23] constructed different hyperedges through sliding windows of different sizes and extracted user intent through hypergraph attention for session-based recommender systems. MHGNN [27] uses multi-hypergraph neural networks to explore the latent correlation among multiple physiological signals and the relationship among different subjects. Inspired by these works, we treat dialogue as a hypergraph and solve the ERC task using a hypergraph neural network.

## 3  Methodology

### 3.1  Hypergraph Definition

Hypergraph is defined as: $HG = (V, E)$, where $V = \{v_1, v_2, \ldots, v_N\}$ is a node-set, $E = \{HE_1, HE_2, \ldots, HE_N\}$ is a collection of hyperedges. The node-set belonging to hyperedge $HE_n$ is a subset of $V$. The structure of a hypergraph HG can also be represented by an incidence matrix A, with entries defined as:

$$A_{ij} = \begin{cases} 0, & v_i \notin HE_j, \\ 1, & v_i \in HE_j \end{cases} \tag{1}$$

We use $X = \{x_1, x_2, \ldots, x_N\}$ to denote the attribute vector of nodes in the hypergraph. So the hypergraph can also be represented by $HG = (A, X)$. In this paper, we use matrix M to store the relative position weight of the utterance in the hypergraph. The structure of matrix M is similar to the incidence matrix A. Each row in M corresponds to a hyperedge, and the non-zero items in each row represent the utterance node in this hyperedge. The size of the non-zero items is related to the position between nodes in the hyperedge. In the following, we use $HG = (M, X)$ to represent the hypergraph.

**Vertices.** Each utterance in a conversation is represented as a node $v_i \in V$. Each node $v_i$ is initialized with the utterance embeddings $h_i$. We update the embedding representations of vertices via hypergraph convolution.

**Hyperedge.** Since each hyperedge is generated based on a specific current utterance, we need to calculate the influence of other utterances on the current utterance, and these influences will be weakened according to the relative position between the utterances. We set the position weight of the current utterance to 1, and the position weight of the remaining utterances gradually decreases with the relative distance. See the Algorithm 1 for the specific process of hypergraph and hyperedge construction. We design two kinds of hypergraphs, one is speaker-specific hypergraph (SSHG), and the other is non-speaker-specific hypergraph (NSHG). The hyperedges in SSHG are speaker-specific hyperedges (SSHE). We select some utterances in the context window to add to SSHE, and the speaker of these utterances are the same as the speaker of the current utterance. The

**Algorithm 1.** Constructing Hypergraph

---

**Input:** the dialogue $\{h_1, h_2, \ldots, h_N\}$, speaker identity $p(\cdot)$, context window $w$.
**Output:** $SSHG, NSHG$.
1: $X, M_{SSHG}, M_{NSHG} \leftarrow \{h_1, h_2, \ldots, h_N\}, \emptyset, \emptyset$
2: **for all** $i \in [1, N]$ **do**
3:     $M^i_{SSHG}, M^i_{NSHG} \leftarrow \{0, 0, \ldots, 0\}, \{0, 0, \ldots, 0\}$ // N zero in total
4:     $w_p, w_f, count, M^i_{NSHG}[i] \leftarrow i - w, i + w, 0, 1$ // $w_p, w_f \in [1, N]$
5:     **for** $j = w_p; j <= w_f; j++$ **do**
6:       **if** $p(h_i) = p(h_j)$ **then**
7:         $M^i_{SSHG}[j] \leftarrow 1/(1 + \mathbf{abs}(i - j))$
8:         count++
9:       **else if** $p(h_i)! = p(h_j)$ and $count = 0$ **then**
10:        $M^i_{NSHG}[j] \leftarrow 1/(1 + \mathbf{abs}(i - j))$
11:       **end if**
12:     **end for**
13: **end for**
14: **return** $SSHG = (M_{SSHG}, X), NSHG = (M_{NSHG}, X)$

---

hyperedges in NSHG are non-speaker-specific hyperedges (NSHE). We take the past utterance of the speaker of the current utterance as a selective constraint, and select some utterances in the context window to add to NSHE. The speakers of these utterance are different from speaker of the current utterance.

### 3.2 Problem Definition

Given an input sequence containing N utterances $\{u_1, u_2, \ldots, u_N\}$, which is annotated with a sequence of labels $\{y_1, y_2, \ldots, y_N\}$. Utterance $u_i$ spoken by $p(u_i)$. The task of ERC aims to predict the emotion label $y_i$ for each utterance $u_i$.

### 3.3 Model

An overview of our proposed model is shown in Fig. 2, which consists of the Feature Extraction module, the Hypergraph Convolution Layer module, and the Emotion Classification module. Hyperedges are generated according to the third, fourth and fifth utterances.

**Utterance Feature Extraction.** Following COSMIC [4], we employ RoBERTa-Large [14] as feature extractor. The pre-trained model is firstly fine-tuned on each ERC dataset, and its parameters are then frozen while training our model. More specifically, a special token [CLS] is appended at the beginning of the utterance to create the input sequence for the model. Then, we use the [CLS]'s pooled embedding at the last layer as the feature representation $h_i$ of $u_i$.

**Hypergraph Convolution (HGC) Layer.** We utilize the two hypergraphs to perform hypergraph convolutions separately, and then obtain different utterance representations. The process of performing hypergraph convolution for each graph can be divided into the following three steps.

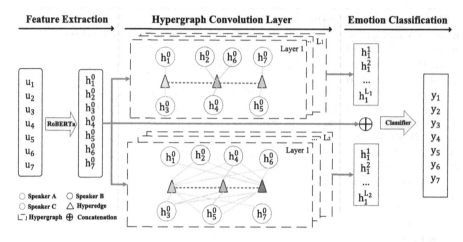

**Fig. 2.** Overview of our proposed model. In the Hypergraph Convolutional Layer module, the red dotted line represents the information transfer between the hyperedges. (Color figure online)

**Node to Edge Aggregation.** The first step is the aggregation from nodes to hyperedges. Here, we use the position weight $m_j^i$ to calculate the weight $\alpha_{ji}^{pos}$ of the weighted average aggregation. Since some nodes on a hyperedge are informative but others may not be, we should pay varying attention to the information from these nodes while aggregating them together. We utilize the attention mechanism to model the significance of different nodes. Here, we use a function $S(\cdot, \cdot)$ to calculate attention weights $\alpha_{ji}^{ATT}$. Function $S(\cdot, \cdot)$ is derived from the Scaled Dot-Product Attention formula [22]. Then, the obtained weight $\alpha_{ji}^{pos}$, attention weight $\alpha_{ji}^{ATT}$ and node information $h_i^{l-1}$ are aggregated to obtain hyperedge feature $f_j^l$. The specific formula is as follows:

$$\alpha_{ji}^{pos} = \frac{m_j^i}{\sum_{k|v_k \in HE_j} m_j^k} \tag{2}$$

$$\alpha_{ji}^{att} = \frac{S(W_1 h_i^{l-1}, t^l)}{\sum_{f|v_f \in HE_j} S(W_1 h_f^{l-1}, t^l)} \tag{3}$$

$$f_j^l = \sigma\left(\sum_{v_i \in HE_j} \alpha_{ji}^{pos} \alpha_{ji}^{att} W_2 h_i^{l-1}\right) \tag{4}$$

$$S(a, b) = \frac{S(a^T b)}{\sqrt{D}} \tag{5}$$

where $HE_j$ is the j-th hyperedge, resulting from the j-th utterance. $m_j^i$ is stored in the association matrix M, which represents the size of the position weight of the i-th node in the j-th hyperedge. $h_i^{l-1}$ represents the features of the utterance node. $t^l$ represents a trainable node-level context vector for the l-th HGC layer. $W_1$ and $W_2$ is a trainable parameter matrices. D is the dimension size.

**Edge to Edge Aggregation.** The second step is to transfer information between hyperedges. In order to make the current utterance have better interaction with the context, we use the hyperedge generated by each utterance to model the conversation context. We use BiLSTM to complete the information transfer.

$$q_j^l, hidden_j = \overleftrightarrow{LSTM}^c(f_j^l, hidden_{j-1}) \tag{6}$$

where $hidden_j$ is the j-th hidden state of the LSTM, $q_j^l$ represents the hyperedge feature obtained after the information passed by the hyperedge.

**Edge to Node Aggregation.** To update the feature for a node, we need to aggregate the information from all its connected hyperedges. We also use $S(\cdot, \cdot)$ to calculate the similarity between the node and hyperedge features.

$$h_i^l = \sigma\left(\sum_{HE_j \in E_i} \beta_{ij} W_3 q_j^l\right) \tag{7}$$

$$\beta_{ij} = \frac{S(W_4 q_j^l, W_1 h_i^{l-1})}{\sum_{HE_p \in E_i} S(W_4 q_p^l, W_5 h_i^{l-1})} \tag{8}$$

where $E_i$ is the set of hyperedges containing the i-th node. $W_3$, $W_4$, and $W_5$ denote trainable parameters, and $S(\cdot, \cdot)$ is same as Eq. 6.

## 3.4   Classifier

We concatenate the hidden states of the two hypergraphs in all HGC layers and pass it through a feedforward neural network to get the predicted emotion:

$$H_i^{HG} = \|_{l=1}^{L_{HG}} (h_{HG})_i^l \tag{9}$$

$$P_i = softmax(W_{smax}[h_i^0 : H_i^{SSHG} : H_i^{NSHG}] + b_{smax}) \tag{10}$$

$$\hat{y}_i = argmax_k(P_i[k]) \tag{11}$$

where $H_i^{HG}$ represents the result of hypergraph convolution performed on the hypergraph, $HG$ can be SSHG and NSHG, and $L_{HG}$ is the number of layers for hypergraph convolution of the corresponding hypergraph.

## 4   Experimental Setting

### 4.1   Datasets

We evaluate our model on two ERC datasets. The statistics of them are shown in Table 1. They are all multimodal datasets, but our task mainly focuses on textual modality to conduct our experiments.

**MELD** [17] is derived from the Friends TV series. The utterances are annotated with one of seven labels, namely neutral, joy, surprise, sadness, anger, disgust, and fear. The dataset consists of multi-party conversations and involves too many plot backgrounds. Non-neutral emotions account for 53%.

**Table 1.** The statistics of datasets. 'Avg.' denotes the average number of utterances. 'WA-F1' denotes the Weighted-average F1.

|          | #Dial.(Train/Dev./Test) | #Utt.(Train/Dev./Test) | Avg. | Classes | Metrics |
|----------|-------------------------|------------------------|------|---------|---------|
| MELD     | 1432(1038/114/280)      | 13708(9989/1109/2610)  | 9.57 | 7       | WA-F1   |
| EmoryNLP | 897(713/99/85)          | 12606(9934/1344/1328)  | 14.05| 7       | WA-F1   |

**EmoryNLP** [26] is also collected from Friends' TV scripts, but varies from MELD in the choice of scenes and emotion labels. The emotion labels include neutral, sad, mad, scared, powerful, peaceful, and joyful.

### 4.2  Compared Methods

For a comprehensive evaluation of our proposed ERMC-MHGNN, we compare it with the following baseline methods:

**Recurrence-based Models:** DialogueRNN [15], COSMIC [4], DialogueCRN [6], TODKAT [28], DialogXL [19], VAE-ERC [16], DialogueRNN-RoBERTa [4], CoG-BART [12], and EmpCaps [13].

**Graph-based Models:** DialogueGCN [5], RGAT [8], RGAT-RoBERTa [20], DialogueGCN-RoBERTa [20], SKAIG [10], ERMC-DisGCN [21], MMGCN [7], TUCORE-GCN [9], and DAG-ERC [20].

### 4.3  Implementation Details

We conducted experiments on a Windows10 using NVIDIA GeForce GTX 1650 GPU with 4 GB of memory. We used PyTorch 1.7.0 and CUDA toolkit 11.0. We adopt AdamW as the optimizer. Table 2 is hyperparameter settings. For feature dimension, the utterance feature dimension extracted by RoBERTA extractor is 1024, and after linear layer, the utterance feature dimension becomes 100.

**Table 2.** Hyperparameter settings.

| #        | Batchsize | Dropout | Lr     | Window | $Layer_{SSHG}$ | $Layer_{NSHG}$ |
|----------|-----------|---------|--------|--------|----------------|----------------|
| MELD     | 32        | 0.1     | 0.001  | 1      | 1              | 1              |
| EmoryNLP | 16        | 0.4     | 0.0009 | 4      | 4              | 6              |

## 5  Results and Discussions

### 5.1  Overall Performance

Table 3 shows the performance of different models on the MELD and EmoryNLP test sets. We can see that our model outperforms all baselines, which demonstrates the effectiveness of our proposed model. At the same time, we find that

**Table 3.** Overall performance on the two datasets. '-' signifies that no results were reported for the given dataset. 'CSK' stands for model that introduces commonsense knowledge. '*' represents the results of the model in the text-only modality.

| Model | CSK | MELD | EmoryNLP |
|---|---|---|---|
| RoBERTa | × | 62.88 | 37.78 |
| DialogueRNN | × | 57.03 | - |
| +RoBERTa | × | 63.61 | 37.44 |
| DialogueCRN | × | 58.39 | - |
| VAE-ERC | × | 65.34 | - |
| DialogXL | × | 62.4 | 34.73 |
| BERT+MTL | × | 61.90 | 35.92 |
| CoG-BART | × | 64.81 | 39.04 |
| COSMIC | √ | 65.21 | 38.11 |
| TODKAT | √ | 65.47 | 38.69 |
| EmoCaps* | × | 63.51 | - |
| DialogueGCN | × | 58.10 | - |
| +RoBERTa | × | 63.02 | 38.10 |
| RGAT | × | 60.91 | 34.42 |
| +RoBERTa | × | 62.80 | 37.89 |
| TUCORE-GCN | × | 62.47 | 36.01 |
| +RoBERTa | × | 65.36 | 39.24 |
| DAG-ERC | × | 63.65 | 39.02 |
| ERMC-DisGCN | × | 64.22 | 36.38 |
| SKAIG | √ | 65.18 | 38.88 |
| MMGCN* | × | 57.72 | - |
| ERMC-MHGNN | × | **66.4** | **40.1** |

models using CSK on MELD generally perform better, while our model achieves good results without relying on external knowledge. In this paper, we focus on modeling the two kinds of dependencies among speakers by building multiple hypergraphs, so we do not incorporate external knowledge. On the EmoryNLP dataset, we found that models using large-scale pre-trained models to extract features have better results. For example, both DAG-ERC and TUCORE-GCN use RoBERTa as feature extractors. These models can achieve over 39% on EmoryNLP. Our model also uses RoBERTa as a feature extractor and achieves relatively better results by separately modeling the two speaker dependencies.

## 5.2   Ablation Study

To investigate the impact of various modules in the model, we evaluate our model by separately removing two weights in the node-to-edge aggregation process in hypergraph convolution. In addition, we also conduct experiments on hypergraph convolution with a single hypergraph. The results are shown in Table 4.

**Table 4.** Results of ablation study.

| Method | MELD | EmoryNLP |
|---|---|---|
| full model | **66.4** | **40.1** |
| w/o $\alpha^{pos}$ | 65.61 ($\downarrow$0.79) | 39.15 ($\downarrow$0.95) |
| w/o $\alpha^{att}$ | 65.64 ($\downarrow$0.76) | 39.05 ($\downarrow$1.05) |
| w/o $SSHG$ | 65.3 ($\downarrow$1.1) | 38.93 ($\downarrow$1.17) |
| w/o $NSHG$ | 65.19 ($\downarrow$1.21) | 38.9 ($\downarrow$1.2) |

As shown in the Table 4, we can see that after removing the weights $\alpha^{att}$, there is a relatively large drop in performance on both datasets. Through the attention function, the surrounding utterances can be given different weights, so that the current utterance can better receive information from other utterances. Therefore, the use of attention weights $\alpha^{att}$ is beneficial to the aggregation of node information. When we remove the $\alpha^{pos}$ weights, both datasets also have relatively large drops. The distance between utterances may affect the interaction between two utterances. Appropriately reducing the influence of surrounding utterances according to the relative distance can also make the model better aggregate node features to a certain extent.

When we use one hypergraph and remove other hypergraphs, we only perform hypergraph convolution of one hypergraph. From the results in Table 4, we can see that the performance of the model is degraded no matter which hypergraph is removed. Among them, the model will also have a relatively large performance drop after removing the NSHG, which also shows that the method of modeling for non-specific speakers is feasible. In multi-party dialogues, the influences of utterances from other speakers should be considered in a targeted manner.

## 5.3    Effect of Depths of GNN and Window Sizes

We explore the relationship between model performance and the depth of ERMC-MHGNN on EmoryNLP datasets. From Fig. 3, the best $\{L^{SSHG}, L^{NSHG}\}$ is $\{4, 6\}$ on EmoryNLP datasets, which obtain 40.1% Weighted-average F1. Note that the convolution on EmoryNLP requires more NSHG layers. This may be related to the number of labels in the conversation and the length of the conversation. The proportion of each label in EmoryNLP is more balanced than MELD, the proportion of emotional shift is relatively larger, and the conversation length is also larger. Therefore, EmoryNLP needs more NSHG layers for convolution to deal with inter-dependency.

We also experimented with both datasets by increasing the window size of the past and future. The experimental results are shown in Fig. 4. From the figure, we can see that the window size of the context has a relatively small effect on the two datasets, but the context window size for obtaining relatively good results for different datasets is not the same. On MELD, there are a relatively certain number of conversations with less than three utterances, while on EmoryNLP,

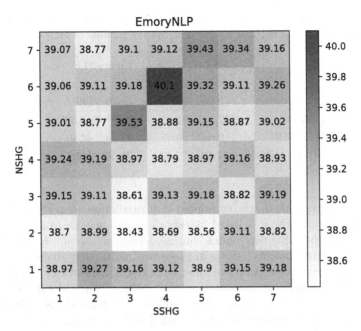

**Fig. 3.** Effect of depths of GNN. We report the Weighted-F1 score on the EmoryNLP. The darker the color, the better the performance.

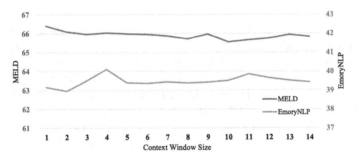

**Fig. 4.** Effect of window sizes.

the length of conversations is generally greater than five utterances. Therefore, MELD gets better results when the window size of the past and future is 1, while EmoryNLP requires a relatively large context window.

## 5.4   Error Analysis

We study the emotional shift issue, which means the emotions of two consecutive utterances from the same speaker are different. Since DialogXL does not provide the corresponding emotional shift prediction accuracy on MELD and EmoryNLP, we reproduce it. The weighted average f1 of DialogXL on MELD and EmoryNLP is 62.67% and 35.0%, respectively, both higher than the results

**Table 5.** Test accuracy of ERMC-MHGNN and partial baseline models on samples with emotional shift and without it. '()' indicates the number of samples.

| # | MELD | | EmoryNLP | |
|---|---|---|---|---|
| | shift (1003) | w/o shift (861) | shift (673) | w/o shift (361) |
| DialogXL | 57.33 | 71.43 | 33.88 | **43.77** |
| DAG-ERC | 59.02 | 69.45 | 37.29 | 42.10 |
| ERMC-MHGNN | **62.01** | **72.36** | **38.93** | 41.83 |

in the paper. Among them, the emotional shift prediction accuracy of DialogXL on MELD and EmoryNLP is listed in Table 5. It can be seen from Table 5 that compared with the other two models, our model has greatly improved the accuracy of identifying emotional shifts in these two multi-party dialogue datasets. However, improving the accuracy of identifying emotional shifts can easily reduce the accuracy of identifying without emotional shifts. Compared with other models, we can improve the accuracy of recognizing emotional shifts while keeping the accuracy of recognizing without emotional shifts at a high level.

## 6    Conclusion

This paper constructs two different hypergraphs for the two speaker dependencies, and designs a multi-hypergraph neural network for multi-party conversation emotion recognition, namely ERMC-MHGNN, to better handle speaker dependencies. The experimental results show that ERMC-MHGNN has good performance. Furthermore, through comprehensive evaluation and ablation studies, we confirm the advantages of ERMC-MHGNN and the impact of its modules on performance. Several conclusions can be drawn from the experimental results. First, our approach to non-speaker-specific modeling of utterances from other speakers is feasible. Second, combining average aggregation with attention aggregation can obtain better hyperedge features. Finally, our model has achieved relatively good results on emotional shift issue.

## References

1. Chatterjee, A., Narahari, K.N., Joshi, M., Agrawal, P.: SemEval-2019 task 3: Emo-Context contextual emotion detection in text. In: Proceedings of SemEval, pp. 39–48. Association for Computational Linguistics, Minneapolis, Minnesota, USA (Jun 2019). https://doi.org/10.18653/v1/S19-2005
2. Ding, K., Wang, J., Li, J., Li, D., Liu, H.: Be more with less: Hypergraph attention networks for inductive text classification. In: Proceedings of EMNLP, pp. 4927–4936. Association for Computational Linguistics, Online (Nov 2020). https://doi.org/10.18653/v1/2020.emnlp-main.399

3. Feng, Y., You, H., Zhang, Z., Ji, R., Gao, Y.: Hypergraph neural networks. In: Proceedings of AAAI, pp. 3558–3565. AAAI Press (2019). https://doi.org/10.1609/aaai.v33i01.33013558

4. Ghosal, D., Majumder, N., Gelbukh, A., Mihalcea, R., Poria, S.: COSMIC: COmmonSense knowledge for eMotion identification in conversations. In: Proc. of EMNLP Findings, pp. 2470–2481. Association for Computational Linguistics, Online (Nov 2020). https://doi.org/10.18653/v1/2020.findings-emnlp.224

5. Ghosal, D., Majumder, N., Poria, S., Chhaya, N., Gelbukh, A.: DialogueGCN: A graph convolutional neural network for emotion recognition in conversation. In: Proc. of EMNLP, pp. 154–164. Association for Computational Linguistics, Hong Kong, China (Nov 2019). https://doi.org/10.18653/v1/D19-1015

6. Hu, D., Wei, L., Huai, X.: DialogueCRN: Contextual reasoning networks for emotion recognition in conversations. In: Proceedings of ACL, pp. 7042–7052. Association for Computational Linguistics, Online (Aug 2021). https://doi.org/10.18653/v1/2021.acl-long.547

7. Hu, J., Liu, Y., Zhao, J., Jin, Q.: MMGCN: Multimodal fusion via deep graph convolution network for emotion recognition in conversation. In: Proceedings of ACL, pp. 5666–5675. Association for Computational Linguistics, Online (Aug 2021). https://doi.org/10.18653/v1/2021.acl-long.440

8. Ishiwatari, T., Yasuda, Y., Miyazaki, T., Goto, J.: Relation-aware graph attention networks with relational position encodings for emotion recognition in conversations. In: Proceedings of EMNLP, pp. 7360–7370. Association for Computational Linguistics, Online (Nov 2020). https://doi.org/10.18653/v1/2020.emnlp-main.597

9. Lee, B., Choi, Y.S.: Graph based network with contextualized representations of turns in dialogue. In: Proceedins of EMNLP, pp. 443–455. Association for Computational Linguistics, Online and Punta Cana, Dominican Republic (Nov 2021). https://doi.org/10.18653/v1/2021.emnlp-main.36

10. Li, J., Lin, Z., Fu, P., Wang, W.: Past, present, and future: Conversational emotion recognition through structural modeling of psychological knowledge. In: Proceedings of EMNLP Findings, pp. 1204–1214. Association for Computational Linguistics, Punta Cana, Dominican Republic (Nov 2021). https://doi.org/10.18653/v1/2021.findings-emnlp.104

11. Li, J., Zhang, M., Ji, D., Liu, Y.: Multi-task learning with auxiliary speaker identification for conversational emotion recognition. CoRR abs/2003.01478 (2020)

12. Li, S., Yan, H., Qiu, X.: Contrast and generation make BART a good dialogue emotion recognizer. In: Proceedings of AAAI, pp. 11002–11010. AAAI Press (2022)

13. Li, Z., Tang, F., Zhao, M., Zhu, Y.: Emocaps: Emotion capsule based model for conversational emotion recognition. In: Muresan, S., Nakov, P., Villavicencio, A. (eds.) Proc. of ACL Findings. pp. 1610–1618. Association for Computational Linguistics (2022). 10.18653/v1/2022.findings-acl.126

14. Liu, Y.,et al.: Roberta: A robustly optimized BERT pretraining approach. CoRR abs/1907.11692 (2019)

15. Majumder, N., Poria, S., Hazarika, D., Mihalcea, R., Gelbukh, A., Cambria, E.: Dialoguernn: An attentive rnn for emotion detection in conversations. In: Proceedings of AAAI. AAAI'19/IAAI'19/EAAI'19, AAAI Press (2019). https://doi.org/10.1609/aaai.v33i01.33016818

16. Ong, D., et al.: Is discourse role important for emotion recognition in conversation? In: Proceedings of AAAI, pp. 11121–11129. AAAI Press (2022)

17. Poria, S., Hazarika, D., Majumder, N., Naik, G., Cambria, E., Mihalcea, R.: MELD: A multimodal multi-party dataset for emotion recognition in conversations. In: Proceedings of ACL, pp. 527–536. Association for Computational Linguistics, Florence, Italy (Jul 2019). https://doi.org/10.18653/v1/P19-1050

18. Poria, S., Majumder, N., Mihalcea, R., Hovy, E.: Emotion recognition in conversation: Research challenges, datasets, and recent advances. IEEE Access **7**, 100943–100953 (2019). https://doi.org/10.1109/ACCESS.2019.2929050

19. Shen, W., Chen, J., Quan, X., Xie, Z.: Dialogxl: All-in-one xlnet for multi-party conversation emotion recognition. In: Proceedings of AAAI (2021)

20. Shen, W., Wu, S., Yang, Y., Quan, X.: Directed acyclic graph network for conversational emotion recognition. In: Proceedings of ACL, pp. 1551–1560. Association for Computational Linguistics (2021). https://doi.org/10.18653/v1/2021.acl-long.123

21. Sun, Y., Yu, N., Fu, G.: A discourse-aware graph neural network for emotion recognition in multi-party conversation. In: Proceedings off EMNLP Findings,D pp. 2949–2958. Association for Computational Linguistics, Punta Cana, Dominican Republic (Nov 2021). https://doi.org/10.18653/v1/2021.findings-emnlp.252

22. Vaswani, A., et al.: Attention is all you need. In: Proceedings of NeurIPS (2017)

23. Wang, J., Ding, K., Zhu, Z., Caverlee, J.: Session-based recommendation with hypergraph attention networks. In: Demeniconi, C., Davidson, I. (eds.) Proceedings of SDM. pp. 82–90. SIAM (2021). https://doi.org/10.1137/1.9781611976700.10

24. Yang, Z., Dai, Z., Yang, Y., Carbonell, J., Salakhutdinov, R., Le, Q.V.: XLNet: Generalized Autoregressive Pretraining for Language Understanding. Curran Associates Inc., Red Hook, NY, USA (2019)

25. Yi, J., Park, J.: Hypergraph convolutional recurrent neural network. In: Gupta, R., Liu, Y., Tang, J., Prakash, B.A. (eds.) Proceedings of KDD, pp. 3366–3376. ACM (2020). https://doi.org/10.1145/3394486.3403389

26. Zahiri, S.M., Choi, J.D.: Emotion detection on TV show transcripts with sequence-based convolutional neural networks. In: Proceedings of AAAI. AAAI Technical Report, vol. WS-18, pp. 44–52. AAAI Press (2018)

27. Zhu, J., Zhao, X., Hu, H., Gao, Y.: Emotion recognition from physiological signals using multi-hypergraph neural networks. In: Proceedings of ICME (2019). https://doi.org/10.1109/ICME.2019.00111

28. Zhu, L., Pergola, G., Gui, L., Zhou, D., He, Y.: Topic-driven and knowledge-aware transformer for dialogue emotion detection. In: Proceedings of ACL, pp. 1571–1582. Association for Computational Linguistics, Online (Aug 2021). https://doi.org/10.18653/v1/2021.acl-long.125

# Using Emoji as an Emotion Modality in Text-Based Depression Detection

Pingyue Zhang, Mengyue Wu[✉], and Kai Yu[✉]

MoE Key Lab of Artificial Intelligence, X-LANCE Lab, Department of Computer
Science and Engineering, Shanghai Jiao Tong University, Shanghai, China
{williamzhangsjtu,mengyuewu,kai.yu}@sjtu.edu.cn

**Abstract.** Text-based depression detection has long been investigated
by exploring useful handcrafted linguistic features and word embeddings.
This paper focuses on utilizing emoji as an emotional modality to detect
whether a subject is depressed or not based on text. In particular, we
propose to extract sentence-level emotional information with model pre-
trained to predict emoji of text on social media and semantic information with widely used embedding model. The embeddings are then input
to the classification model to predict one's mental state. Experiments
are conducted on user-generated posts from three datasets and clinical
conversational data from DAIC-WOZ. Results on social media data indicate emojis' superior performance in general, with further enhancement
derived from modality fusion. Furthermore, emoji outperforms contextual text embeddings in sparse scenarios like clinical interview dialogues.
We also provide a detailed analysis showing that the emojis extracted
from healthy and depressed subjects are significantly different, suggest-
ing that emoji can be a reliable emotion representation in such implicit
yet complex sentiment analysis settings.

**Keywords:** Depression detection · Emotion detection · Deep learning

## 1 Introduction

Depression is an illness that affects, knowingly or unknowingly, millions of people
worldwide. Efficient and effective automatic depression diagnosis can be of sub-
stantial benefit. However, this is an arduous task since a variety of complicated
symptoms are reported while publicly available data is limited. Primarily, two
kinds of datasets are broadly used in text-based depression detection: 1) user-
generated data collected from social media and written texts; 2) dialogue recording transcriptions during clinical interviews with professionally rated labels from
doctors. The second data type is more scarce, but the labels that whether one is
depressed are more accurate, since social media determines one's mental state on
a self-report basis. Both data sources are imbalanced and contain more healthy
subjects. Accordingly, a major challenge for text-based depression detection is
to extract meaningful features that can be used to distinguish depressed patients
from healthy participants.

ⓒ The Author(s), under exclusive license to Springer Nature Singapore Pte Ltd. 2023
L. Zhenhua et al. (Eds.): NCMMSC 2022, CCIS 1765, pp. 59–67, 2023.
https://doi.org/10.1007/978-981-99-2401-1_5

In particular, different text features have been investigated, ranging from hand-crafted feature types such as n-grams, Bag of Words (BoW), Linguistic Inquiry and Word Count (LIWC) [17] to neural word embeddings like Word2Vec [13], fastText [3,11], as well as global vectors (GloVe) [14]. In addition, linguistic metadata including word and grammar use, readability, and keywords summary has exhibited great performance in depression detection [17].

Psychology studies suggest that depressed mood can directly influence an individual's emotional expression and perception [10]. Emotion and sentiment analysis is hence a useful tool in detecting depression. [2] proposes Bag of Sub-Emotions (BoSE) and suggests that depression detection benefits from emotional information. In [4], they use eight basic emotions (anger, fear, happiness, etc.) as features to identify the risk of depression on Twitter.

Recently a few studies predict emoji representations from text [7,9,19]. Emoji, a rich emotional representation, is thus found effective in multiple tasks, including sentiment analysis, emotion detection, and sarcasm detection [7]. Emoji exhibits certain characteristics compared with traditional sentiment/emotion analysis where a fixed label (i.e. positive/negative, happy, angry, etc.): has more categories than the traditional binary or five emotion categories. In this way, the emotional projection space is more continuous rather than discrete quadrants. Such a mapping mechanism is more natural to human emotion expressions that we do not always have a clear definition to our emotional state.

Therefore, we propose to use emoji representations as an emotional modality in depression detection, which can be utilized on its own or in combination with the existing textual features.

## 2    Emoji Extraction and Depression Detection

This work proposes to generate emoji from text sequences as an additional emotion modality, which is used to determine whether a subject is depressed or not. We firstly introduce our method of obtaining Emoji representations from null-emoji depression data and then present our model for text-based depression detection.

### 2.1    Emotion and Semantic Features

Emoji can be seen as a tool for emotion expression, which makes it more suitable for text classification tasks involving emotions. Therefore, depression detection, which highly correlates with emotion, can surely benefit from it. By its nature, training text and emojis is entitled to large publicly available data since emojis can be regarded as noisy labels for text as they are often used in tandem in one sentence. Pretrained models like DeepMoji [7] take advantage of 1.6 billion Tweets that incorporate emojis in each post. As a result, 64 different types of emojis are embedded as the classification output. By utilizing DeepMoji, we obtained highly emotion-correlated emoji representations from the text.

In addition, we also utilize task-agnostic pretraining textual embeddings to represent semantic features, since the use of pretrained features can be seen as one possible approach to ameliorate the data scarcity problem. BERT [6] is used to extract high-level context-sensitive features as a semantic modality.

A fusion of BERT and Emoji of combining the semantic and emotion modalities are also investigated.

## 2.2 Depression Detection Model

Our approach for depression detection models each subjects' text as a sequence of sentences [1], seen in Fig. 1.

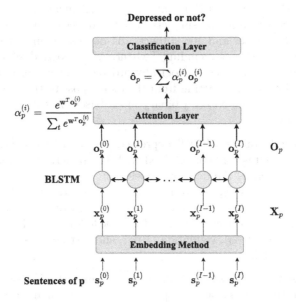

**Fig. 1.** Detection framework. Input feature can be sentence-level text features, emoji features, or the combination of both.

Modeling sequential data is commonly done using recurrent neural networks (RNN). Specifically, we use a bidirectional long short-term memory (BLSTM) network. The network receives a sequence of vectorized features $\mathbf{X}_p = [\mathbf{x}_p^{(1)}, \mathbf{x}_p^{(2)}, \ldots, \mathbf{x}_p^{(I)}]$ obtained from text of patient $p$ with length $I$. The sequence $\mathbf{X}_p$ is then fed into a BLSTM network, producing $\mathbf{O}_p = \text{BLSTM}(\mathbf{X}_p) = [\mathbf{o}_p^{(1)}, \ldots, \mathbf{o}_p^{(I)}]$. Since depression state labels are only given once for each dialog, a time pooling layer is required to remove all time variability. Here, attention is used as the preferred time pooling method. For each time step $i$, we calculate weight based on the output $\mathbf{o}_p^{(i)}$ of the BLSTM:

$$\alpha_p^{(i)} = \frac{e^{\mathbf{w}^T \mathbf{o}_p^{(i)}}}{\sum_t e^{\mathbf{w}^T \mathbf{o}_p^{(t)}}}, \tag{1}$$

where $\mathbf{w}$ is a learnable attention parameter. The attention output $\hat{\mathbf{o}}_p = \sum_i \alpha_p^{(i)} \mathbf{o}_p^{(i)}$ is then fed into a final classification layer which predicts the mental state.

## 3    Experiments

*Datasets* We experiment with two dataset types for text-based depression detection. One concerns user-generated social media posts (the eRisk [12] and Topic-restricted [18]), and the other collects clinical conversation transcriptions (the DAIC-WOZ dataset [5]).

**eRisk.** The eRisk Challenge 2017 Dataset (eRisk) [12] includes 531,453 posts and comments (sentences) from 892 subjects (137 Depressed), with 486 subjects (83 depressed) in training set and 406 subjects (54 depressed) in test set, collected from social media platforms including Twitter, ATL and Reddit. For each subject, the dataset contains a chronologically ordered sequence of writings, which is divided into ten chunks, starting from the oldest post to the newest one. The depression state is evaluated on a binary level, based on the subjects' website visiting frequency.

**Topic-Restricted.** consists of 4,947 depressed users and 7,159 control users [18]. Since the original dataset is not released with the paper, we follow the implementation of [8] to crawl the dataset ourselves in the same way, which yields a dataset containing 5,564 depressed users and 7,556 control ones. The depressed users are those who started a thread in the depression subreddit while the control users are those who started a thread in AskReddit subreddit. The dataset is split into the train (80%), validation (10%), and test (10%) set with the label distribution preserved.

**DAIC.** The DAIC-WOZ (DAIC) [5] dataset contains multiple modalities including visual (face-landmarks), audio (speech) and text (human transcribed) collected from clinical conversations from a total of 142 subjects (42 depressed). The training set contains 16895 instances from 107 subjects (30 depressed) while the development set 6674 written sentences from 35 subjects (12 depressed). And test set is only available for people attended the challenge. Though the DAIC dataset contains less data than eRisk, its quality of labels, as well as transcriptions, is higher.

*Feature Extraction.* Two text-based features, pretrained on large datasets, are investigated in this work. Both features are extracted on the sentence level. DeepMoji [7], is used as our feature extractor to extract 2304-dimension **Emoji** features. Regarding text embeddings, we use the publicly available model (*bert-base-uncased*), obtaining a 768-dimension **BERT** feature [6]. **Fusion** is done on the feature level, whereas the previous two embeddings are concatenated, obtaining a single 3,072 dimensional feature.

*Experimental Setup.* Our proposed approach can be viewed as a sequential modeling approach, which processes each patients' response in succession. For eRisk and topic-restricted, contents from both title and text domain of a post are concatenated. The provided text corpus of eRisk was firstly preprocessed by removing URLs, tags, punctuation, and digit number and lowercasing all the words. For DAIC, we only consider each patient's response.

*Training Details.* For eRisk dataset, the BLSTM model consists of 3 layers, with a hidden size of 256 for a single Emoji and BERT feature and 512 for their fusion ones, and a dropout rate of 0.1. It is trained using cross-entropy as the default criterion, running with a batch size of 4. The model is optimized using Adam, with an initial learning rate of $4 \times 10^{-5}$.

For TopicRestricted dataset, the BLSTM model has 2 layers, with a hidden size of 256 and a dropout rate of 0.1 for all three kinds of features. It is also trained using cross-entropy as the default criterion, running with a batch size of 32. The model is optimized using Adam, with an initial learning rate of $1 \times 10^{-4}$.

For eRisk dataset, we run the training processes for each feature multiple times with the same settings except for the seed and pick 5 results with the highest $F_1$ scores on the test set. Then we report the mean score with a standard deviation along with the maximum one we have achieved. And for TopicRestricted dataset, we run the training processes for 10 times with different seeds and report the mean score with a standard deviation.

*Evaluation Metrics.* Automatic depression detection can be assessed via classification performance using $F_1$ score. If not specified, $F_1$ stands for average over the positive case (binary).

## 4   Results

Our results are two-fold: we first compare the results of using emojis and BERT on two dataset types: social media and dialogue transcription text. For the social media dataset eRisk, our results are compared with the best systems in the challenge and the state-of-the-art by far. Regarding the dialogue text DAIC, our baselines include the work that has individually tested textual modality.

### 4.1   Depression Detection on Social Media Text

Experiment results on eRisk can be seen in Table 1, with reference to previously published best results. BERT features perform more robust than Emoji embeddings with a smaller std value and a slightly higher mean $F_1$ score, while the best result of Emoji outperforms BERT. By further fusing semantic and emoji embeddings, additional performance gains can be observed. We achieve the most robust results in terms of classification (mean $F_1 = 0.65$, max $F_1 = 0.66$) by using the Fusion feature. The results illustrate that Emoji model, which is pretrained by a simple emoji classification task, can perform similarly with BERT

**Table 1.** Comparison of $F_1$ results on eRisk. 1st-3rd Places indicate the results from eRisk Challenge. Best results highlighted in bold.

| Method | Feature | $F_1$ | Max $F_1$ |
|---|---|---|---|
| [2] | Bag of Sub-Emotions | 0.64 | – |
| [17] (1st Place) | 5 Feature Combinations | 0.64 | – |
| [15] (2nd Place) | DepEmbed + DepWords + Metamap | 0.51 | – |
| [15] (3rd Place) | DepWords + Metamap | 0.44 | – |
| Ours | Emoji | 0.63±0.013 | 0.65 |
| | BERT | 0.64±0.003 | 0.64 |
| | Fusion | **0.65±0.009** | **0.66** |

**Table 2.** Comparison on $F_1$ scores on Topic Restricted and DAIC dataset

(a) Comparison on $F_1$ score on Topic-Restricted. Best results highlighted in bold.

| Method | Feature | $F_1$ |
|---|---|---|
| | Emoji | 0.764 ± 0.009 |
| Ours | BERT | 0.744 ± 0.009 |
| | Fusion | **0.781 ± 0.011** |

(b) Comparison on $F_1$ score on DAIC. Best results highlighted in bold.

| Method | Feature | $F_1$ macro |
|---|---|---|
| [1] | Word2Vec | 0.67 |
| [1] | Doc2Vec | 0.67 |
| | Emoji | **0.76** |
| Ours | BERT | 0.69 |
| | Fusion | 0.69 |

on eRisk task, demonstrating that emotional text information is beneficial to the depression detection task.

The results on the Topic-Restricted dataset in Table 2a show Emoji outperforms BERT and Fusion performs the best.

## 4.2   Depression Detection on Dialogue Text

Table 2b presents our results on the smaller dataset of clinical interviews. It can be seen that the use of DeepMoji outperforms other text embeddings, including Word2Vec, and BERT. By contrast, BERT exhibited similar performance to DeepMoji on eRisk as it is collected from social media platforms while DAIC is more specifically purposed at mental state assessment, leading to a great benefit from the use of emotion modality. For the Topic-Restricted dataset, the depressed users are those who started a thread in the depression subreddit, which is more topic restricted and not so general like the eRisk dataset, hence Emoji also outperforms BERT on it.

**Table 3.** Top 10 most possible emojis for Healthy and Depressed subjects, predicted from our approach

| Mental state | eRisk | DAIC |
|---|---|---|
| Depressed | 💔 😔 😓 😒 😪 😓 😖 😁 😫 😓 | 😒 😇 😌 •• 😌 😁 😌 😔 😪 😓 |
| Healthy | 😳 •• 😁 😄 😄 🎤 🎂 😄 🍡 😎 | 😄 😳 🎵 🎶 😄 •• 😌 😳 😌 😁 |

## 5 Analysis

*Emoji Representations.* We analyzed the emoji representations extracted for depressed and healthy subjects, shown in Table 3. It can be seen that emojis representing *depressed* and *healthy* subjects varied greatly from each other, with quite a little overlap across the two groups. For depressed subjects, their emoji representations extracted via DeepMoji from their textual embeddings are shown more negative, e.g. heartbroken and crying emojis are more likely to occur. In contrast, healthy subjects' emoji representations are more positive-directed, i.e. music and smiling faces tend to appear.

Quantitative analysis, as shown in Table 4, is calculated via the polarity score allocated to each emoji, with an integer between $-4$ (most negative) and 4 (most positive)[1]. For the eRisk dataset, the sums of polarity score of the top 20 emojis for depressed and healthy subjects are $-19$ and 3; while for the DAIC dataset, the sums are $-2$ and 7 respectively.

**Table 4.** The Sum of 20 Emoji Representations' Polarity Score for Depressed and Healthy subjects

| Polarity Score | eRisk | DAIC |
|---|---|---|
| Depressed | −19 | −2 |
| Healthy | 3 | 7 |

*LIWC Analysis.* LIWC [17] has also been applied on eRisk for Pearson's correlation between linguistic use and one's mental state. Results indicated that the most positive correlated language use feature is the number of 1st person singular "I" ($\rho = 0.43$), personal pronouns (0.42), and Authenticity (0.42) while the most negative correlated linguistic uses are the analytical thinking ($\rho = -0.45$), clout ($-0.32$) and articles ($-0.28$). This is in line with previous research that depressed subjects are more inclined to express personal feelings hence their language usage concerns more first-person pronouns [16]. Though textual embeddings can help differentiate depressed and healthy subjects' expressions, their most distinguishable characteristics lie in sentiment/emotion divergence.

---

[1] https://github.com/words/emoji-emotion.

Therefore, emoji can hence be an additional input to the handcrafted metadata and further enhance depression detection performance.

To sum up, Emojis exhibit a helpful way in sentiment analysis due to their extensive categories. When applied to depression detection, it is powerful in capturing implicit and complex sentiments expressed by depressed subjects. Further, it not only enhances detection accuracy but more impressively, can deliver a practical way to visualize and explain the differences in depressed and healthy subjects' semantic variation.

## 6   Conclusion

This paper proposes to use emoji as an emotional modality in tandem with semantic information in depression detection. Pretrained features DeepMoji and BERT are incorporated on three text-based depression detection datasets. Results indicate that emojis can work as a robust modality on their own in detecting depression states from the text. In sparse data scenarios such as clinical interviews, emoji features outperform text embeddings by large. Future work can treat emojis as an individual modality for multi-modal fusion depression detection i.e., including semantic, emoji, visual and auditory cues.

**Acknowledgements.** This work has been supported by National Natural Science Foundation of China (No. 61901265, 92048205), Major Program of National Social Science Foundation of China (No. 18ZDA293), and Shanghai Municipal Science and Technology Major Project (2021SHZDZX0102). Experiments have been carried out on the PI supercomputer at Shanghai Jiao Tong University.

## References

1. Al Hanai, T., Ghassemi, M., Glass, J.: Detecting depression with audio/text sequence modeling of interviews. In: Proceedings of Interspeech 2018, pp. 1716–1720 (2018). 10.21437/Interspeech. 2018–2522

2. Aragón, M.E., López-Monroy, A.P., González-Gurrola, L.C., Montes, M.: Detecting depression in social media using fine-grained emotions. In: Proceedings of the 2019 Conference of the North American Chapter of the Association for Computational Linguistics: Human Language Technologies, vol. 1 (Long and Short Papers), pp. 1481–1486 (2019)

3. Bojanowski, P., Grave, E., Joulin, A., Mikolov, T.: Enriching word vectors with subword information. Trans. Assoc. Comput. Linguist. **5**, 135–146 (2017)

4. Chen, X., Sykora, M.D., Jackson, T.W., Elayan, S.: What about mood swings: Identifying depression on twitter with temporal measures of emotions. In: Companion Proceedings of the The Web Conference 2018, pp. 1653–1660 (2018)

5. DeVault, D., et al.: Simsensei kiosk: A virtual human interviewer for healthcare decision support. In: Proceedings of the 2014 International Conference on Autonomous Agents and Multi-agent Systems, AAMAS 2014, pp. 1061–1068. International Foundation for Autonomous Agents and Multiagent Systems, Richland, SC (2014). http://dl.acm.org/citation.cfm?id=2615731.2617415

6. Devlin, J., Chang, M.W., Lee, K., Toutanova, K.: Bert: Pre-training of deep bidirectional transformers for language understanding. arXiv preprint arXiv:1810.04805 (2018)
7. Felbo, B., Mislove, A., Søgaard, A., Rahwan, I., Lehmann, S.: Using millions of emoji occurrences to learn any-domain representations for detecting sentiment, emotion and sarcasm. arXiv preprint arXiv:1708.00524 (2017)
8. Harrigian, K., Aguirre, C., Dredze, M.: Do models of mental health based on social media data generalize? In: Proceedings of the 2020 Conference On Empirical Methods In Natural Language Processing: Findings, pp. 3774–3788 (2020)
9. Hayati, S.A., Muis, A.O.: Analyzing incorporation of emotion in emoji prediction. In: Proceedings of the Tenth Workshop on Computational Approaches to Subjectivity, Sentiment and Social Media Analysis, pp. 91–99 (2019)
10. J. Joormann, M.Q.: Cognitive processes and emotion regulation in depression. Depression Anxiety **31** (2014). https://doi.org/10.1002/da.22264
11. Joulin, A., Grave, E., Bojanowski, P., Mikolov, T.: Bag of tricks for efficient text classification. arXiv preprint arXiv:1607.01759 (2016)
12. Losada, D.E., Crestani, F., Parapar, J.: eRISK 2017: CLEF Lab on early risk prediction on the internet: experimental foundations. In: Jones, G.J.F., et al. (eds.) CLEF 2017. LNCS, vol. 10456, pp. 346–360. Springer, Cham (2017). https://doi.org/10.1007/978-3-319-65813-1_30
13. Mikolov, T., Chen, K., Corrado, G., Dean, J.: Efficient estimation of word representations in vector space. In: Bengio, Y., LeCun, Y. (eds.) 1st International Conference on Learning Representations, ICLR 2013, Scottsdale, Arizona, USA, 2–4 May 2013, Workshop Track Proceedings (2013). http://arxiv.org/abs/1301.3781
14. Pennington, J., Socher, R., Manning, C.: Glove: Global vectors for word representation. In: Proceedings of the 2014 conference on empirical methods in natural language processing (EMNLP), pp. 1532–1543 (2014)
15. Sadeque, F., Xu, D., Bethard, S.: Measuring the latency of depression detection in social media. In: Proceedings of the Eleventh ACM International Conference on Web Search and Data Mining, pp. 495–503 (2018)
16. Trotzek, M., Koitka, S., Friedrich, C.M.: Linguistic metadata augmented classifiers at the clef 2017 task for early detection of depression. In: CLEF (Working Notes) (2017)
17. Trotzek, M., Koitka, S., Friedrich, C.M.: Word embeddings and linguistic metadata at the CLEF 2018 tasks for early detection of depression and anorexia. In: CEUR Workshop Proceedings, vol. 2125 (2018). http://www.reddit.com/r/depression
18. Wolohan, J., Hiraga, M., Mukherjee, A., Sayyed, Z.A., Millard, M.: Detecting linguistic traces of depression in topic-restricted text: Attending to self-stigmatized depression with nlp. In: Proceedings of the First International Workshop on Language Cognition and Computational Models, pp. 11–21 (2018)
19. Zanzotto, F.M., Santilli, A.: Syntnn at semeval-2018 task 2: is syntax useful for emoji prediction? embedding syntactic trees in multi layer perceptrons. In: Proceedings of The 12th International Workshop on Semantic Evaluation, pp. 477–481 (2018)

# Source-Filter-Based Generative Adversarial Neural Vocoder for High Fidelity Speech Synthesis

Ye-Xin Lu, Yang Ai[✉], and Zhen-Hua Ling

National Engineering Research Center of Speech and Language Information
Processing, University of Science and Technology of China,
Hefei, People's Republic of China
yxlu0102@mail.ustc.edu.cn, {yangai,zhling}@ustc.edu.cn

**Abstract.** This paper proposes a source-filter-based generative adversarial neural vocoder named SF-GAN, which achieves high-fidelity waveform generation from input acoustic features by introducing F0-based source excitation signals to a neural filter framework. The SF-GAN vocoder is composed of a source module and a resolution-wise conditional filter module and is trained based on generative adversarial strategies. The source module produces an excitation signal from the F0 information, then the resolution-wise convolutional filter module combines the excitation signal with processed acoustic features at various temporal resolutions and finally reconstructs the raw waveform. The experimental results show that our proposed SF-GAN vocoder outperforms the state-of-the-art HiFi-GAN and Fre-GAN in both analysis-synthesis (AS) and text-to-speech (TTS) tasks, and the synthesized speech quality of SF-GAN is comparable to the ground-truth audio.

**Keywords:** Neural vocoder · Source-filter model · Generative adversarial networks · Speech synthesis

## 1 Introduction

Speech synthesis, a technology that converts text information into human-like speech, has been actively studied by researchers to chase the high naturalness, intelligibility, and expressiveness of synthesized speech. In recent years, statistical parametric speech synthesis (SPSS) [1] has become the dominant speech synthesis method due to its small system size, high robustness, and flexibility. SPSS generally consists of two stages: *text to acoustic feature* and *acoustic feature to speech*, which are implemented by acoustic models and vocoders, respectively.

For years, conventional vocoders such as STRAIGHT [2] and WORLD [3] have been widely used in the SPSS framework. Because of their source-filter architecture, these vocoders tend to have good controllability of acoustic components. While subject to the signal-processing mechanism, some deficiencies exist, such as the loss of spectral details and phase information, which lead to the degradation of speech quality.

Thanks to the developments of deep learning and neural networks, neural vocoders rapidly emerged and made great progress. Initially, autoregressive (AR) neural waveform generation models symbolized by WaveNet [4], SampleRNN [5], and WaveRNN [6] were proposed to build neural vocoders and achieved breakthroughs in synthesized speech quality compared to the conventional methods. However, due to the AR structure, they struggle in inference efficiency when synthesizing high temporal resolution waveforms. Subsequently, to address this problem, knowledge-distilling-based models (e.g., Parallel WaveNet [7], and ClariNet [8]), flow-based models (e.g., WaveGlow [9], and WaveFlow [10]) were proposed. Although the inference efficiency has made significant progress, the computational complexity is still considerable, which limits their applications in resource-constrained scenarios.

Recently, researchers have been prone to pay more attention to waveform generation models without AR or flow-like structures. The neural source-filter (NSF) model [11] combines speech production mechanisms with neural networks and realizes the prediction of speech waveform from explicit F0 and mel-spectrogram. Generative adversarial networks (GANs) based models [12–17] , simultaneously improved the synthesized speech quality with GANs and inference efficiency with parallelizable inference process. Among them, HiFi-GAN has achieved both high-fidelity and efficient speech synthesis. For the generative process, HiFi-GAN cascades multiple upsampling layers and multi-receptive field fusion (MRF) modules that consist of parallel residual blocks to gradually upsample the input mel-spectrogram to the temporal resolution of the final waveform while performing convolution operations. And for the discriminative process, HiFi-GAN adopts adversarial training with a multi-period discriminator (MPD) and a multi-scale discriminator (MSD) to ensure high-fidelity waveform generation. Based on it, Fre-GAN [18] further improved the speech quality in frequency space by adopting a resolution-connected generator and resolution-wise discriminators to capture various levels of spectral distributions over multiple frequency bands. However, due to their entirely data-driven manner, compared to source-filter-based vocoders, they tend to lack the controllability of speech components and robustness. Lately, there sprung some generative adversarial neural vocoders based on the source-filter model, such as quasi-periodic Parallel WaveGAN (QPPWG) [19], unified source-filter GAN (uSFGAN) [20], and harmonic plus noise uSFGAN (HN-uSFGAN) [21]. Although they successfully achieved high-controllability speech generation, there still exists a gap in speech quality between synthesized and ground-truth audios.

To achieve high-fidelity speech synthesis, we propose SF-GAN, which integrates the source excitation into a GAN-based neural waveform model. Firstly, we design a source module through which the voiced and unvoiced segments of the excitation signal are generated from the upsampled F0 and gaussian noise processed by a deep neural network (DNN), respectively. Secondly, we propose a resolution-wise conditional filter module to reconstruct speech waveform from the excitation signal and mel-spectrogram. Inspired by the layer-wise upsampling architecture in HiFi-GAN, we subsample the excitation signal to various resolu-

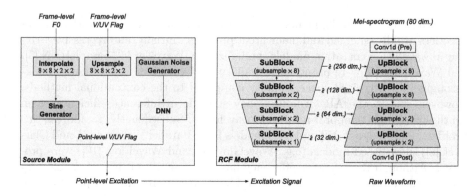

**Fig. 1.** The architecture of SF-GAN. In the source module, the V/UV flag is the abbreviation of the Voiced/Unvoiced flag. In the resolution-wise convolutional filter module, SubBlock denotes the subsampling residual block, while UpBlock denotes the concatenation of a transposed convolution and a parallel convolutional residual block.

tions and subsequently condition them to each upsampling layer of HiFi-GAN. And then, we redesign the residual blocks in the MRF module by using parallel convolutions to combine the transformed excitation signals with the hierarchical intermediate features processed from mel-spectrogram. Additionally, to further apply our model to a text-to-speech (TTS) system, we design an F0 predictor to predict F0 from mel-spectrogram generated by an acoustic model. We use both the predicted F0 and mel-spectrogram as our model inputs. Both objective and subjective evaluations demonstrate that our proposed SF-GAN vocoder outperforms HiFi-GAN and Fre-GAN in aspect of speech quality.

## 2    Proposed Method

### 2.1    Overview

In this paper, we propose the SF-GAN vocoder, which applies a source excitation access method to the HiFi-GAN architecture. As illustrated in Fig. 1, the proposed SF-GAN vocoder consists of a source module and a filter module, and it takes frame-level F0 and mel-spectrogram as inputs and outputs a speech waveform. The source module converts frame-level F0 into point-level excitation signal, while the filter module reconstructs raw waveform from this excitation along with the mel-spectrogram. Our proposed model adopts the GAN-based training strategy, and we use the MSD and MPD in HiFi-GAN to capture consecutive and periodic patterns.

### 2.2    Source Module

Given the frame-level F0 sequence $f_{1:L}$ as input, where $L$ denotes the number of frames, the source module firstly extracts the voiced/unvoiced (V/UV) flag sequence $v_{1:L}$ from it. Subsequently, $f_{1:L}$ is interpolated $N = T/L$ times to match the temporal resolution of the raw waveform, where $T$ denotes the number of waveform sample points and $v_{1:L}$ is also upsampled $N$ times by repeating V/UV flag values in each frame. The interpolated F0 sequence $f_{1:T}$ with V/UV flags sequence $v_{1:T}$ are converted to an excitation signal $e_{1:T}$, which is a sine-based signal for voiced segments and a DNN-transformed Gaussian white noise for unvoiced segments. By assuming the F0 value and V/UV flag value at the $t$-th time step are $f_t$ and $v_t$, the $t$-th value of excitation signal $e_t$ can be defined mathematically as

$$e_t = \begin{cases} \alpha \sin\left( \sum_{k=1}^{t} 2\pi \frac{f_k}{N_s} + \phi \right) + n_t, & v_t = 1 \\ g\left(\frac{1}{3\sigma}n_t\right), & v_t = 0 \end{cases}, \tag{1}$$

where $v_t = 0$ or 1 denotes that $e_t$ belongs to unvoiced or voiced segment, $\alpha$ and $\sigma$ are hyperparameters, $g(\cdot)$ represents a DNN-based transformation, $n_t \sim \mathcal{N}(0, \sigma^2)$ is a Gaussian noise, $\phi \in (-\pi, \pi]$ is a random initial phase, and $N_s$ denotes the waveform sampling rate. In the source module, only the DNN is trainable.

### 2.3   Resolution-Wise Conditional Filter Module

Given the mel-spectrogram and excitation signal as inputs, the filter module aims to reconstruct raw waveform from them. Our filter module is a combination of excitation subsampling blocks and the HiFi-GAN generator. The HiFi-GAN generator implements layer-wise upsampling on the frame-level mel-spectrogram to gradually match the temporal resolution of the speech waveform. In each upsampling layer, a transposed convolution is used for upsampling the intermediate feature, and an MRF module is used to observe patterns of various lengths in parallel. To condition the excitation signal to each layer of the HiFi-GAN generator, we propose the resolution-wise conditional filter (RCF) module. Specifically, inspired by the residual blocks (ResBlocks) in MRF, as shown in Fig. 2(a), we design the subsampling blocks (SubBlocks), which apply layer-wise subsampling operations on the excitation signal to match the resolutions of corresponding intermediate features. And then, to combine the subsampled excitation signals with the corresponding intermediate features, we design parallel convolutional residual blocks (PC-ResBlocks) in the MRF module and conclude the redesigned MRF module together with the transposed convolution as an UpBlock. As illustrated in Fig. 2(b), in each layer of the PC-ResBlock, a transformed excitation signal $\hat{e}$ and an intermediate feature $\hat{c}$ are converted to the next-level intermediate feature $\hat{c}'$. To achieve this, we propose a feature to feature mapping

$$f_{k,d}(\boldsymbol{x}, \boldsymbol{y}) = LReLU(\boldsymbol{W}_{k,d} * \boldsymbol{x} + \boldsymbol{W}_{k,d} * \boldsymbol{y}), \tag{2}$$

where $*$ denotes a convolution operator, $+$ denotes an element-wise addition operator, $LReLU$ is a leaky rectified linear unit (LReLU) [22] activation function,

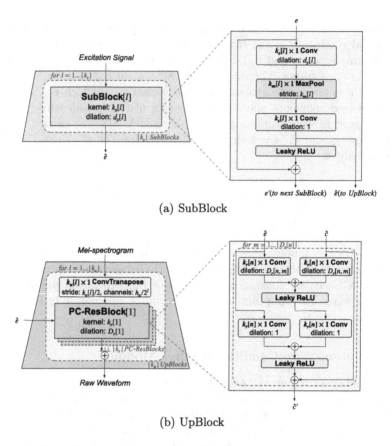

(a) SubBlock

(b) UpBlock

**Fig. 2.** (a) The SubBlocks subsample the excitation signal $|k_s|$ times to match the resolution of intermediate features in each UpBlock with max-pooling. The $l$-th SubBlock with kernel size $k_s[l]$ and dilation rates $d_s[l]$ is depicted. (b) The UpBlocks upsample the mel-spectrogram $|k_u|$ times with the transposed convolutions and combine the upsampled features with the corresponding subsampled excitation signals in the PC-ResBlocks. The $n$-th PC-ResBlock with kernel size $k_r[n]$ and dilation rates $D_r[n]$ in the $l$-th UpBlock is depicted.

and $W_{k,d}$ is a 1D-convolution with kernel size $k$ and dilation rate $d$. With this mapping function, the generative process of $\hat{c}'$ can be defined as

$$\hat{c}' = f_{k,1}(\hat{e}, f_{k,d}(\hat{e}, \hat{c})) + \hat{c}. \tag{3}$$

Our proposed RCF module has the following advantages: Firstly, by feeding the excitation signal to a filter module, speech waveforms can be reconstructed under the direction of explicit fundamental frequencies for better frequency modeling. Secondly, with the resolution-wise condition architecture, which combines the input excitation signal and mel-spectrogram at different resolutions, our model can capture fundamental frequency information and acoustic properties across multiple frequency bands. Thirdly, through the MRF module with PC-Resblocks, our model can observe various receptive field patterns of excitation signals and acoustic features in parallel. In a way, the receptive fields are extended.

## 3   Experiments

### 3.1   Experimental Setup

To compare our model with others on seen data, we conducted experiments on the LJSpeech dataset [23], which consists of 13,100 audio clips of a single English female speaker and is of about 24 h. The audio sampling rate is 22.05 kHz with a format of 16-bit PCM. We randomly divided the dataset into training, validation, and test sets in a ratio of 8:1:1. Our proposed model was compared against the official implementation of HiFi-GAN[1] and an open source implementation of Fre-GAN[2]. All the models were trained until 2.5M steps (about 3800 epochs).

To evaluate the generalization ability of our proposed model in a speaker-unseen scenario, we conducted multi-speaker experiments on the VCTK corpus [24]. The VCTK dataset consists of 44,257 audio clips from 109 native English speakers with various accounts, and the total length is about 44 h. The audio sampling rate is 44 kHz with a format of 16-bit PCM, and we subsampled all the audio clips to 22.05 kHz. At the training stage, we randomly excluded nine speakers from the training set. In the rest 100 speakers, we randomly selected nine audio clips from each speaker for validation and trained the model with all the rest audio clips. At the generation stage, we randomly selected 100 audio clips from each unseen speaker to generate samples. We also trained our proposed SF-GAN together with HiFi-GAN and Fre-GAN until 2.5M steps.

Most of the above experiments were conducted based on two generator variations (i.e.,$V1$ and $V2$), while for the Fre-GAN, we just used its $V1$ version, which was adequate to reveal the performance differences between models. In the source module, we set $\alpha = 0.1$ and $\sigma = 0.003$. As the RCF module shown in Fig. 2, for $V1$, we set $h_u = 512$, $k_u = [16, 16, 4, 4]$, $k_r = [3, 7, 11]$, and $D_r = [[1, 3, 5] \times 3]$ for the UpBlocks while $k_m = [1, 2, 2, 8]$, $k_s = [15, 11, 7, 3]$, and $d_s = [7, 5, 3, 1]$ for the SubBlocks. The $V2$ version reduces the hidden dimensions $h_u$ to 128 but with the same receptive fields as the $V1$ version. We used F0 and 80-dimensional mel-spectrogram as input conditions, and both their hanning window size and hop size were set to 1024 and 256, respectively. Additionally, for the mel-spectrogram, the FFT point number was set to 1024. At the training stage, we adopted the AdamW optimizer [25], and all the hyperparameters were set in accordance with HiFi-GAN. The synthesized audio samples are available at the demo website[3].

---

[1] https://github.com/jik876/hifi-gan.

[2] https://github.com/rishikksh20/Fre-GAN-pytorch.

[3] https://yxlu-0102.github.io/SF-GAN-Demo.

**Table 1.** Objective evaluation results for analysis-synthesis experiments.

| Dataset | Model | SNR (dB)↑ | LAS-RMSE (dB)↓ | MCD (dB)↓ | F0-RMSE (cent)↓ | V/UV error (%)↓ |
|---|---|---|---|---|---|---|
| LJSpeech | HiFi-GAN $V1$ | 4.3186 | 6.3555 | 1.5340 | 42.3113 | 5.0804 |
| | Fre-GAN $V1$ | 4.4277 | 6.2462 | 1.5478 | 40.3120 | 4.9076 |
| | SF-GAN $V1$ | **4.6161** | **6.1812** | **1.4229** | **33.7797** | **4.8340** |
| | HiFi-GAN $V2$ | 3.6501 | 6.9994 | 1.9805 | 48.3591 | 5.8613 |
| | SF-GAN $V2$ | **3.7774** | **6.6931** | **1.7629** | **44.9624** | **5.7700** |
| VCTK | HiFi-GAN $V1$ | 2.1841 | 6.8098 | 2.3486 | 45.8586 | 8.3386 |
| | Fre-GAN $V1$ | 2.3693 | 6.5795 | 2.2559 | 38.7379 | 8.0487 |
| | SF-GAN $V1$ | **2.5788** | **6.4566** | **2.1518** | **34.8631** | **7.3406** |
| | HiFi-GAN $V2$ | 1.7438 | 7.3316 | 2.7787 | 48.0714 | 9.2443 |
| | SF-GAN $V2$ | **1.9405** | **7.2284** | **2.5440** | **43.5458** | **8.1595** |

### 3.2   Comparison Among Neural Vocoders

**Evaluations on Analysis-Synthesis Tasks.** We implemented both objective and subjective evaluations on the LJSpeech and VCTK dataset to evaluate our proposed SF-GAN vocoder with HiFi-GAN and Fre-GAN in terms of synthesized speech quality. For objective evaluation, we adopted five metrics from our previous work [26], including signal-to-noise ratio (SNR), root MSE (RMSE) of log amplitude spectra (LAS-RMSE), mel-cepstrum distortion (MCD), MSE of F0 (F0-RMSE), and V/UV error. For subjective evaluation, we conducted the ABX preference tests on the Amazon Mechanical Turk platform[4] to compare the differences between two comparative systems. In each ABX test, 20 utterances synthesized by two comparative systems were randomly selected from the test set and evaluated by at least 30 native English listeners. The listeners were asked to judge which utterance in each pair had better speech quality or whether there was no preference. In order to calculate the average preference scores, the $p$-value of a $t$-test was used to measure the significance of the difference between two systems.

For experiments on the LJSpeech dataset, the objective results are presented in the top half of Table 1. Obviously, our proposed SF-GAN vocoder outperformed HiFi-GAN and Fre-GAN among all the objective metrics, demonstrating the distinct advantages of the proposed model in synthesized speech quality. The subjective ABX test results are illustrated in the top part of Fig. 3, which shows that the performance of our proposed SF-GAN $V1$ was comparable with the ground-truth natural speech ($p = 0.25$), and significantly better than HiFi-GAN $V1$ ($p < 0.01$) and Fre-GAN $V1$ ($p < 0.05$). In addition, the performance of SF-GAN $V2$ was slightly better than HiFi-GAN $V2$ ($p$ is slightly higher than 0.05). These above experimental results verified the advantage of our model in improving synthesized speech quality.

---

[4] https://www.mturk.com.

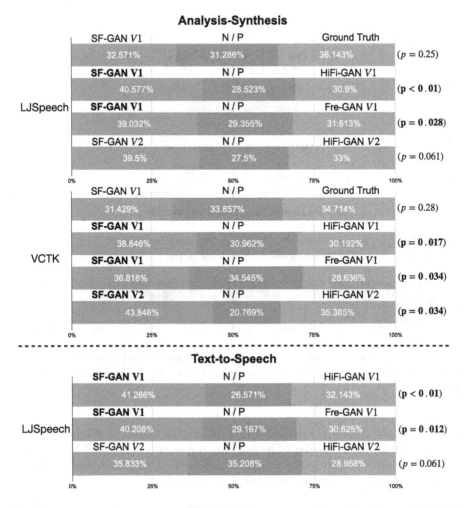

**Fig. 3.** Average preference scores (%) of ABX tests on speech quality between SF-GAN and other three systems (i.e., Ground Truth, HiFi-GAN, and Fre-GAN), where N/P stands for "no preference" and $p$ denotes the $p$-value of a $t$-test between two systems.

For experiments on the VCTK datasets, we used utterances of nine unseen speakers excluded from the training set for the objective evaluation, and randomly selected 20 utterances from them for the ABX test as above. The objective results are presented in the bottom half of Table 1, and among all the metrics, our proposed SF-GAN surpassed HiFi-GAN and Fre-GAN. The subjective ABX test results are illustrated in the middle part of Fig. 3, which also shows that the performance of our proposed SF-GAN was better than that of HiFi-GAN in both

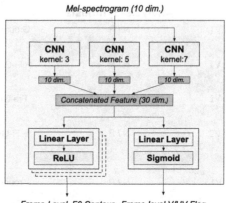

**Fig. 4.** The architecture of the F0 predictor. We use predicted F0 and synthesized mel-spectrogram as the inputs of the F0 predictor in the TTS task.

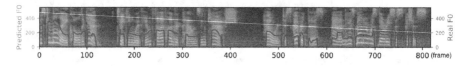

**Fig. 5.** An example of comparison between a predicted F0 and a real F0. The red curve denotes the predicted F0 and the purple curve denotes the real F0. (Color figure online)

$V1$ and $V2$ ($p < 0.05$) and Fre-GAN in $V1$ ($p < 0.05$). Moreover, it is noteworthy that even with unseen speakers, the performance of SF-GAN was still comparable with ground-truth audios ($p = 0.28$), which verified the generalization ability of our proposed model in the speaker-unseen scenario.

**Evaluations on TTS Tasks.** Since our proposed SF-GAN needs an additional F0 input, which is distinct from other mel-spectrogram vocoders. To apply our proposed model to the TTS task, we proposed an F0 predictor to predict F0 from the first ten dimensions of the mel-spectrogram, which contains sufficient F0 information. The architecture of the F0 predictor is illustrated in Fig. 4. We first input the first 10-dimensional mel-spectrogram into three convolutional neural networks (CNNs) with different kernel sizes and get three 10-dimensional intermediate features. We then input the concatenated 30-dimensional feature to two linear layers with different activation functions (i.e., ReLU and Sigmoid) to get the F0 contour and V/UV flag, respectively. We used the LJSpeech dataset to train and test the F0 predictor with the same data division as in the analysis-synthesis experiment, and we evaluated the synthesized F0 contours and V/UV flags from real mel-spectrograms with two objective metrics, including F0-RMSE and V/UV error. The F0-RMSE and V/UV errors on the test set are 115.7072

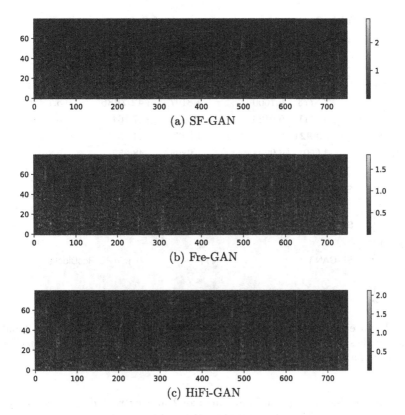

(a) SF-GAN

(b) Fre-GAN

(c) HiFi-GAN

**Fig. 6.** Pixel-wise difference between a mel-spectrogram from Tacotron2 and mel-spectrograms from waveforms generated by SF-GAN, Fre-GAN, and HiFi-GAN.

cents and 4.3901%, respectively. A close examination found that most of the errors came from the inaccuracy of F0 extraction. An example is shown in Fig. 5, where we can see the F0 extraction errors occurred between the 560-th frame and the 700-th frame.

We herein predicted the mel-spectrograms from texts by adopting the most popular implementation of Tacotron2[5] [27] with the provided pre-trained weights on the LJSpeech dataset. We fed the mel-spectrogram predicted by Tacotron2 as the input condition to HiFi-GAN and Fre-GAN. Then we predicted the F0 contour and V/UV flag from the predicted mel-spectrogram and fed all of them as input conditions to our proposed SF-GAN. We also conducted ABX preference tests between them and the results are illustrated in the bottom part of Fig. 3, which demonstrates our proposed SF-GAN $V1$ significantly outperformed HiFi-GAN $V1$ ($p < 0.01$) and Fre-GAN $V1$ ($p < 0.05$), while SF-GAN $V2$ was slightly better than HiFi-GAN $V2$ ($p$ is slightly higher than 0.05). In addition, when we investigated the pixel-wise mel-spectrogram difference between

---

[5] https://github.com/NVIDIA/tacotron2.

**Table 2.** Objective evaluation results for ablation studies.

| Model | SNR (dB)↑ | LAS-RMSE (dB)↓ | MCD (dB)↓ | F0-RMSE (cent)↓ | V/UV error (%)↓ |
|-------|-----------|----------------|-----------|-----------------|------------------|
| SF-GAN $V2$ | 3.7774 | **6.6931** | **1.7629** | 44.9624 | **5.6931** |
| w/o DNN | 3.8773 | 6.7600 | 1.8197 | **42.7686** | 5.7700 |
| w/o SubBlock | 3.7111 | 6.9126 | 1.9078 | 47.1645 | 6.0300 |
| w/o PC-ResBlock | **3.8921** | 6.8701 | 1.8472 | 44.7955 | 5.7856 |
| HiFi-GAN $V2$ | 3.6501 | 6.9994 | 1.9805 | 48.3591 | 5.8613 |

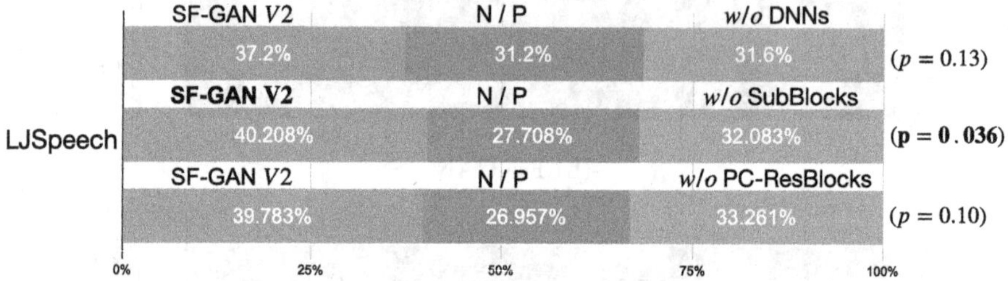

**Fig. 7.** The average preference scores (%) of ABX tests between SF-GAN $V2$ and its ablation models in speech quality, where N/P stands for "no preference" and $p$ denotes the $p$-value of a $t$-test between two systems.

a generated mel-spectrogram from Tacotron2 and extracted mel-spectrograms from waveforms generated by SF-GAN, HiFiGAN, and Fre-GAN, as presented in Fig. 6, it's apparently that the pixel-wise difference of our model is less than those of HiFi-GAN and Fre-GAN. To sum up, we conclude that our SF-GAN can be well applied to the TTS task.

### 3.3   Ablation Studies

We implemented ablation studies on the DNN, SubBlock, and PC-ResBlock used in the SF-GAN vocoder to verify the effectiveness of each component in terms of synthesized speech quality in the analysis-synthesis task, where ablating DNN denotes removing the DNN in the source module, ablating SubBlock denotes directly subsampling the excitation signal to four resolutions instead of using SubBlocks, and ablating PC-ResBlock denotes simply adding the subsampled excitation signals with the upsampled intermediate features followed by the ResBlocks instead of combining them in the PC-ResBlocks. The $V2$ version with fewer hidden dimensions was used as the generator for the ablation studies, and all the ablation models were trained until 2.5M steps.

As objective results presented in Table 2, ablating SubBlock demonstrates distinct degradation among all the metrics while ablating DNN and PC-ResBlock show slight degradation in partial metrics. To further verify the effect of these components subjectively, we also conducted ABX tests. The subjective results are illustrated in Fig. 7, which demonstrates that all the three components contribute to the improvement. In accordance with the objective results, SubBlock remarkably contributes to the synthesized speech quality ($p < 0.05$), while ablating DNN ($p = 0.13$) or PC-ResBlock ($p = 0.10$) cause slight but insignificant quality degradations.

# 4  Conclusion

In this work, we proposed the SF-GAN vocoder, which can synthesize high-quality speech from input F0 and mel-spectrogram. We took the inspiration from the layer-wise upsampling architecture of HiFi-GAN, applied a resolution-wise source-filter access method to the HiFi-GAN framework where excitation signal generated by a source module and mel-spectrogram are combined at various resolutions, which significantly contributes to the quality of synthesized speech audios according to the ablation studies. Above all, the proposed SF-GAN vocoder outperforms the state-of-the-art HiFi-GAN and Fre-GAN in both analysis-synthesis and TTS tasks of speech synthesis, and the speech quality of synthesized audios by SF-GAN is even comparable with that of natural ones. It is noteworthy that the SF-GAN vocoder shows strong generalization ability in the speaker-unseen scenario and the ability to be applied to the TTS task. In our future work, we will apply our proposed method to other GAN-based vocoders.

**Acknowledgment.** This work was partially funded by the National Nature Science Foundation of China under Grant 61871358.

# References

1. Zen, H., Tokuda, K., Black, A.W.: Statistical parametric speech synthesis. Speech Commun. **51**(11), 1039–1064 (2009)
2. Kawahara, H., Masuda-Katsuse, I., De Cheveigne, A.: Restructuring speech representations using a pitch-adaptive time-frequency smoothing and an instantaneous-frequency-based f0 extraction: Possible role of a repetitive structure in sounds. Speech Commun. **27**(3–4), 187–207 (1999)
3. Morise, M., Yokomori, F., Ozawa, K.: WORLD: A vocoder-based high-quality speech synthesis system for real-time applications. IEICE Trans. Inf. Syst. **99**(7), 1877–1884 (2016)
4. Oord, A.v.d., et al.: WaveNet: A generative model for raw audio. In: 9th ISCA Speech Synthesis Workshop, pp. 125–125 (2016)
5. Mehri, S., et al.: SampleRNN: An unconditional end-to-end neural audio generation model. In: Proceedings of ICLR (2017)
6. Kalchbrenner, N., et al.: Efficient neural audio synthesis. In: Proceedings of ICML, pp. 2410–2419 (2018)
7. Oord, A.v.d., et al.: Parallel WaveNet: Fast high-fidelity speech synthesis. In: Proceedings of ICML, pp. 3918–3926 (2018)
8. Ping, W., Peng, K., Chen, J.: ClariNet: Parallel wave generation in end-to-end text-to-speech. In: Proceedings of ICLR (2019)

9. Prenger, R., Valle, R., Catanzaro, B.: WaveGlow: A flow-based generative network for speech synthesis. In: Proceedings of ICASSP, pp. 3617–3621 (2019)

10. Ping, W., Peng, K., Zhao, K., Song, Z.: WaveFlow: A compact flow-based model for raw audio. In: Proceedings of ICML, pp. 7706–7716 (2020)

11. Wang, X., Takaki, S., Yamagishi, J.: Neural source-filter-based waveform model for statistical parametric speech synthesis. In: Proceedings of ICASSP, pp. 5916–5920 (2019)

12. Goodfellow, I., et al.: Generative adversarial nets. In: Proceedings of NeurIPS, pp. 2672–2680 (2014)

13. Donahue, C., McAuley, J., Puckette, M.: Adversarial audio synthesis. In: Proceedings of ICLR (2018)

14. Bińkowski, M., et al.: High fidelity speech synthesis with adversarial networks. In: Proceedings of ICLR (2019)

15. Kumar, K., et al.: MelGAN: Generative adversarial networks for conditional waveform synthesis. In: Advances In Neural Information Processing Systems 32 (2019)

16. Yamamoto, R., Song, E., Kim, J.M.: Parallel WaveGAN: A fast waveform generation model based on generative adversarial networks with multi-resolution spectrogram. In: Proceedings of ICASSP, pp. 6199–6203 (2020)

17. Kong, J., Kim, J., Bae, J.: HiFi-GAN: Generative adversarial networks for efficient and high fidelity speech synthesis. Adv. Neural. Inf. Process. Syst. **33**, 17022–17033 (2020)

18. Kim, J.H., Lee, S.H., Lee, J.H., Lee, S.W.: Fre-GAN: Adversarial frequency-consistent audio synthesis. In: Proceedings of InterSpeech 2021, pp. 3246–3250 (2021)

19. Wu, Y.C., Hayashi, T., Okamoto, T., Kawai, H., Toda, T.: Quasi-periodic Parallel WaveGAN: A non-autoregressive raw waveform generative model with pitch-dependent dilated convolution neural network. IEEE/ACM Trans. Audio Speech Lang. Process. **29**, 792–806 (2021)

20. Yoneyama, R., Wu, Y.C., Toda, T.: Unified source-filter GAN: Unified source-filter network based on factorization of quasi-periodic parallel wavegan. In: Proceedings of InterSpeech 2021, pp. 2187–2191 (2021)

21. Yoneyama, R., Wu, Y.C., Toda, T.: Unified source-filter GAN with harmonic-plus-noise source excitation generation. arXiv preprint arXiv:2205.06053 (2022)

22. Maas, A.L., Hannun, A.Y., Ng, A.Y.: Rectifier nonlinearities improve neural network acoustic models. In: Proceedings of ICML, vol. 30, p. 3 (2013)

23. Ito, K., Johnson, L.: The lj speech dataset (2017). https://keithito.com/LJ-Speech-Dataset/

24. Veaux, C., Yamagishi, J., MacDonald, K., et al.: CSTR VCTK corpus: English multi-speaker corpus for CSTR voice cloning toolkit. University of Edinburgh, The Centre for Speech Technology Research (CSTR) (2017)

25. Loshchilov, I., Hutter, F.: Decoupled weight decay regularization. In: International Conference on Learning Representations (2018)

26. Ai, Y., Ling, Z.H.: A neural vocoder with hierarchical generation of amplitude and phase spectra for statistical parametric speech synthesis. IEEE/ACM Trans. Audio Speech Lang. Process. **28**, 839–851 (2020)

27. Shen, J., et al.: Natural tts synthesis by conditioning wavenet on mel spectrogram predictions. In: 2018 IEEE International Conference On Acoustics, Speech and Signal Processing (ICASSP), pp. 4779–4783. IEEE (2018)

# Semantic Enhancement Framework for Robust Speech Recognition

Baochen Yang[1,2] and Kai Yu[1,2(✉)]

[1] X-LANCE Lab, Department of Computer Science and Engineering, MoE Key Lab
of Artificial Intelligence, AI Institute, Shanghai Jiao Tong University,
Shanghai, China
[2] State Key Laboratory of Media Convergence Production Technology and Systems,
Shanghai, China
`kai.yu@sjtu.edu.cn`

**Abstract.** Auto speech recognition (ASR) has been widely used in dialogue systems of various domains, performing as a crucial part of technology. Since the output of the ASR system will provide input to the subsequent system, the semantic intelligibility problem of the recognition results draws wide attention, yet remains unsolved. We propose a semantic enhancement framework to extract global semantic information from the audio to guide the recognition results. We evaluate our method on the Wall Street Journal (WSJ) dataset. The proposed framework gain relative 5.9% and 9.1% improvement of the WER on dev93 set and eval92 set compared to the baseline model.

**Keywords:** Semantic enhancement · Speech recognition

## 1 Introduction

The ASR aims to transfer the continuous audio input to human readable text output. It is an essential technology for many artificial intelligence application. A typical ASR system consists of many modules including acoustic, lexicon, language models (LM) that apply fusion during decoding. Over the past decades, with the rapid development of neural network, the performance of large vocabulary continuous speech recognition has attained impressive advancement. However, there still remains many problems for researchers to eliminate.

- **Domain problem**: The output of ASR system mainly focus on the acoustic feature. The system significantly degenerate when facing low-quality audio input or another domain audio input.
- **Semantic intelligibility problem**: Due to the fineness of the modeling symbols such as char, phoneme, byte pair encoding which are designed using linguistic knowledge, the ASR system are subject to semantic intelligibility problem. A complicated post-processing process which use the semantic information is required between the output of the model and the final decoded output to fix the semantic intelligibility problem.

L. Zhenhua et al. (Eds.): NCMMSC 2022, CCIS 1765, pp. 81–88, 2023.
https://doi.org/10.1007/978-981-99-2401-1_7

- **Incoherence in optimization**: ASR system normally need a language model during inference stage to reduce the word error rate. However the language model and acoustic model need to be trained separately with different objectives that need to be balanced and may cause mismatches between modules during application.

Consequently, it is quite important to couple the semantic information into ASR system training to address the above issues. In this paper, we propose a semantic enhancement framework to utilize the semantic information to enhance the ASR system. We characterize the main contribution of our method as follows:

- We introduce a semantic enhancement framework which extract the semantic information to provide the global semantic information to other parts of system to reduce the semantic intelligibility problem.
- We utilize the pre-train language model which is trained on a large text dataset to develop our semantic information extractor, which established a connection between pre-train language models and speech recognition systems.

## 2   Related Work

In this part, we concisely overview about recent effort relevant to our proposed work.

### 2.1   Contextual Method

Contextual methods aim to bias the results towards tokens, generally proper nouns or rare words or jargon, which are thought likely to be produced given the context of an audio signal. Correct transcription of these tokens might have an outsized impact on the value of the output, and incorrect transcription might otherwise be likely. These models normally accepts additional input about contextual information. These works can be categorized into deep contextualization [1,2,8,11] and external contextualization [6,9,10,17]. Deep contextualization integrate the contextual module into the end-to-end (E2E) deep neural module. External contextualization apply external modules such as language models, error correction models, and weighted finite-state transducers to the output hypotheses of ASR systems.

### 2.2   Adaptive Method

Adaptive methods usually utilize information or knowledge from other task models with ASR systems to make the models adaptive in multiple domains. The basic hypothesis of these methods is that the acoustic environment remains the same while deviations exist in the textual domains. Models for other tasks are trained using domain-specific data, or have been learned for specific knowledge which will help ASR systems to be more robust. The general methods is LM fusion or learn from external LM into the ASR systems. LM fusion use an external LM in decoding. Depending on the specific method, it can be further divided into shallow fusion [5,7] , deep fusion (DF) [5,14] and cold fusion (CF) [13,14]. Shallow fusion is also called joint-score decoding. DF and CF both concatenate hidden states from LM and ASR while CF train the ASR system under the guidance of LM. Learn from external LM methods [3,12] tuning part of the ASR systems under the external LM to adapt to different domain transformations.

## 3   Method

In this section we describe the proposed semantic enhancement framework in detail. The general framework is shown in Fig. 1, which consists of 2 major components: 1) The semantic module for predicting semantic information from the acoustic feature, the pre-train language model is used during training, 2) A hybrid connectionist temporal classification (CTC) module and attention-based decoder to eliminate irregular alignments between acoustic features and recognition symbols. It is important to note that they both use the shared Encoder which is responsible for processing acoustic features. First we explain the hybrid CTC/attention architecture in Sect. 3.1, and introduce the pre-trained language model in Sect. 3.2. In Sect. 3.3, we describe our semantic enhancement framework.

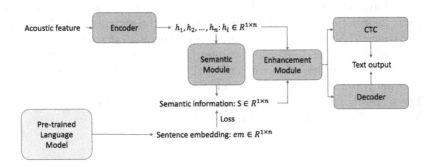

**Fig. 1.** The overall architecture of the semantic enhancement framework

### 3.1   Hybrid CTC/Attention Architecture

With the respect to the input waveform $X$, the audio speech recognition aims to produce the corresponding token sequence $C$. The hybrid CTC/attention architecture utilizes both benefits of CTC and attention during the training and decoding steps in ASR. The The CTC module and attention module share the same encoder which is trained by multi-objective learning(MOL). The CTC objective function serve as an auxiliary task to train the encoder. So the overall objective to be maxmized is a linear combination of the CTC and attention:

$$L_{MOL} = \lambda \log p_{ctc}(C|X) + (1 - \lambda) \log p_{att}(C|X). \tag{1}$$

where the $\lambda$, satisfies $0 \leq \lambda \leq 1$, serve as a hyper-parameter in this framework.

## 3.2  Pre-train Language Model

Language model pre-training has been proven to be effective for learning general language representations. BERT, along with its variants has reached state-of-the-art results in most downstream natural language understanding (NLU) tasks. Pre-train LM can be trained on the large unlabeled dataset and then fine-tuned on the limited labeled dataset. Contrastive learning is a self-supervised learning method used to learn general presentation $em$ by letting the model learn the similarity and differences between data from the unlabeled dataset. The process of contrastive learning can be decomposed into three steps: data augmentation, encoding, and computing loss. Through contrastive learning, the model can learn more efficient high-dimensional representations of the data.

## 3.3  Semantic Enhancement Framework

The hybrid CTC/attention architecture works well in many speech recognition scenarios except for some semantic intelligibility problem. Due to the lack of global semantic information, the model may be less effective facing low-quality audio. Semantic enhancement framework utilize the global semantic information to address these problems. The proposed semantic module predicts extra semantic information $S$ from the audio. Moreover, the sentence embedding from the pre-trained LM as the ground truth of the semantic information to improve the performance. After that, the semantic information is used to enhance the acoustic features. In this way, our framework is robust to low-quality audio and can provide reasonable recognition results.

The semantic task serves as an auxiliary task to raw hybrid CTC/attention architecture. The overall loss function is as follows:

$$L = L_{MOL} + \beta L_{sem}. \tag{2}$$

where $L_{MOL}$ is the loss function of hybrid CTC/attetion architecture defined in Eq. (1), and $L_{sem}$ is to measure the similarity of the predicted semantic information respect to the sentence embedding of the transcription label from the pre-train LM. We try the Mean-Squared-Error (MSE) loss and cosine embedding loss in our experiment. $\beta$ is an important hyper-parameter to control the importance of the semantic embedding loss and will be evaluated in our experiment. That is to say, the loss function of the $L_{sem}$ is as follows.

$$\begin{aligned} L_{sem} &= MSE(S, em). \\ L_{sem} &= 1 - cos(S, em). \end{aligned} \tag{3}$$

At training time, the ratio of use sentence embedding from pre-train LM or directly use predicted semantic information for semantic enhancement is a nonnegligible problem. Balancing the ratio of the two in the training phase can effectively help the model learn better semantic information and utilize the semantic information better. We evaluate the different ratios and choose 0.8 as the final ratio in our experiment. That is to say, the model will use the sentence embedding during the training phase with probability 0.8.

### 3.4   Evaluation Metrics

In addition to the most commonly used word error rate (WER) metrics. We introduce the keyword error rate (KeywordER) metrics which indicate the keywords difference between recognize text and ground truth text. The keyword is extracted by the keybert [4] tool which select keywords according to the cosine similarity between words embedding and sentence embedding. The keyword sequence is distilled from the original sentence containing the significant semantic information.

For example, the keyword sequence of the sentence "Mary, it is rainy outside, remember to take your umbrella." will be "Mary, rainy, umbrella". We arrange the keywords in the order of the original sequence, and calculate the KeywordER on the sequence to evaluate the difference in semantic information between the two sentences. Thus, KeywordER can be considered as a metric of semantic intelligibility.

## 4   Experiment

### 4.1   Dataset

The WSJ dataset is used to evaluate the effectiveness of the proposed framework. We use 40-dimension fbank features with 25 ms window and 10 ms stride. The recognition modeling units for our experiment are characters. There are total 71 modeling units including characters, non-language symbols and special symbols.

### 4.2   Configuration

We use ESPNet[16] toolkit to build both the vanilla hybrid CTC/attention based E2E ASR baselines and our proposed semantic enhancement ASR models. Moreover, we use an external LM for shallow fusion in our decoding part.

1) CTC/attention Baseline: The encoder consists of 12 transformer [15] blocks. We use 2048 hidden units in transformer feed forward layers, and 256 hidden units in the multi-head self-attention layers with 4 heads. The decoder is composed of a stack of 6 Transformer layers with 2048 hidden units in the multi-head self-attention layers. The CTC branch contains an additional fully connected layer to predict labels.

2) Semantic enhancement module: We use a fully connected layer to make encoder and pre-train LM the same output size. Then, a mean pooling and $l_2$ normalization is performed to get the semantic embedding. As for the enhancement module, we tried three enhancements methods including just

learning to extract semantic information without using it, directly concatenate the semantic embedding with the encoder output and use the semantic embedding as the initial state of the recurrent neural network. The last enhancement method achieved the best results and will be further analyzed by us in the following sections.

3) The two pre-train LM we used including BERT-base which is a 12 layers transformer with 768 hidden units. And SimCSE which is trained by contrastive loss with the same architecture of BERT.

### 4.3  Impact of Losses

In this part, we evaluate the impact of different loss measure methods and different hyper parameters in our proposed framework. From the Table 1, we can see that although cosine embedding loss perform poorly in the WER, it is relatively better in KeywordER, meaning that cosine embedding loss is more efficient for improving the comprehensibility of recognition results.

**Table 1.** Different losses results

| Loss | weight | WER (%) | | KeywordER (%) | |
|---|---|---|---|---|---|
| – | – | dev93 | eval92 | dev93 | eval92 |
| Baseline | – | 6.8 | 4.8 | 15.7 | 12.3 |
| MSE | 1e2 | 7.1 | 4.9 | 16.3 | 12.0 |
| | 1e1 | **6.4** | **4.4** | 15.6 | 11.6 |
| | 1e0 | 6.9 | 4.6 | 16.4 | 12.2 |
| COS | 1e0 | 7.1 | 4.8 | 15.8 | 12.0 |
| | 1e-1 | 6.8 | 4.6 | 15.5 | **11.4** |
| | 1e-2 | 7.0 | 4.6 | **15.3** | 11.5 |

### 4.4  Results

From the Table 2, our proposed framework gains relative 5.9% and 9.1% improvement of the WER on dev93 set and eval92 set compared to the baseline model. Furthermore, our proposed framework also achieve a decrease in the keywords error rate meaning that the recognition results is much more understandable.

**Table 2.** Main Result of the method

| Method | WER (%) | | KeywordER (%) | |
|---|---|---|---|---|
| – | dev93 | eval92 | dev93 | eval92 |
| Baseline | 6.8 | 4.8 | 15.7 | 12.3 |
| Sem-enh | 6.4 | 4.4 | 15.6 | 11.6 |

For a case in eval92 set, baseline model recognize an audio to "driven that system to its observed extreme" while our method correctly recognized it to "driven that system to its abused extreme". Apparently our method incorporates semantic information into the recognition result and corrects the error.

## 5    Conclusions

In this paper, we propose a semantic enhancement framework to reduce the semantic intelligibility problem. We evaluate our framework on the WSJ dataset and gain some improvements. However, the modules of our proposed framework still need further refinement and modification to achieve greater improvement. In future work, we will explore more implementation to achieve more robust speech recognition.

**Acknowledgements.** This study was supported by State Key Laboratory of Media Convergence Production Technology and Systems Project (No.S KLMCPT S2020003), Major Program of National Social Science Foundation of China (No.18ZDA293) and Shanghai Municipal Science and Technology Major Project (2021SHZDZX0102).

## References

1. Bruguier, A., Prabhavalkar, R., Pundak, G., Sainath, T.N.: Phoebe: Pronunciation-aware contextualization for end-to-end speech recognition. In: ICASSP 2019–2019 IEEE International Conference on Acoustics, Speech and Signal Processing (ICASSP), pp. 6171–6175. IEEE (2019)
2. Chang, F.J., et al.: Context-aware transformer transducer for speech recognition. In: 2021 IEEE Automatic Speech Recognition and Understanding Workshop (ASRU), pp. 503–510. IEEE (2021)
3. Deng, K., Cheng, G., Yang, R., Yan, Y.: Alleviating asr long-tailed problem by decoupling the learning of representation and classification. IEEE/ACM Trans. Audio Speech Lang. Process. **30**, 340–354 (2021)
4. Grootendorst, M.: Keybert: Minimal keyword extraction with bert (2020). https://doi.org/10.5281/zenodo.4461265
5. Gulcehre, C., et al.: On using monolingual corpora in neural machine translation. arXiv preprint arXiv:1503.03535 (2015)
6. Jung, N., Kim, G., Chung, J.S.: Spell my name: keyword boosted speech recognition. In: ICASSP 2022–2022 IEEE International Conference on Acoustics, Speech and Signal Processing (ICASSP), pp. 6642–6646. IEEE (2022)
7. Kannan, A., Wu, Y., Nguyen, P., Sainath, T.N., Chen, Z., Prabhavalkar, R.: An analysis of incorporating an external language model into a sequence-to-sequence model. In: 2018 IEEE International Conference on Acoustics, Speech and Signal Processing (ICASSP), pp. 1–5828. IEEE (2018)
8. Le, D., Keren, G., Chan, J., Mahadeokar, J., Fuegen, C., Seltzer, M.L.: Deep shallow fusion for rnn-t personalization. In: 2021 IEEE Spoken Language Technology Workshop (SLT), pp. 251–257. IEEE (2021)
9. Liu, D.R., Liu, C., Zhang, F., Synnaeve, G., Saraf, Y., Zweig, G.: Contextualizing asr lattice rescoring with hybrid pointer network language model. arXiv preprint arXiv:2005.07394 (2020)

10. Michaely, A.H., Zhang, X., Simko, G., Parada, C., Aleksic, P.: Keyword spotting for google assistant using contextual speech recognition. In: 2017 IEEE Automatic Speech Recognition and Understanding Workshop (ASRU), pp. 272–278. IEEE (2017)
11. Pundak, G., Sainath, T.N., Prabhavalkar, R., Kannan, A., Zhao, D.: Deep context: end-to-end contextual speech recognition. In: 2018 IEEE Spoken Language Technology Workshop (SLT), pp. 418–425. IEEE (2018)
12. Pylkkönen, J., Ukkonen, A., Kilpikoski, J., Tamminen, S., Heikinheimo, H.: Fast text-only domain adaptation of rnn-transducer prediction network. arXiv preprint arXiv:2104.11127 (2021)
13. Sriram, A., Jun, H., Satheesh, S., Coates, A.: Cold fusion: Training seq2seq models together with language models. arXiv preprint arXiv:1708.06426 (2017)
14. Toshniwal, S., Kannan, A., Chiu, C.C., Wu, Y., Sainath, T.N., Livescu, K.: A comparison of techniques for language model integration in encoder-decoder speech recognition. In: 2018 IEEE Spoken Language Technology Workshop (SLT), pp. 369–375. IEEE (2018)
15. Vaswani, A., et al.: Attention is all you need. In: Advances in Neural Information Processing Systems 30 (2017)
16. Watanabe, S., et al.: ESPnet: End-to-end speech processing toolkit. In: Proceedings of Interspeech, pp. 2207–2211 (2018). https://doi.org/10.21437/Interspeech.2018-1456
17. Zhao, D., et al.: Shallow-fusion end-to-end contextual biasing. In: Interspeech, pp. 1418–1422 (2019)

# Achieving Timestamp Prediction While Recognizing with Non-autoregressive End-to-End ASR Model

Xian Shi[✉], Yanni Chen, Shiliang Zhang, and Zhijie Yan

Speech Lab, Alibaba Group, Hangzhou, China
{shixian.shi,cyn244124,sly.zsl,zhijie.yzj}@alibaba-inc.com

**Abstract.** Conventional ASR systems use frame-level phoneme posterior to conduct force-alignment (FA) and provide timestamps, while end-to-end ASR systems especially AED based ones are short of such ability. This paper proposes to perform timestamp prediction (TP) while recognizing by utilizing continuous integrate-and-fire (CIF) mechanism in non-autoregressive ASR model - Paraformer. Foucing on the fire place bias issue of CIF, we conduct post-processing strategies including fire-delay and silence insertion. Besides, we propose to use scaled-CIF to smooth the weights of CIF output, which is proved beneficial for both ASR and TP task. Accumulated averaging shift (AAS) and diarization error rate (DER) are adopted to measure the quality of timestamps and we compare these metrics of proposed system and conventional hybrid force-alignment system. The experiment results over manually-marked timestamps testset show that the proposed optimization methods significantly improve the accuracy of CIF timestamps, reducing 66.7% and 82.1% of AAS and DER respectively. Comparing to Kaldi force-alignment trained with the same data, optimized CIF timestamps achieved 12.3% relative AAS reduction.

**Keywords:** end-to-end ASR · non-autoregressive ASR · timestamp · force alignment

## 1 Introduction

Timestamp prediction is one of the most important and widely used subtasks of automatic speech recognition (ASR). Kinds of speech related tasks (text-to-speech, key-word spotting) [1,2], speech/language analysis [3–5] and ASR training strategies [6] can be conducted with a reliable timestamp predicting system. At present, conventional hybrid HMM-GMM or HMM-DNN systems are mostly used to conduct force-alignment (FA) - the given transcriptions are expanded to phoneme sequences for processing Viterbi decoding in WFST (weighted finite state transducers) composed by acoustic model, lexicon and language model. Recent years have seen the rapid growth of end-to-end (E2E) ASR models [7–10], which skip the complex training process and preparation of language

© The Author(s), under exclusive license to Springer Nature Singapore Pte Ltd. 2023
L. Zhenhua et al. (Eds.): NCMMSC 2022, CCIS 1765, pp. 89–100, 2023.
https://doi.org/10.1007/978-981-99-2401-1_8

related expert knowledge and convert speech to text with a single neural network. Generally, E2E ASR models are classified into two categorise: Time-synchronous models including CTC and RNN-Transducer, token-synchronous models including listen-attend-and-spell (LAS) and Transformer based AED models. The former utilize CTC-like criterion over frame-level encoder output and predict posterior probabilities of tokens together with blank while the later conduct cross-attention between acoustic information and character information to achieve soft alignment. These models have shown strong competitiveness and have replaced conventional ASR model in many scenarios. At the same time, however, due to the inherent deficiency of timestamp prediction ability of these models, some ASR systems have to use an additional conventional ASR model to predict the timestamp of recognition results, which introduce computation overhead and training difficulty.

In this paper, we propose to achieve timestamp prediction while recognizing with non-autoregressive E2E ASR model Paraformer [11]. Continuous integrate-and-fire (CIF) [12] is a soft and monotonic alignment mechanism proposed for E2E ASR which is adopted by Paraformer as predictor, it predicts the number of output tokens by integrating frame-level weights, once the accumulated weights exceed the fire threshold, the encoder output of these frames will be summed up to one step of acoustic embedding. One of the core ideas of achieving non-autoregressive decoding in Paraformer is to generate character embedding which has the same length as output sequence. The modeling characteristic of Parafirner CIF delights us to conduct timestamp prediction basing on CIF output. Focusing on the distribution of original CIF inside Paraformer, we propose scaled-CIF training strategies and three post-processing methods to achieve timestamp prediction of high quality and also explore to measure the timestamp prediction systems with AAS and DER metrics. The following part of this paper is organized as below. Timestamp prediction and FA related works are introduced in Sect. 2. We briefly introduce CIF and Paraformer and look into original CIF distribution in Sect. 3. Section 4 comes about the proposed methods including scaled-CIF and post-processing strategies while Sect. 5 describe out experiments and results in detail. Section 6 ends the paper with our conclusion.

**Our Contributions Are:**

- From the aspects of timestamp quality, the proposed scaled-CIF and post-processing strategies improve the accuracy of timestamp and outperforms the conventional hybrid model trained with the same data.
- This paper propose to predict timestamps naturally while recognizing with Paraformer, such system can predict accurate timestamp of recognition results and reduce computation overhead which is of value in commercial usage.

## 2   Related Works

In this section we briefly introduce the mechanism of sequence-to-sequence modeling and discuss about the recent works related to timestamp predicton and force-alignment. Time-synchronous models and token-synchronous solve unequal length sequence prediction in different ways. Transducer [8] performs forward and backward algorithm as shown in 1 (a), it is allowed to move in time axis or label axis to establish connection between token sequence and time sequence of unequal length. However, it turns out that a well trained CTC [7] or Transducer model tends to predict posterior probabilities with sharp peaks (single frame with extremely high probability for a token except blank) [13], and the position of the peak can not reflect the real time of the token, especially when the modeling units have long duration. AED based E2E models like Transformer conduct cross-attention between encoder and decoder, the score matrix inside cross-attention can be regarded as alignment but it is soft and nonmonotonic, so it's hard to conduct timestamp prediction in origin AED-based models like Transformer. CIF is a time-synchronous method which is also adopted by AED based models [11,21]. It generates monotonic alignment by predicting integrate weights in frame level which can be naturally treated as timestamps.

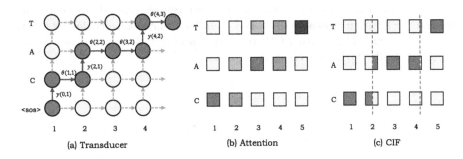

**Fig. 1.** Illustration of alignment of transducer, attention and CIF.

Recently, neural network based timestamp prediction and force-alignment strategies without conventional hybrid systems are explored. Kürzinger et al. [14] proposed CTC-segmentation for German speech recognition, which determines the alignment through forward and backward probabilities of CTC model. The proposed alignment system beats the conventional ones including HTK [17] and Kaldi [18] in German task. Li et al. [15] proposed NEUFA to conduct force-alignment, which deploys bidirectional attention mechanism to achieve bidirectional relation learning for parallel text and speech data. The proposed boundary detector takes the attention weights from both ASR and TTS directions as inputs to predict the left and right boundary signals for each phoneme. Their system achieves better accuracies at different tolerances comparing to MFA [19]. Besides, systems like ITSE [16] also achieves an accurate, lightweight text-to-speech alignment module implemented without expertise such as pronunciation lexica.

These works, however, conduct force-alignment outside of ASR models to get timestamps. For an ASR model which is required to obtain timestamp prediction ability, such models introduce additional computation overhead, which might be unacceptable for ASR systems in commercial usage. The excepted timestamp prediction models inside ASR system is supposed to introduce less computation overhead and predict accuracy timestamps naturally.

## 3    Preliminaries

### 3.1    Continuous Integrate-and-Fire

Continuous integrate-and-fire (CIF) is a soft and monotonic alignment mechanism for E2E ASR different from time-synchronized models and label-synchronized models. CIF performs integrate process over the output of encoder $e_{1:T}$ and predicts frame-level weights $\alpha_{1:T}$, once the accumulated weights exceed the fire threshold, these frames will sum up to acoustic embedding $\mathbf{E}_{1:L'}$ which is synchronized with output tokens. Such process is illustrated in Fig. 2. In ASR models, the modeling character of CIF is of great value. The frame-level weights actually conduct alignment between acoustic representation and output tokens, and the fire location indicates the token boundary. Such capacity delights us to explore the feasibility of using CIF to predict accurate timestamps of decoded tokens naturally in the process of recognition.

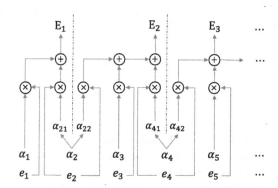

**Fig. 2.** Illustration of integrate-and-fire on encoder output $e_{1:T}$ with predicted weights $\alpha = (0.3, 0.9, 0.4, 0.4, 0.3)$. The integrated acoustic embedding $\mathbf{E}_1 = 0.3 \times e_1 + 0.7 \times e_2$, $\mathbf{E}_2 = 0.2 \times e_2 + 0.4 \times e_3 + 0.4 \times e_4$. The sum of weights $\alpha$ is $L'$ - the length of prediction sequence.

### 3.2    Paraformer

We adopt Paraformer - a novel NAR ASR model which achieves non-autoregressive decoding capacity by utilizing CIF and two-pass training strategy inside an AED backbone. Avoiding the massive computation overhead introduced by autoregressive decoding and beam-search, Paraformer gains more than 10x speedup with even lower error rate comparing to Conformer baseline. The overall framework of Paraformer is illustrated in Fig. 3.

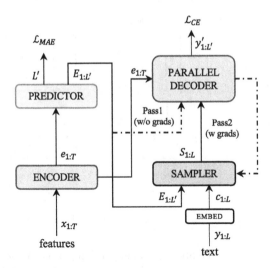

**Fig. 3.** Illustration of Paraformer structure.

Paraformer contains three modules, namely encoder, predictor and parallel decoder (sampler interpolates acoustic embeding and char embedding without parameters). Encoder is same as AR encoders of Conformer which contains self-attention, convolution and feed-forward networks (FFN) layers to generate acoustic representation $\mathbf{e}_{1:T}$ from down-sampled Fbank features. Predictor uses CIF to predict the number of output tokens $L'$ and generate acoustic embedding $\mathbf{E}_{1:L'}$. Parallel decoder and sampler conduct two-pass training with the vectors above: $\mathbf{E}_{1:L'}$ is directly sent to decoder to calculate cross-attention with $\mathbf{e}_{1:T}$, and decode all tokens $y'_{1:L'}$ in NAR form at once. Then sampler interpolates $\mathbf{E}_{1:L'}$ and char embedding $\mathbf{c}_{1:L}$ according to the edit distance between hypothesis $y'_{1:L'}$ and $y_{1:L}$. The interpolated vector (named semantic embedding, noted by $S_{1:L'}$) is sent to parallel decoder again to conduct the same calculation, which makes up the second pass. Note that the forward process of the first pass is gradient-free, cross-entropy (CE) loss is calculated between output of the second pass and ground truth. In the training process, as the accuracy of first pass decoding raises, $\mathbf{E}_{1:L'}$ makes up increasing proportion of $S_{1:L'}$. In the inference stage, Paraformer use the first pass to decode.

CIF is of vital importance in Paraformer as it predicts the number of output tokens and generates monotonic token boundaries implicitly. Figure 4 shows the origin fire place comparing to the timestamp generated by kaldi force-alignment system. Weights $\alpha$ shows that the pattern of CIF is regard less of the real length of tokens, for an ASR model with 4-times down-sampling layer (60 ms each time step), the integrate process for each token finishes in around 4 frames, which leads to a large offset for end point timestamp of character with long duration.

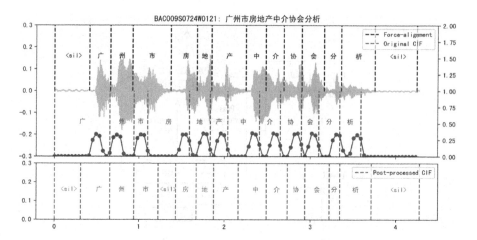

**Fig. 4.** CIF fire places and weights $\alpha$ of demo utterance from Aishell-1. Axis of $\alpha$ is in the right. The subfigure above shows the timestamps of origin CIF comparing to FA system, the CIF fire place roughly indicate the corresponding token place but it is always inaccurate, especially the end of integrate is usually later. The subfigure at bottom shows the effect of weights post-processing strategies introduced in Sect. 4.2.

## 4     Methods

Considering of the pattern of CIF weights and the characteristic of Paraformer, we optimize the timestamp prediction strategy from two aspects. In the training process, we propose to scale the CIF weights after sigmoid function in order to alleviate the sharpness of weights and also cut off the gradient towards encoder of irrelevant position. Then we adopt several post-processing strategies for weights to achieve more precise timestamp prediction.

### 4.1     Scaled-CIF Training

Original CIF calculates $\alpha$ with sigmoid activation function after feed-forward network, sharp spikes and glitches can be observed from the curve in Fig. 4. We propose to scale and smooth the CIF weights with the operation below:

$$\alpha' = \gamma \cdot ReLU\big((sigmoid(x) - \beta)\big) \tag{1}$$

First, ReLU function and $\beta$ smooth the glitches of $\alpha$, cutting off the gradient towards encoder of irrelevant position, which is supposed to be beneficial for ASR task. Besides, scaling the output of ReLU with $\gamma$ relieves the spikes of the curve, trying to achieve level and smooth weights.

### 4.2     Weights Post-processing

Another optimization comes about from the aspect of post-processing. According to the distribution of alphas observed in Fig. 4, most of the weights begins to accumulate from the exact frame but the accumulation ends in fixed steps (4 frames in the figure), which delights us to conduct the following 4 processing strategies:

**Begin/end Silence.** As the begging place of CIF is always precise, the beginning frames with weights under a threshold $\theta_s$ is considered as silence. So as to the frames at the end but we set an interval of 3 frames.

**Fire Delay.** Original CIF timestamps are always unreliable for the end point prediction for long-lasting tokens. We propose to conduct fire delay operation when frames with low weights are observed, such frames with be grouped to the previous token except the last frame.

**Silence Insertion.** When the low-weight frames last longer than $L_s$, we insert a silence token in between.

**Weight Averaging.**[1] For ASR models with higher down-sampling rate, the process of integrate finishes even faster. When using 6-times low frame rate features, the CIF tends to output weights around 1.0 or 0.0, which makes CIF weights no longer stand for the accumulation procedure. Then we proposed to weaken the weight spikes.

$\theta_s$ and $L_s$ are model related hyper-parameters, for Paraformer model with original CIF predictor and 4-times down-sampling encoder embedder, we set $\theta_s = 0.05$ and $L_s = 3$, the timestamp thus generated of the demo utterance above is shown in Fig. 4. Subjectively, the proposed post-processing strategies optimize the quality of CIF timestamps in a simple but efficient way.

### 4.3  Evaluation Metrics

In addition to the visual analysis of timestamps, we propose to use accumulated averaging shift (AAS) and diarization error rate (DER) as the evaluation metrics of timestamp accuracy.

**AAS.** The first metrics measures the averaging time shift of each token. The time shift of begging timestamp and end timestamp are summed up and averaged over the entire testset. Formally, we calculate the metrics between timestamp $i$ and timestamp $j$ as

$$AAS = \frac{\sum_{k=1}^{K} |start_{k,i} - start_{k,j}| + |end_{k,i} - end_{k,j}|}{2K}, \tag{2}$$

where $K$ is for the number of aligned token pairs[2].

**DER.** Speaker diarization systems are evaluated by DER, which calculates the proportion of frames which are classified correctly. DER is the sum of three

---

[1] In this paper, we conduct experiments with 4-times down-sampling encoder, this methods is thus not validated.

[2] Considering the transcriptions of two timestamps might differ, only paired tokens according to edit distance are included in the calculation.

different error types: False alarm of speech, missed detection of speech and confusion between speaker labels. Treating tokens as speakers, we introduce this metrics to measure the quality of timestamp.

$$DER = \frac{False\ Alarm + Missed + Speaker\ Confusion}{Total\ Duration\ of\ Time}. \tag{3}$$

## 5 Experiments and Results

### 5.1 Datasets

Two datasets are used in our experiments. First, Aishell-1 is used for training the Paraformer model and visualizing timestamps. Aishell-1 contains 178 h speech data with transcription, which is a widely used open-source Mandarin ASR corpus. Besides, we use a TTS dataset called M7 for evaluation, which contains 5550 Mandarin utterances. Except transcriptions, M7 also contains manually marked timestamps in token level, which is regarded as reference in the calculation of AAS and DER.

### 5.2 Experiment Setup

The Paraformer model is trained from scratch using ESPnet toolkit with the following setups. The model contains 12-layer Conformer encoder (implemented as with kernel size 15) and 6-layer Transformer decoder with attention dimension $d_{attn} = 256$ and feed-forward network dimension $d_{linear} = 2048$. The input layer of encoder conducts 4-times down-sampling for Fbank features, one step of encoder output thus stands for 40ms. The model is jointly trained with CTC loss ($\alpha = 0.3$) and we set dropout rate $r_{dropout} = 0.1$ for entire model. 3-times speed perturbation is adopted for Aishell-1 data and we apply spec-augment with 2 frequence masks range in $[0, 30]$ and 2 time masks range in $[0, 40]$ for each utterance. We use dynamic batch size ($numel = 2000k$) and $n_{epoch} = 50$. No language model is used in the inference stage. For scaled-CIF, we use $\gamma = 0.8$ and $\beta = 0.05^3$.

We prepare the force-alignment system with Kaldi toolkit as baseline timestamp, the training setup and model configuration is exactly same as open-source recipe Aishell-1, from flat-start to HMM-GMM. Timestamps of phonemes are extracted from lattice using model $tri5a$, and then converted to characters timestamps.

### 5.3 Quality of Timestamp

In this section we evaluate the quality of the timestamps and analyse the effect of the proposed methods. Table 1 shows the AAS and DER metircs of the timestamps from different models over testset M7. First we test the force-alignment

---

[3] For ASR models of different frame rate and encoder down-sampling rate, we fine it better to adjust scaling coefficients $\gamma$ and $\beta$, so as to the post-processing hyper-parameters $\theta_s$ and $L_s$ to achieve better performance.

system with ground truth transcription and Paraformer recognition results. It turns out that using these two kinds of transcription lead to slight difference of DER and almost the same AAS, which is because the majority error of Paraformer's recognition results is substitution error, reference and hypothesis have nearly the same expanded phoneme sequences, thus AAS of FA-GT and FA-HYP differs little.

**Table 1.** Accuracy of timestamp from force-alignment system and Paraformer CIF.

|  | Exp | Sys | AAS (sec) | DER (%) |
|---|---|---|---|---|
| force-alignment systems | FA-GT | FA with groundtruth | 0.080 | 6.34 |
|  | FA-HYP | FA with decoded trans | 0.081 | 7.50 |
| CIF timestamps | CIF-0 | origin CIF timestamp | 0.213 | 45.39 |
|  | CIF-1 | +begin/end silence | 0.161 | – |
|  | CIF-2 | +fire-delay | 0.124 | – |
|  | CIF-3 | +silence insertion | 0.112 | 17.09 |
| scaled-CIF timestamps | SCIF-0 | scaled-CIF timestamp | 0.143 | 29.75 |
|  | SCIF-1 | +begin/end silence | 0.098 | – |
|  | SCIF-2 | +fire-delay | 0.080 | – |
|  | SCIF-3 | +silence insertion | 0.071 | 8.11 |

Comparing CIF-0 and FA-HYP, it is obvious that original CIF weights as timestamps is of unacceptable quality. With the addition of post-processing strategies, the accuracy of CIF timestamps has been improved step by step, CIF-3 achieves 47.4% AAS reduction. Among the three post-processing methods, fire delay is the most effective and also the most tricky one. Scaled-CIF brings further help to timestamps prediction, 32.9% AAS reduction and 34.5% DER reduction are observed without any post-processing strategies. In SCIF-3, CIF timestamps outperforms force-alignment systems in AAS (11.3% relatively), but DER is still higher.

Figure 5 shows comparison of manually marked timestamps, original CIF timestamps and optimized timestamps. Comparing the blue curve and the purple curve, it can be observed that the peak of the curve is weakened to some extent, and the glitches in the curve disappear (not obvious in the figure). Considering the start and ending time of each token, the effect of fire delay is obvious, especially when the token is followed by low weight frames (like character '家' in the demo).

## 5.4   ASR Results

Results in Sect. 5.3 shows that scale-CIF significantly improves timestamp prediction accuracy, and we find it is also benificial for ASR. Comparing to vanilla Conformer AR model, Paraformer achieves better recognition accuracy and even

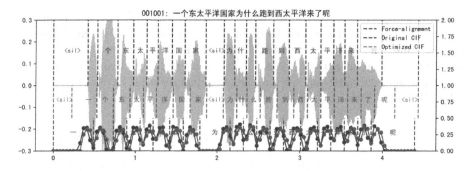

**Fig. 5.** Demo for utterance 001001 in M7. It can be observed that the peak in the $\alpha$ is weakened and the optimized timestamps are more accurate.

lower real time factor (RTF) [11]. On Aishell-1 task, Paraformer gets 4.6%, 5.2% CER over dev and test set and the RTF is 0.0065 (RTF of Conformer is 0.1800 under the same test environment). In our experiments, Paraformer with scaled-CIF outperforms the baseline, CER on Aishell-1 dev and test set are 4.5% and 5.2% respectively. On testset M7, Paraformer gets 11.70% CER with origin CIF and 11.26% with scaled CIF (3.76% relative CER reduction). Such results prove that smoothing the weights of CIF and cutting off the gradient towards encoder of irrelevant frames are beneficial for recognition task of Paraformer.

## 6    Conclusion

As the inherent deficiency of predicting timestamps with AED based models, we propose to predict timestamps according to CIF weights while recognition with Paraformer. In this paper, we first explore the characteristic of CIF in Paraformer for Mandarin. It turns out that the integrate of CIF weights tends to start from the right place but the process ends in fixed number of steps, regardless of tokens' real length. Besides, sharp peaks and glitches are observed in CIF weights. Such behavior delights us to improve the accuracy of CIF timestamp with scaled-CIF and post-processing strategies. We compare the CIF timestamp with force-alignment results of conventional HMM-GMM systems, and evaluate the quality of timestamps with AAS and DER metrics. The results in Table 1 show that with the help of the proposed methods, CIF timestamps achieve comparable performance as the FA baseline, 12.3% relative AAS reduction is observed while DER is a little worse. Comparing CIF timestamps before and after optimization, 66.7% AAS reduction and 82.1% DER reduction is achieved. In summarize, the proposed scaled-CIF and post-processing strategies improve the accuracy of timestamp and out system outperforms the conventional hybrid model trained with the same data, and such system reduces computation overhead which might be of value in commercial usage.

In the future, we will explore the CIF timestamps of different languages and different frame-rate, and modify the silence insertion strategy with dynamic silence length threshold according to the phoneme.

# References

1. Yang, Z., et al.: CaTT-KWS: A Multi-stage Customized Keyword Spotting Framework based on Cascaded Transducer-Transformer. arXiv preprint arXiv:2207.01267 (2022)
2. Ren, Y., et al.: Fastspeech: Fast, robust and controllable text to speech. In: Advances in Neural Information Processing Systems 32 (2019)
3. Labov, W., Rosenfelder, I., Fruehwald, J.: One hundred years of sound change in Philadelphia: Linear incrementation, reversal, and reanalysis. Language, pp. 30–65 (2013)
4. DiCanio, C., Nam, H., Whalen, D.H., Timothy Bunnell, H., Amith, J.D., García, R.C.: Using automatic alignment to analyze endangered language data: Testing the viability of untrained alignment. J. Acoustical Soc. Am. $134$(3), 2235–2246 (2013)
5. Yuan, J., Liberman, M., Cieri, C.: Towards an integrated understanding of speaking rate in conversation. In: Ninth International Conference on Spoken Language Processing (2006)
6. Fu, L., et al.: SCaLa: Supervised Contrastive Learning for End-to-End Automatic Speech Recognition. arXiv preprint arXiv:2110.04187 (2021)
7. Graves, A., Fernández, S., Gomez, F., Schmidhuber, J.: Connectionist temporal classification: labelling unsegmented sequence data with recurrent neural networks. In: Proceedings of the 23rd International Conference on Machine Learning, pp. 369–376 (June 2006)
8. Graves, A.: Sequence transduction with recurrent neural networks. arXiv preprint arXiv:1211.3711 (2012)
9. Vaswani, A., et al.: Attention is all you need. In: Advances In Neural Information Processing Systems, 30 (2017)
10. Chan, W., Jaitly, N., Le, Q., Vinyals, O.: Listen, attend and spell: A neural network for large vocabulary conversational speech recognition. In 2016 IEEE international conference on acoustics, speech and signal processing (ICASSP), pp. 4960–4964. IEEE (March 2016)
11. Gao, Z., Zhang, S., McLoughlin, I., Yan, Z.: Paraformer: Fast and Accurate Parallel Transformer for Non-autoregressive End-to-End Speech Recognition. arXiv preprint arXiv:2206.08317 (2022)
12. Dong, L., Xu, B.: Cif: Continuous integrate-and-fire for end-to-end speech recognition. In: ICASSP 2020–2020 IEEE International Conference on Acoustics, Speech and Signal Processing (ICASSP), pp. 6079–6083. IEEE (May 2020)
13. Zeyer, A., Schlüter, R., Ney, H.: Why does CTC result in peaky behavior?. arXiv preprint arXiv:2105.14849 (2021)
14. Kürzinger, L., Winkelbauer, D., Li, L., Watzel, T., Rigoll, G.: CTC-segmentation of large corpora for german end-to-end speech recognition. In: Karpov, A., Potapova, R. (eds.) SPECOM 2020. LNCS (LNAI), vol. 12335, pp. 267–278. Springer, Cham (2020). https://doi.org/10.1007/978-3-030-60276-5_27

15. Li, J., et al.: Neufa: neural network based end-to-end forced alignment with bidirectional attention mechanism. In ICASSP 2022–2022 IEEE International Conference on Acoustics, Speech and Signal Processing (ICASSP), pp. 8007–8011. IEEE (May 2022)

16. Yang, R., Cheng, G., Zhang, P., Yan, Y.: An E2E-ASR-based iteratively-trained timestamp estimator. IEEE Signal Process. Lett. **29**, 1654–1658 (2022)

17. Young, S. J., Young, S.: The HTK hidden Markov model toolkit: Design and philosophy (1993)

18. Povey, D.: The Kaldi speech recognition toolkit. In: IEEE 2011 Workshop On Automatic Speech Recognition and Understanding. IEEE (2011)

19. McAuliffe, M., Socolof, M., Mihuc, S., Wagner, M., Sonderegger, M.: Montreal forced aligner: trainable text-speech alignment using kaldi. In: Interspeech, vol. 2017, pp. 498–502 (August 2017)

20. Watanabe, S., et al.: Espnet: End-to-end speech processing toolkit. arXiv preprint arXiv:1804.00015 (2018)

21. Yu, F., et al.: Boundary and context aware training for CIF-based non-autoregressive end-to-end ASR. In: 2021 IEEE Automatic Speech Recognition and Understanding Workshop (ASRU), pp. 328–334. IEEE (December 2021)

# Predictive AutoEncoders Are Context-Aware Unsupervised Anomalous Sound Detectors

Xiao-Min Zeng[1], Yan Song[1(✉)], Li-Rong Dai[1], and Lin Liu[2]

[1] National Engineering Research Center of Speech and Language Information Processing, University of Science and Technology of China, Hefei, China
ZengXiaoMin@mail.ustc.edu.cn, {songy,lrdai}@ustc.edu.cn
[2] iFLYTEK Research, iFLYTEK CO., LTD., Hefei, China
liulin@iflytek.com

**Abstract.** In this paper, we propose a Predictive AutoEncoder (PAE) capable of exploiting context information for unsupervised anomalous sound detection (ASD). The conventional unsupervised ASD approaches mainly employ the straightforward deep neural network (DNN) to detect abnormal sounds. However, this model fails to consider the utilization of the relationship between frames, resulting in limited performance and constrained input length. Recently, context information has been proven to be valid for sequence data processing. In our method, the PAE consisting of transformer blocks is proposed to predict unseen frames by remaining available inputs. Based on the self-attention mechanism, our model captures not only content information within the frame but also context information between frames to improve ASD performance. Moreover, our method extends the input length of AE-based models due to its outstanding capability of long-range sequence modeling. The extensive experiments conducted on the DCASE2020 Task2 development dataset demonstrate that our method outperforms the state-of-the-art AE-based methods and verify the effectiveness and stability of our proposed method for long-range temporal inputs.

**Keywords:** Self-Attention · Context Information · Unsupervised Learning · Anomalous Sound Detection

## 1 Introduction

Anomalous sound detection (ASD) is an acoustic task to detect whether a clip of sound emitted from a given machine is abnormal or not. Recently, ASD has gained more and more attention since this technology can be widely applied to the automatic monitoring of industrial machines. However, anomalous events are diverse but rare, resulting in high costs for collecting enough anomalous sound data. Generally, only normal data is provided to develop an ASD system, which

L. Zhenhua et al. (Eds.): NCMMSC 2022, CCIS 1765, pp. 101–113, 2023.
https://doi.org/10.1007/978-981-99-2401-1_9

is one of the reasons why ASD is exceptionally challenging. Since 2020, DCASE[1] challenge has arranged an unsupervised ASD task to attract further research.

The ASD systems are trained to model the distribution of normal data by unsupervised learning. The probability that a sound belongs to this distribution is regarded as a distinguishable criterion that can detect anomalies from plenty of normal data. Various ASD methods have been proposed, which can be categorized into discriminative [3,9,14,15,24] and generative approaches [7,9–11,20,29,31,32]. The discriminative methods convert the binary classification task of anomaly detection into a machine-IDs identification problem. Benefiting from the representation learning capability of neural networks, these methods achieve superior performance. The generative approaches mainly calculate the errors between the generated output and the input features to detect anomalies, including GAN [10,22,33] and Flow [7,17]. AutoEncoder [20] is also one of the unsupervised generative approaches. Based on its framework, Interpolation Deep Neural Network (IDNN) [29], Conditional AutoEncoder (CAE) [11,18], and Attentive Neural Processes (ANP) [31] are proposed. These AE-based methods pay more attention to local details and mainly depend on content information within the frame.

However, the discriminative methods [3,9,14,24] are only suitable for some anomaly detection tasks as the presence of data without machine-ID labels. Although AE-based generative approaches [11,18,20,29,31] are universal for unsupervised ASD, they break the relationship between frames due to the limitations of deep neural networks (DNN). Furthermore, these methods restrict the length of the input, enabling insufficient ability to model long-range temporal inputs. The temporal association has been demonstrated to be effective for various acoustic tasks [2,4,6,16,26,28]. Therefore, the relationship between frames should be reasonably utilized in ASD.

In this paper, we concentrate on the universal AE-based methods. Go beyond previous methods, we introduce Transformer [30] into unsupervised anomalous sound detection and propose Predictive AutoEncoder (PAE) as a context-aware detector. Applying Transformer to ASD, we find that the temporal association of each frame can be obtained from the self-attention mechanism [30], which presents as a distribution of its association weights to all frames. The distribution can provide a more informative description of the context, so it is named context information in acoustic applications. Based on the transformer blocks, our method exploits context and content information to detect abnormal sounds. Furthermore, due to the transformer's ability to process sequence data, our PAE extends the input length.

We experimentally evaluate our method on the DCASE2020 Challenge Task2 development dataset. Compared with the previous AE-based methods, our PAE achieves significant improvement in ASD performance. Meanwhile, we conduct ablation experiments to verify the effectiveness of our methods for long-range temporal inputs.

---

[1] DCASE: Detection and Classification of Acoustic Scenes and Events, https://dcase.community.

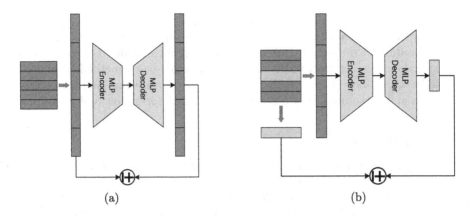

**Fig. 1.** The AE-based methods. (a) The overview of AutoEncoder (AE) [20]. (b) The framework of IDNN [29].

## 2  Related Work

### 2.1  Unsupervised Anomalous Sound Detection

As practical application requirements, unsupervised anomalous sound detection has been widely explored. Categorizing by training manner, the paradigms of ASD roughly include discriminative and generative methods.

The discriminative approaches address the machine-ID identification problems as new ways to accomplish anomalous detection tasks according to the idea of outlier exposure [14]. These methods are based on CNN models to distinguish the machine-ID labels, like ResNet [3,13,15], MobileNetV2 [9,27], MobileFaceNet [5,24]. Moreover, CRNN [23] is utilized to model short-range dependencies between frames to identify machine-ID labels. The classification-based approaches train descriptive decision boundaries by squeezing between different machine-IDs. These methods require careful and delicate design on labels so that they may be unavailable for some unsupervised anomalous sound detection tasks.

The key idea of the generative approach is to reconstruct the information of normal samples better than anomalies. The dominant frameworks of these methods include GAN [10], Flow [7], and AutoEncoder [20]. These methods train the model according to the principle of minimizing errors between the input and output, forcing the model to pay attention to details on the frames. Following the framework of AutoEncoder, IDNN [29] is proposed to predict unseen frames by available information, and CAE [11] integrates category labels into AE-based methods. However, the structure of conventional AE-based methods is a simple DNN, which requires the input features to be reshaped into a one-dimensional vector, as illustrated in Fig. 1(a)(b). This operation explicitly breaks the relationship between frames and causes most context information to be discarded. ANP [31] proposes to predict unseen frames and frequency bands using the attentive association between coordinate positions and magnitudes. Nevertheless, this method views the acoustic spectrum as an image, so it also does not take advantage of the time-frequency relationship of the audio. Due to the lack of association between frames, the performance of AE-based methods is limited. To address the above issues, PAE is proposed to capture the inter-frame relationships and make full use of context and content information to detect sound anomalies.

## 2.2 Transformer

Recently, the powerful ability of Transformer [30] has been shown in sequential data processing, such as natural language processing [6,26], audio processing [2, 4,16,28] and computer vision [8,12]. With the benefit of the Transformer's self-attention mechanism, long-range temporal relationships are available for audio detection tasks. Inspired by it, we enable the well-designed detector to exploit the association along the time dimension to detect anomalous sounds. Unlike Masked AutoEncoder in computer vision [12], our model focuses on exploiting the relationship between frames and the representations within the frame to predict unseen frames, which is more matching for acoustic anomalous detection applications.

## 3    Proposed Method

This section introduces the self-attention mechanism applied to anomalous sound detection in the beginning. Then, the architecture of our proposed Predictive AutoEncoder (PAE) will be elaborated. Lastly, we describe the training objective in brief.

### 3.1    Self-attention Mechanism

The conventional methods are based on the DNN that consists of several superficial fully-connect layers. This simple structure restricts the relationship between frames to be utilized for sound applications. Given the limitation of the DNN module for anomaly sound detection, we renovate this module to the Transformer. The transformer block is shown in Fig. 2.

The transformer blocks are characterized by stacking the Multi-Head Attention (MHA) and Feed-Forward Network (FFN) alternately.

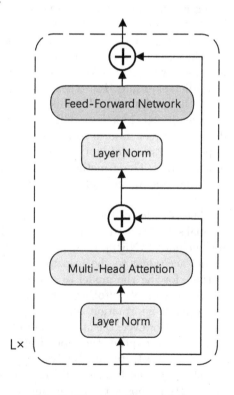

**Fig. 2.** The transformer block.

This stacking structure is favorable for learning the informative association of frames from deep multi-level features. Suppose the model contains $L$ transformer blocks with length-$T$ input sequence $\mathbf{x}$. The overall equations of the $l$-th layer are formalized as:

$$\mathbf{z}^l = \text{MHA}\left(\text{LN}\left(\mathbf{x}^{l-1}\right)\right) + \mathbf{x}^{l-1}$$
$$\mathbf{x}^l = \text{FFN}\left(\text{LN}\left(\mathbf{z}^l\right)\right) + \mathbf{z}^l \tag{1}$$

where $\mathbf{x}^l$ denotes the output of the $l$-th transformer layer, $\mathbf{z}^l$ is the $l$-th layer's hidden representation. LN represents the Layer-Norm module [1].

The Multi-Head Attention (MHA) mechanism adaptively finds the most effective relationship from raw frames. It obtains the query $\mathcal{Q}$, key $\mathcal{K}$, and value $\mathcal{V}$ of the input sequence by a linear transformation and calculates the attention weights $\mathcal{A}$ between frames of the input sequence with the query $\mathcal{Q}$ and key $\mathcal{K}$. The Multi-Head Attention in the $l$-th is:

$$\mathcal{Q}, \ \mathcal{K}, \ \mathcal{V} = \mathbf{x}_{\text{LN}}^{l-1} W_{\mathcal{Q}}^l, \ \mathbf{x}_{\text{LN}}^{l-1} W_{\mathcal{K}}^l, \ \mathbf{x}_{\text{LN}}^{l-1} W_{\mathcal{V}}^l$$
$$\mathcal{A} = \text{Softmax}\left(\frac{\mathcal{Q}\mathcal{K}^\top}{\sqrt{d_m}}\right) \tag{2}$$
$$\hat{\mathbf{z}}^l = \mathcal{A}\mathcal{V}$$

where $\mathbf{x}_{\text{LN}}^{l-1} = \text{LN}\left(\mathbf{x}^{l-1}\right)$ and $\hat{\mathbf{z}}^l$ is the output of Multi-Head Attention module. $W_{\mathcal{Q}}^l$, $W_{\mathcal{K}}^l$, $W_{\mathcal{V}}^l \in \mathbb{R}^{d_m \times d_m}$ represent the parameter matrices for $\mathcal{Q}$, $\mathcal{K}$, $\mathcal{V}$ in the $l$-th layer respectively, and $d_m$ is the dimension of the $l$-th transformer block. The relationship of frames is calculated by adaptive learning since $W_{\mathcal{Q}}^l$, $W_{\mathcal{K}}^l$, $W_{\mathcal{V}}^l$ are learnable parameters of the model. In the multi-head version that we use, the dimension of $\mathcal{Q}_m$, $\mathcal{K}_m$, $\mathcal{V}_m$ is $\frac{d_m}{M}$. The block concatenates the output $\left\{\hat{z_m}^l\right\}_{1 \leq m \leq M}$ from the multiple heads and gets the final result $\hat{\mathbf{z}}^l$. Based on the Multi-Head Attention mechanism, the context information between frames can be captured entirely and applied to reconstruct or predict the output frames.

The role of the Feed Forward Network (FFN) is to fulfill non-linear transformation, similar to the conventional DNN. This module will obtain content information within the frame through several fully-connect layers.

We introduce the transformer blocks into AE-based anomaly sound detection methods and encourage the model equipped with the self-attention mechanism to capture the relationship between frames. On the basis of the previous intra-frame content information being utilized, our model also takes full advantage of inter-frame context information to assist in reconstructing or predicting the outputs. Furthermore, our model extends the frame length of the input features since the self-attention mechanism has a solid capability to measure long-range association.

## 3.2   The Architecture of Predictive AutoEncoder

Predictive AutoEncoder (PAE) is an unsupervised generative model based on the self-attention mechanism, and the overview architecture of PAE is shown in Fig. 3.

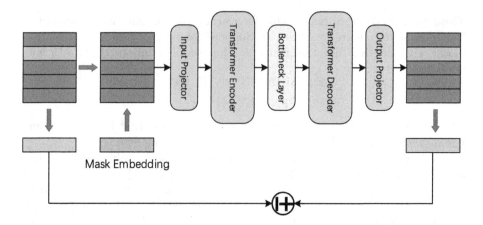

**Fig. 3.** The architecture of our Predictive AutoEncoder (PAE). The input acoustic features will be reshaped into a one-dimensional vector in Fig. 1(a)(b). Instead, PAE maintains the original time-frequency structure.

Let $\mathbf{x} \in \mathbb{R}^{F \times T}$ be the input original acoustic features, where $F$ denotes the dimension of the spectrum and $T$ represents the number of frames. The random subset of the input features will be replaced by mask embedding, which is one of the parameters of the PAE model. We define the masked input as $\mathbf{x}_m$. The PAE model embeds $\mathbf{x}_m$ by a linear projection with added positional embedding, that is:

$$\mathbf{x}_{E\text{-}in} = FC\left(\mathbf{x}_m\right) + PE \tag{3}$$

where $\mathbf{x}_{E\text{-}in} \in \mathbb{R}^{d_m \times T}$ represents the input of encoder, $FC$ is the linear projection layer, and $PE$ is the positional embedding which is a learnable parameter matrix in our model.

After input projecting, $\mathbf{x}_{E\text{-}in}$ is fed into the encoder $E_\theta$ that consists of multiple transformer blocks. The encoder learns latent representations of the input sequence. Moreover, we set up a decoder $D_\phi$ with another series of transformer blocks to perform reconstruction from latent representations.

In the conventional AE-based methods, the input features are compressed to a low-dimensional vector, which is like a filter to select helpful information related to normal data since only normal data is used to train the model. Therefore, an additional bottleneck layer $L_w$ composed of MLPs is designed for dimensional transformation and feature compression, that is:

$$L_w\left(\cdot\right) = FC\left(\text{ReLu}\left(FC\left(\cdot\right)\right)\right) \tag{4}$$

The bottleneck layer $L_w$ is placed between the encoder $E_\theta$ and decoder $D_\phi$. In total, the processing of the encoder and decoder is expressed as follows:

$$\mathbf{x}_{D\text{-}out} = D_\phi\left(L_w\left(E_\theta\left(\mathbf{x}_{E\text{-}in}\right)\right)\right) \tag{5}$$

where $\mathbf{x}_{D\text{-}out} \in \mathbb{R}^{d_m \times T}$ is the output of decoder. Due to the self-attention mechanism and FFN, both context and content information are extracted to model the distribution of normal data.

Finally, a non-linear transformation output projector is applied to reconstruct the original acoustic feature, that is:

$$\tilde{\mathbf{x}} = FC\left(\text{ReLu}\left(FC\left(\mathbf{x}_{D\text{-}out}\right)\right)\right) \tag{6}$$

where $\tilde{\mathbf{x}} \in \mathbb{R}^{F \times T}$ is the output of the entire model, used to calculate the training loss and anomaly score.

### 3.3 Training Strategy

Similar to IDNN [29] and ANP [31], our method also employs the idea of masking and prediction strategy to train the PAE model for anomaly detection. We randomly select partial frames to be substituted by mask embedding and treat these selected frames as the target outputs. Compared to IDNN [29], our method allows random frames to be selected as unseen frames, enabling the PAE model to accept more diverse inputs and be more manageable.

Following the previous AE-based approaches, we adopt the mean squared error (MSE) as the loss function. We train the whole PAE model end-to-end with minimizing training objective, that is:

$$\min \left\| \text{M} \odot \mathbf{x} - \text{M} \odot \tilde{\mathbf{x}} \right\|_2^2 \tag{7}$$

where M is a binary mask matrix representing the position of random mask frames, and $\odot$ is an element-wise multiplication. The test utterance's anomaly scores are obtained by calculating the MSE between the output frames and target frames.

## 4 Experiments and Results

### 4.1 Experimental Setup

**Dataset.** We evaluate Predictive AutoEncoder (PAE) on the DCASE2020 challenge task2 development dataset. This dataset comprises part of ToyADMOS dataset [21] and MIMII dataset [25] and contains six machine types, including ToyCar, ToyConveyor, Fan, Pump, Slider, and Valve. Each machine type has either three or four different machines, respectively, and these different machines are marked with machine-IDs. In our experiments, the training set with only normal sounds is used to train the model, and the test set, which contains both normal and abnormal sounds, is employed for evaluation.

**Table 1.** Average AUC (%) and pAUC (%) of each machine type in different AE-based methods. Average means the average AUC (%) and pAUC (%) of all machine types.

| methods | | AE [20] | IDNN [29] | ANP [31] | PAE |
|---|---|---|---|---|---|
| ToyCar | AUC | **80.90** | 80.19 | 72.50 | 75.35 |
| | pAUC | 69.90 | **71.87** | 67.30 | 69.70 |
| ToyConveyor | AUC | 73.40 | 75.74 | 67.00 | **77.58** |
| | pAUC | 61.10 | 61.26 | 54.50 | **61.37** |
| Fan | AUC | 66.20 | 69.15 | 69.20 | **72.94** |
| | pAUC | 53.20 | 53.53 | **54.40** | 54.37 |
| Pump | AUC | 72.90 | 74.06 | 72.80 | **74.27** |
| | pAUC | 60.30 | 61.26 | 61.80 | **62.01** |
| Slider | AUC | 85.50 | 88.32 | 90.70 | **91.92** |
| | pAUC | 67.80 | 69.07 | 74.20 | **74.39** |
| Valve | AUC | 66.30 | 88.31 | 86.90 | **95.41** |
| | pAUC | 51.20 | 65.67 | 70.70 | **81.24** |
| Average | AUC | 74.20 | 79.30 | 76.52 | **81.25** |
| | pAUC | 60.58 | 63.78 | 63.82 | **67.18** |

**Implementation.** We apply STFT on sound recording with a Hanning window size of 1024 and hop length of 512. We extract the log-Mel spectrogram with Mel filter banks of 128 as the input features of our system. In this case, the dimension of input $F$ is equal to 128.

We train the PAE models for each machine type, *i.e.* six individual models are trained using the training data for the corresponding machine type. We crop $T$ frames as the input, and the stride of cropping is set to 1. Specifically, assuming a sound clip is converted to a log-Mel spectrogram with $N$ frames, we will crop it to $N - T + 1$ input segments. As the number of cropped frames increases, the input length of model will be expanded, but the actual number of segments available for training will decrease. During the input phase, $c\%$ frames are replaced by the mask embedding, and these frames are considered as targets for computing the MSE loss. Unless otherwise stated, both encoder $E_\theta$ and decoder $D_\phi$ consist of two transformer blocks, and the $d_m$ of them are 512 and 256, and the number of multi-head are 8 and 4, respectively. Besides, the dimension of bottleneck layer $L_w$ is 64 by default. PAE is optimized using Adam optimizer [19] with $\beta = \{0.9, 0.999\}$. The batchsize is set to 512 for 60 epochs with the initial learning rate of 1e-3, and the learning rate will decline to 5e-4 and 1e-4 at the 30th and 50th epochs, respectively.

**Evaluation Metrics.** The performance is evaluated by using the area under the receiver operating characteristic curve (AUC) and the partial-AUC (pAUC), which is calculated as the AUC over a low false-positive-rate (FPR) range $[0, p]$, and $p = 0.1$. We calculate the AUC and pAUC for each machine-ID data and

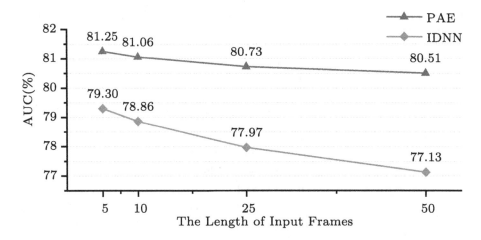

**Fig. 4.** The average AUC (%) for various input length in PAE and IDNN [29].

average these results to get the performance metrics of each machine type. Eventually, we calculate the average AUC and pAUC across all machine types to obtain total AUC and pAUC results.

## 4.2   Results

We compare the Predictive AutoEncoder (PAE) with the previous AE-based generative models and present the performance metrics of each machine type in Table 1. Note that since all the previous methods input five frames into the model, we also set the length of input $T$ to 5 in our method. The mask ratio $c\%$ is set to 20% in PAE, as both IDNN and ANP have 20% of input data unseen when training the model.

The performance of PAE is superior to other AE-based methods for most machine types, especially valve. We believe the primary reason is that our model is an enhancement of IDNN [29] that is extremely favorable for detecting anomalies in non-stationary sounds. With the collaborative improvement of context and content information, our PAE shows the highest average AUC and pAUC, which is persuasive for the advantage of context information in anomalous sound detection. It suggests that our model is more likely to be valuable in real-world applications.

Due to the application of the self-attention mechanism, unseen frames can be predicted by longer temporal relationships. We conduct ablation experiments on the length of input frames $T$ and present the AUC results in Fig. 4. Moreover, to prove the improvement of PAE in extending the input length, we also compare the AUC results of IDNN [29] when the input data is longer. It is not hard to observe that no matter how long the input length is, the performance of PAE is superior to IDNN. Meanwhile, IDNN shows worse sensitivity to the length of input frames $T$, as the simple DNN constrains the utilization of context information.

**Table 2.** Average AUC (%) and pAUC (%) of ablation experiments with different numbers of transformer blocks in encoder $E_\theta$ and decoder $D_\phi$

| $E_\theta$ | $D_\phi$ | AUC | pAUC |
|---|---|---|---|
| 1 | 1 | 79.56 | 66.04 |
| 2 | 1 | 80.15 | 66.88 |
| 2 | 2 | **81.25** | 67.18 |
| 4 | 2 | 80.60 | 67.32 |
| 4 | 4 | 80.91 | 67.35 |
| 6 | 2 | 80.85 | **68.42** |
| 6 | 4 | 81.12 | 67.58 |
| 8 | 2 | 80.47 | 67.69 |
| 8 | 4 | 80.47 | 66.64 |
| 8 | 6 | 80.67 | 66.21 |

We also notice a slight decrease in PAE performance as the input length $T$ increases. That is because longer inputs lead to a broader global attention range, causing the model to pay less attention to local details. In addition, extending the input length reduces the number of segments for training, which is also a potential reason for performance decrease.

In Table 2, we study the influences of the number of transformer blocks of encoder $E_\theta$ and decoder $D_\phi$ on performance. It is surprising that encoder $E_\theta$ and decoder $D_\phi$ with only one layer of transformer blocks can get acceptable performance. This result suggests the significant advantages of the relationship between frames for anomalous sound detection. As the number of parameters in the model increases, the model reaches performance saturation due to the limited amount of training data and shows slight variations in AUC and pAUC because of the randomness in the training process. Considering the balance between runtime and performance, we design encoder $E_\theta$ and decoder $D_\phi$ with two layers of transformer blocks in other ablation experiments.

Furthermore, we conduct masking-prediction experiments with various mask ratios. In this experiment, a lower mask ratio means that more local information is available. To avoid the effects of limited input, we adjust the length $T$ to 25

**Table 3.** Average AUC (%) and pAUC (%) for different mask ratios $c\%$

| mask ratio | AUC | pAUC |
|---|---|---|
| 20% | **80.73** | **67.90** |
| 40% | 80.57 | 67.37 |
| 60% | 80.01 | 67.08 |
| 80% | 79.01 | 65.86 |

frames in this experiment. Table 3 shows the results on different mask ratios $c\%$. It can be found that the more frames that are masked, the worse performance. On the contrary, increasing the number of available input frames is conducive to the improvement of ASD performance. The results shown in Fig. 4 and Table 3 further indicate that the AE-based generative approaches concentrate on the local details of frames. Nevertheless, this does not affect the improvement of context information for anomalous sound detection since the context information contributes to reconstructing or predicting the local details of frames. Therefore, the global attention from the transformer blocks and the local characterization need to be traded off in the practical application.

## 5  Conclusion

This paper studies the unsupervised anomalous sound detection problem. Unlike the previous AE-based methods, our proposed Predictive AutoEncoder (PAE) effectively learns the more promising context information through the self-attention mechanism. With the combination of content and context information, our PAE significantly outperforms other AE-based models. Furthermore, our model overcomes the limitation of the input length and still maintains competitive performance for the long-range temporal input. On the basis of our research, the AE-based generative methods can be more appropriately applied to utterance-level representation learning tasks in future works.

**Acknowledgements.** This work was supported by the Leading Plan of CAS (XDC08030200)

## References

1. Ba, J.L., Kiros, J.R., Hinton, G.E.: Layer normalization. arXiv preprint arXiv:1607.06450 (2016)
2. Baevski, A., Zhou, Y., Mohamed, A., Auli, M.: wav2vec 2.0: A framework for self-supervised learning of speech representations. In: Advances in Neural Information Processing Systems 33, pp. 12449–12460 (2020)
3. Chen, H., Song, Y., Dai, L.R., McLoughlin, I., Liu, L.: Self-supervised representation learning for unsupervised anomalous sound detection under domain shift. In: ICASSP 2022–2022 IEEE International Conference on Acoustics, Speech and Signal Processing (ICASSP), pp. 471–475. IEEE (2022)
4. Chen, S., et al.: Wavlm: Large-scale self-supervised pre-training for full stack speech processing. IEEE J. Selected Topics Signal Process. (2022)
5. Chen, S., Liu, Y., Gao, X., Han, Z.: MobileFaceNets: efficient CNNs for accurate real-time face verification on mobile devices. In: Zhou, J., et al. (eds.) CCBR 2018. LNCS, vol. 10996, pp. 428–438. Springer, Cham (2018). https://doi.org/10.1007/978-3-319-97909-0_46
6. Devlin, J., Chang, M.W., Lee, K., Toutanova, K.: Bert: Pre-training of deep bidirectional transformers for language understanding. arXiv preprint arXiv:1810.04805 (2018)

7. Dohi, K., Endo, T., Purohit, H., Tanabe, R., Kawaguchi, Y.: Flow-based self-supervised density estimation for anomalous sound detection. In: ICASSP 2021–2021 IEEE International Conference on Acoustics, Speech and Signal Processing (ICASSP), pp. 336–340 (2021). https://doi.org/10.1109/ICASSP39728.2021.9414662

8. Dosovitskiy, A., et al.: An image is worth 16x16 words: Transformers for image recognition at scale. arXiv preprint arXiv:2010.11929 (2020)

9. Giri, R., Tenneti, S.V., Cheng, F., Helwani, K., Isik, U., Krishnaswamy, A.: Self-supervised classification for detecting anomalous sounds. In: DCASE, pp. 46–50 (2020)

10. Hatanaka, S., Nishi, H.: Efficient gan - based unsupervised anomaly sound detection for refrigeration units. In: 2021 IEEE 30th International Symposium on Industrial Electronics (ISIE), pp. 1–7 (2021). https://doi.org/10.1109/ISIE45552.2021.9576445

11. Hayashi, T., Yoshimura, T., Adachi, Y.: Conformer-based id-aware autoencoder for unsupervised anomalous sound detection. Tech. rep., DCASE2020 Challenge (July 2020)

12. He, K., Chen, X., Xie, S., Li, Y., Dollár, P., Girshick, R.: Masked autoencoders are scalable vision learners. In: Proceedings of the IEEE/CVF Conference on Computer Vision and Pattern Recognition, pp. 16000–16009 (2022)

13. He, K., Zhang, X., Ren, S., Sun, J.: Deep residual learning for image recognition. In: Proceedings of the IEEE Conference On Computer Vision and Pattern Recognition, pp. 770–778 (2016)

14. Hendrycks, D., Mazeika, M., Dietterich, T.: Deep anomaly detection with outlier exposure. In: International Conference on Learning Representations (2019)

15. Hojjati, H., Armanfard, N.: Self-supervised acoustic anomaly detection via contrastive learning. In: ICASSP 2022–2022 IEEE International Conference on Acoustics, Speech and Signal Processing (ICASSP), pp. 3253–3257. IEEE (2022)

16. Hsu, W.N., Bolte, B., Tsai, Y.H.H., Lakhotia, K., Salakhutdinov, R., Mohamed, A.: Hubert: Self-supervised speech representation learning by masked prediction of hidden units. IEEE/ACM Trans. Audio Speech Lang. Process. **29**, 3451–3460 (2021)

17. Kabir, M.A., Luo, X.: Unsupervised learning for network flow based anomaly detection in the era of deep learning. In: 2020 IEEE Sixth International Conference on Big Data Computing Service and Applications (BigDataService), pp. 165–168. IEEE (2020)

18. Kapka, S.: Id-conditioned auto-encoder for unsupervised anomaly detection. In: Proceedings of the Detection and Classification of Acoustic Scenes and Events 2020 Workshop (DCASE2020), pp. 71–75, Tokyo, Japan (November 2020)

19. Kingma, D.P., Ba, J.: Adam: A method for stochastic optimization. arXiv preprint arXiv:1412.6980 (2014)

20. Koizumi, Y., et al.: Description and discussion on dcase2020 challenge task2: Unsupervised anomalous sound detection for machine condition monitoring. In: Proceedings of the Detection and Classification of Acoustic Scenes and Events 2020 Workshop (DCASE2020), pp. 81–85, Tokyo, Japan (November 2020)

21. Koizumi, Y., Saito, S., Uematsu, H., Harada, N., Imoto, K.: ToyADMOS: A dataset of miniature-machine operating sounds for anomalous sound detection. In: Proceedings of IEEE Workshop on Applications of Signal Processing to Audio and Acoustics (WASPAA), pp. 308–312 (November 2019). https://ieeexplore.ieee.org/document/8937164

22. Li, Y., Peng, X., Zhang, J., Li, Z., Wen, M.: Dct-gan: Dilated convolutional transformer-based gan for time series anomaly detection. IEEE Trans. Knowl. Data Eng. (2021)
23. Liao, W.L., et al.: Dcase 2021 task 2: Anomalous sound detection using conditional autoencoder and convolutional recurrent neural netw. Tech. rep., DCASE2021 Challenge (July 2021)
24. Liu, Y., Guan, J., Zhu, Q., Wang, W.: Anomalous sound detection using spectral-temporal information fusion. In: ICASSP 2022–2022 IEEE International Conference on Acoustics, Speech and Signal Processing (ICASSP), pp. 816–820. IEEE (2022)
25. Purohit, H., et al.: MIMII Dataset: Sound dataset for malfunctioning industrial machine investigation and inspection. In: Proceedings of the Detection and Classification of Acoustic Scenes and Events 2019 Workshop (DCASE2019), pp. 209–213 (November 2019)
26. Radford, A., Narasimhan, K., Salimans, T., Sutskever, I., et al.: Improving language understanding by generative pre-training (2018)
27. Sandler, M., Howard, A., Zhu, M., Zhmoginov, A., Chen, L.C.: Mobilenetv 2: Inverted residuals and linear bottlenecks. In: Proceedings of the IEEE Conference on Computer Vision and Pattern Recognition, pp. 4510–4520 (2018)
28. Schneider, S., Baevski, A., Collobert, R., Auli, M.: wav2vec: Unsupervised pre-training for speech recognition. arXiv preprint arXiv:1904.05862 (2019)
29. Suefusa, K., Nishida, T., Purohit, H., Tanabe, R., Endo, T., Kawaguchi, Y.: Anomalous sound detection based on interpolation deep neural network. In: ICASSP 2020–2020 IEEE International Conference on Acoustics, Speech and Signal Processing (ICASSP), pp. 271–275. IEEE (2020)
30. Vaswani, A., et al.: Attention is all you need. In: Advances in Neural Information Processing Systems 30 (2017)
31. Wichern, G., Chakrabarty, A., Wang, Z.Q., Le Roux, J.: Anomalous sound detection using attentive neural processes. In: 2021 IEEE Workshop on Applications of Signal Processing to Audio and Acoustics (WASPAA), pp. 186–190. IEEE (2021)
32. Xu, J., Wu, H., Wang, J., Long, M.: Anomaly transformer: Time series anomaly detection with association discrepancy. In: International Conference on Learning Representations (2022)
33. Zhou, B., Liu, S., Hooi, B., Cheng, X., Ye, J.: Beatgan: Anomalous rhythm detection using adversarially generated time series. In: IJCAI, pp. 4433–4439 (2019)

# A Pipelined Framework with Serialized Output Training for Overlapping Speech Recognition

Tao Li[1], Lingyan Huang[1], Feng Wang[1], Song Li[2], Qingyang Hong[1(✉)], and Lin Li[2(✉)]

[1] School of Informatics, Xiamen University, Xiamen 361005, China
qyhong@xmu.edu.cn
[2] School of Electronic Science and Engineering, Xiamen University, Xiamen 361005, China
lilin@xmu.edu.cn

**Abstract.** Far-field, noise, reverberation, and overlapping speech make the cocktail party problem one of the greatest challenges in speech recognition. In this paper, we focus on solving the problem of overlapping speech and present a pipelined architecture with serialized output training(SOT). The baseline and the proposed methods are evaluated on the artificially mixed speech datasets generated from the AliMeeting corpus. Experimental results demonstrate that our proposed model outperforms the baseline even with high overlap ratio, which leads to 10.8% and 4.9% relative performance gains in terms of CER for 0.5 overlap ratio and average case, respectively.

**Keywords:** Serialized output training · Multi-talker speech recognition · Overlapping speech · Cocktail party

## 1 Introduction

At a noisy cocktail party, people can ignore the distractions of other people's voices and background noise while communicating with their peers. Humans can do this easily because the brain has special mechanisms for perceiving and paying attention to sound [1], which is still very challenging for computers. Although the best current speech recognition models [2,3] already outperform professional human transcriptionists on ideal single-speaker datasets in terms of accuracy, they hardly work in complex scenarios. Complex scenarios are often composed of multiple factors, including far-field, noise, reverberation, overlapping speech, and so on. The speech overlap problem can be described as the simple superposition of several simultaneously active speech signals.

For overlapping speech recognition, a great deal of work [4–7] followed the approach of "separation before recognition". However, these efforts ignored the inconsistencies in the evaluation metrics of the separation and recognition modules, which cause speech distortion and thus degrade the performance of ASR [8].

L. Zhenhua et al. (Eds.): NCMMSC 2022, CCIS 1765, pp. 114–123, 2023.
https://doi.org/10.1007/978-981-99-2401-1_10

In order to optimize model in the intended direction of ASR, joint optimization was proposed [9,10]. To further alleviate speech distortion, [11,12] adopted different fusion mechanisms to make the most of the information from the original speech signal. Following that, [13] introduced perceptual losses to complement the detailed information of the reconstructed signal. In addition, label ambiguity problem due to models with multiple output layers should also be taken into account, [14,15] proposed a strategy called permutation invariance training (PIT) which calculates the minimum assignment loss for predicted output and label from the frame level and utterance level, respectively.

Based on deep neural networks, an increasing number of studies [16–20] were devoted to the end-to-end model. In contrast to the pipelined models, the end-to-end methods take the form of implicit separation. [16] designed a set of speaker encoders to separate different streams of multiple speakers in feature space and performed speech recognition on these separated features. [20] found that the speaker encoders were not well trained due to the wide variation in characteristics of speakers and energy and therefore proposed to use independent attention modules in the decoder to reduce the burden for the encoder. [17] introduced knowledge distillation to end-to-end model, in which pre-trained single-speaker ASR model plays a role in providing supplementary information.

Although the pipelined and end-to-end models above performed well on some datasets, the common problem is that the models have a fixed number of output layers. In other words, these methods can only work with a fixed number of speakers, the scenario with an unknown number of speakers remains a challenge. To this end, [10] proposed One and Rest PIT (OR-PIT), the idea of which is iterative training, that is, extracting one speaker's utterance from the mixture at a time. [21] proposed a concise but effective method, namely serialized output training (SOT), which introduced a special marker to serially combine the tags of different speakers, allowing the model to learn the dependencies between speakers while learning the speech information. [22] further extended SOT using token-level timestamps to make streaming multi-speaker ASR a reality.

In this paper, we explore the effect of the overlap rate on SOT, and observe that the performance of the baseline significantly decreased at overlap ratio higher than 0.4. The main reason is that the high overlap rate exceeds the upper limit of the ASR model's learning ability. Therefore, we propose to combine the pipelined model with SOT, the separated features contain a priori information about the speaker's identity, which can reduce the burden of the ASR module, in addition, we introduce a fusion module to solve the problem of label ambiguity.

The rest of the paper is organized as follows: In Sect. 2, the adopted methods and the proposed methods are described. In Sect. 3, we show the dataset and the experimental setting. In Sect. 4, we evaluate the proposed approach on the artificially mixed speech datasets generated from the AliMeeting [23] corpus, and the experiments and analysis are given. Finally, the paper is concluded in Sect. 5.

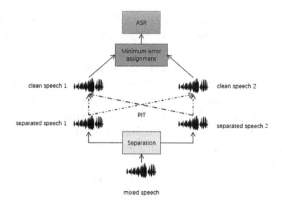

**Fig. 1.** The overall block diagram of pipelined model.

## 2   Overview and the Proposed Methods

### 2.1   Pipelined Model

To simplify the model, the single-channel observed mixture $O \in R^C$ is given by:

$$O = \sum_{i=1}^{N} S_i \tag{1}$$

where the N denotes the number of speakers, $S_i \in R^C$ denotes the speaker's source signal, and the $C$ is the the count of sampling points of the observed mixture. The pipelined architecture aims at separating and recognizing the source signal $S_i, i = 1, ..., N$ from the mixture $O$.

Taking two speakers as an example, the mixture goes through the separation module to get two utterances, then the separated utterances are fed to the recognition module. During the training phase, PIT [15] acts on the separation module to provide the minimum error assignment. The overall block diagram of the model is shown in Fig. 1.

In our expriments, we adopted Conv-TasNet [4] as the separation module. Conv-TasNet is a time-domain speech separation model consisting of an encoder, a separation network, and a decoder.

The encoder and decoder are 1D convolution and 1D transposed convolution, replacing the traditional STFT and iSTFT, respectively, allowing the network to extract features directly in the time domain for mixed speech. 1D convolution maps the observed mixture $O$ to 2D features $\hat{O} \in R^{T \times F}$, where $T$ represents the number of feature frames, whose size mainly depends on the step size of convolution, and $F$ represents the size of feature dimensions, which is corresponding to the number of frequency groups in STFT.

Compared with STFT, data-driven feature extraction can solve the problem of phase mismatch. In most studies on frequency domain models, it is difficult to estimation the phase of each speaker's source signal, and the predicted "clean" signal is reconstructed by synthesizing the masked amplitude spectrum and the original phase of the mixed speech. However, this biased phase estimate determines the upper bound of the model performance. Data-driven feature extraction avoids decoupling the amplitude and phase, maximizing the retention of the consistency of this information.

Then, the feature $\hat{O}$ is fed into the separation network to estimate $N$ mask weight matrices, and the masks are multiplied by $\hat{O}$ to obtain the masked representation of the source signal for each speaker:

$$R_i = mask_i \circ \hat{O}, i = 1, ..., N. \tag{2}$$

where $\circ$ denotes the Hadamard product, $R_i$ and $mask_i \in R^{T \times F}$ represents the masked representation and the mask weight matrix corresponding to the source signal, respectively.

The separation part adopts the temporal convolutional network (TCN) [24], which is a convolutional network for sequence modeling. Unlike RNN, TCN can process sequences in parallel, and thus it's much faster than RNN [25] in terms of computing speed. In addition, the convolutional kernels of each layer of TCN share parameters, which can reduce a large amount of memory occupation. A TCN block consists of multiple convolutional blocks, each of which is a 1D depthwise separable convolution [26] with different dilation rates. This design can greatly reduce the number of parameters, and residual connections are added between the convolutional blocks to alleviate the vanishing gradient problem.

## 2.2   Serialized Output Training

SOT is a training strategy for overlapping speech recognition. It only serializes labels without changing the model structure, and its essence is to distinguish speakers by making use of temporal information.

Let the transcriptions corresponding to the source signal $S_i, i = 1, ...N$ are $U_{s_1e_1}, U_{s_2e_2}, ..., U_{s_Ne_N}$, where the $s_i$ denotes the start time of source signal in the mixed speech, and $e_i$ indicates the end time to check whether overlap occurs. Following the first-in-first-out rule, these tags are rearranged according to the start time, and a special marker <sc> is added between different tags to indicate speaker switching, the reference tags are merged as:

$$U_{merged} = U_{s_1e_1} < sc > U_{s_2e_2} < sc > ... < sc > U_{s_Ne_N} \tag{3}$$

Note that the method is used on the premise that the start time of the source signal is different, if not, then the order of the tags is randomly determined.

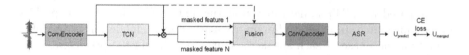

**Fig. 2.** The overall block diagram of proposed model.

In previous work, the number of output layers of models corresponds to the number of speakers, and they are independent of each other, which not only limits the performance of the model in the case of unknown speakers but also causes possible output duplication due to this natural independence [16].

SOT is applied to the single-speaker end-to-end ASR model, so there is only one output layer, and the independence is broken by modeling the dependency relationship between the speakers through special markers. It solves the two difficulties mentioned above at the same time.

However, SOT forces the ASR model to learn the dependencies between speakers through a switch marker while ignoring the limited learning capacity of the model. The more complex the mixed speech, e.g., high overlap rate, low signal-to-noise ratio, etc., the higher the learning pressure on the model. In our experiments of baseline, it is observed that at a high overlap rate, the performance of the model deteriorates substantially, details will be reported in Sect. 4.

### 2.3 Proposed Method

To alleviate the pressure on the ASR model to distinguish speakers, we propose to add a speech separation model before the module. Considering that the features obtained by the separation model are explicitly related to speakers which can further bring some prior information to ASR model.

The proposed model is shown in Fig. 2. The $N$ masked representations from the separator are fed to a fusion module. Since the ASR model only receives the single-channel input, we designed the feature fusion module. Here, we tried three different fusion methods in this module. The first one is to directly add the $N$ features together, and the second one is to concatenate the original features after adding the $N$ features together, as shown in the dashed line. The third method uses cross attention to fuse masked representations with original features:

$$Q = \hat{O}W_Q \in R^{T \times d_k} \tag{4}$$

$$K_i = R_i W_{Ki} \in R^{T \times d_k} \tag{5}$$

$$V_i = R_i W_{Vi} \in R^{T \times d_k} \tag{6}$$

$$F_{inter} = \sum_{i=1}^{N} softmax(\frac{QK_i^{\top}}{\sqrt{d_k}})Vi \tag{7}$$

where $W_Q \in R^{F \times d_k}$, $W_{Ki} \in R^{F \times d_k}$ and $W_{Vi} \in R^{F \times d_k}$, $i = 1, .., N$ are weight matrices, $F_{inter}$ indicates the intermediate feature, in this work, we set $N$ equal to 2.

# 3    Experimental Settings

## 3.1    DataSet

We manually generated single-channel mixed speech of two speakers based on AliMeeting corpus [23]. This corpus contains 120 h of Chinese conference recordings, including far-field data collected by an 8-channel microphone array and near-field data collected by the headphone microphone of each participant.

For each sample, it is generated by mixing the utterances of any two speakers in the AliMeeting dataset, where the overlap rate is obtained by random sampling from the uniform distribution on the interval [0.3, 0.8], and the signal noise ratio (SNR) of one speaker relative to another speaker is obtained on the interval [−5 dB, 5 dB] in the same way.

Following this approach, we generated a 240-hour training set and a 6-hour validation set using near-field data. To facilitate the evaluation of our model, we generated speech with an overlap rate from 0.3 to 0.8 with increments of 0.1 as the test set, the speech with each overlap rate was about 2 h, forming a total of 12 h.

## 3.2    Training and Evaluation Metric

Our models are trained according to SISNR and CER, where SISNR is a signal-to-noise ratio independent of signal variation and is used to guide the training of the front-end separation module, and CER is the Chinese character error rate for the training of back-end speech recognition module. Note that we use PIT to solve the label ambiguity problem when pre-training the separation module. In the testing phase, we only evaluate the performance based on CER.

## 3.3    Model Settings

We adopt the ESPNet-based multi-speaker ASR system (https://github.com/yufan-aslp/AliMeeting/tree/main/asr) from the M2Met challenge as the baseline system. The system uses Conformer [27] as the encoder, which combines the advantages of Transformer and convolution to extract the local and global features of the sequence at the same time.

The separation module adopts Conv-TasNet [4], where the number of channels of encoder and decoder is 256, kernel size is 20, and step size is 10, corresponding to the number of frequency groups, window length, and hop size in STFT, respectively. The separation network has 4 TCN blocks, where each TCN block contains 8 convolution blocks with the bottleneck layer of 256 dimensions and the hidden layer of 512 dimensions. The speech recognition module follows the configuration of the baseline, where the encoder has 12 Conformer blocks with 4 attention heads. The number of units of position-wise feed-forward is 2048, and the decoder has 6 Transformer blocks. The acoustic features are 80-dimensional FBank with frame length of 25 ms and frame shift of 10 ms. For feature augmentation, we employ SpecAugment proposed in [28].

In addition, some tricks such as curriculum learning [29] and warmup [30] are also added to train the model.

## 4   Result Analysis

### 4.1   Baseline Results

First, we evaluated the performance of the baseline model on simulated 2-speaker test mixtures. Results are presented in Table. 1, "average" represents the result over the entire test set. We observe that the performance of the model decreases as the overlap rate gradually increases, and there is a significant drop above 0.4 overlap ratio.

**Table 1.** CER(%) of the baseline model at different overlap ratio.

| Model | overlap ratio | | | | | | average |
|---|---|---|---|---|---|---|---|
| | 0.3 | 0.4 | 0.5 | 0.6 | 0.7 | 0.8 | |
| baseline | 28.5 | 30.2 | 33.1 | 33.3 | 35.2 | 35.3 | 32.7 |

### 4.2   Results of Proposed Method

Then, we trained the proposed model. The initial training was guided entirely by the loss of the ASR module. Surprisingly, the model did not converge, as shown in the results of Model1 in Table 2.

We analyzed the reason and concluded that driving the whole model by the ASR task alone would not allow the front-end separation module to learn effectively as well. To solve this problem, we pre-trained the separation module using clean speech and migrated the trained encoder and separator, freezing the parameters of this part and further training the model.

We implemented three different fusion modules and the results are shown in the rest of Table 2. Model2 represents the direct summation of masked representations, Model3 represents the concatenation of the original features on the basis of Model2, and Model4 represents the fusion of masked representations and original features using cross attention. We observed that Model3 performs the best among all models, with a 10.8% and 4.9% relative improvement compared with the baseline at 0.5 overlap ratio and average case respectively.

**Table 2.** CER(%) of the proposed model at different overlap ratio.

| Model | overlap ratio | | | | | | average |
|---|---|---|---|---|---|---|---|
| | 0.3 | 0.4 | 0.5 | 0.6 | 0.7 | 0.8 | |
| Model1 | 98.7 | - | - | - | - | - | - |
| Model2 | 27.1 | 28.3 | 29.7 | 30.9 | 32.7 | 33.6 | 31.3 |
| Model3 | 26.9 | 28.0 | **29.5** | 30.5 | 32.4 | 33.4 | **31.1** |
| Model4 | 30.5 | 32.3 | 33.6 | 34.1 | 35.4 | 36.8 | 34.2 |

## 5 Conclusion

In this paper, we explored the limitations of the SOT strategy and proposed a fusion method to combine the pipelined model with the SOT. We tried several different feature fusion methods and found that simple addition and concatenation were most effective, which leaded to a relative improvement of 10.8% over the baseline at high overlap rate. In future work, we will explore more complex fusion methods and joint optimization methods. In addition, only single-channel signals are used in this work, and we also consider extending the model to multichannel cases to utilize spatial information.

**Acknowledgements.** This research was funded by the National Natural Science Foundation of China (Grant No. 61876160 and No. 62001405) and in part by the Science and Technology Key Project of Fujian Province, China (Grant No. 2020HZ020005).

## References

1. McDermott, J.H.: The cocktail party problem. Curr. Biol. **19**(22), R1024–R1027 (2009)
2. Xue, B., et al.: Bayesian transformer language models for speech recognition. In: ICASSP 2021–2021 IEEE International Conference on Acoustics, Speech and Signal Processing (ICASSP). IEEE, pp. 7378–7382 (2021)
3. Zeineldeen, M., et al.: Conformer-based hybrid asr system for switchboard dataset. In: ICASSP 2022–2022 IEEE International Conference on Acoustics, Speech and Signal Processing (ICASSP), IEEE, pp. 7437–7441 (2022)
4. Luo, Y., Mesgarani, N.: Conv-tasnet: Surpassing ideal time-frequency magnitude masking for speech separation. IEEE/ACM Trans. Audio, Speech, Lang. Process. **27**(8), 1256–1266 (2019)
5. Luo, Y., Han, C., Mesgarani, N., Ceolini, E., Liu, S.-C.: Fasnet: Low-latency adaptive beamforming for multi-microphone audio processing. In: 2019 IEEE Automatic Speech Recognition and Understanding Workshop (ASRU)
6. Luo, Y., Chen, Z., Yoshioka, T.: Dual-path rnn: efficient long sequence modeling for time-domain single-channel speech separation. In: ICASSP 2020–2020 IEEE International Conference on Acoustics, Speech and Signal Processing (ICASSP). IEEE, 2020, pp. 46–50 (2020)
7. Zeghidour, N., Grangier, D.: Wavesplit: end-to-end speech separation by speaker clustering. IEEE/ACM Trans. Audio, Speech, Lang. Process. **29**, 2840–2849 (2021)

8. Heymann, J., Drude, L., Haeb-Umbach, R.: Neural network based spectral mask estimation for acoustic beamforming. In: IEEE International Conference on Acoustics, Speech and Signal Processing (ICASSP), Shanghai, China, vol. 03, pp. 196–200 (2016)

9. Wang, Z.-Q., Wang, D.: A joint training framework for robust automatic speech recognition. IEEE/ACM Trans. Audio, Speech, Lang. Process. **24**(4), 796–806 (2016)

10. von Neumann, T., et al: Multi-talker asr for an unknown number of sources: Joint training of source counting, separation and asr, arXiv preprint arXiv:2006.02786 (2020)

11. Fan, C., Yi, J., Tao, J., Tian, Z., Liu, B., Wen, Z.: Gated recurrent fusion with joint training framework for robust end-to-end speech recognition. IEEE/ACM Trans. on Audio Speech Lang. Process. **29**, 198–209 (2020)

12. Liu, B., et al.: Jointly adversarial enhancement training for robust end-to-end speech recognition, vol. 09, pp. 491–495 (2019)

13. Zhuang, X., Zhang, L., Zhang, Z., Qian, Y., Wang, M.: Coarse-grained attention fusion with joint training framework for complex speech enhancement and end-to-end speech recognition. In: Interspeech (2022)

14. Yu, D., Kolbæk, M., Tan, Z.-H., Jensen, J.: Permutation invariant training of deep models for speaker-independent multi-talker speech separation. In: 2017 IEEE International Conference on Acoustics, Speech and Signal Processing (ICASSP). IEEE, 2017, pp. 241–245 (2017)

15. Kolbæk, M., Yu, D., Tan, Z.-H., Jensen, J.: Multitalker speech separation with utterance-level permutation invariant training of deep recurrent neural networks. IEEE/ACM Trans. Audio, Speech Lang. Process. **25**(10), 1901–1913 (2017)

16. Seki,H., Hori, T., Watanabe, S., Le Roux, J., Hershey, J.: A purely end-to-end system for multi-speaker speech recognition, vol. 01 2018, pp. 2620–2630 (2018)

17. Zhang, W., Chang, X., Qian, Y.: Knowledge distillation for end-to-end monaural multi-talker asr system, vol. 09 2019, pp. 2633–2637 (2019)

18. Zhang, W., Chang, X., Qian, Y., Watanabe, S.: Improving end-to-end single-channel multi-talker speech recognition. IEEE/ACM Trans. Audio, Speech, Lang. Process. **28**, 1385–1394 (2020)

19. Chang, X., Zhang, W., Qian, Y., Le Roux, J., Watanabe, S.: End-to-end multi-speaker speech recognition with transformer, vol. 02 (2020)

20. Chang, X., Qian, Y., Yu, K., Watanabe, S.: End-to-end monaural multi-speaker asr system without pretraining. In: ICASSP 2019–2019 IEEE International Conference on Acoustics, Speech and Signal Processing (ICASSP), 2019, pp. 6256–6260 (2019)

21. Kanda, N., Gaur, Y., Wang, X., Meng, Z., Yoshioka, T.: Serialized output training for end-to-end overlapped speech recognition, vol. 10, pp. 2797–2801 (2020)

22. N. Kanda, et al.: Streaming multi-talker asr with token-level serialized output training, vol. 02 (2022)

23. Yu, F.: M2met: The icassp 2022 multi-channel multi-party meeting transcription challenge, vol. 10 (2021)

24. Bai, S., Kolter, J.Z., Koltun, V.: An empirical evaluation of generic convolutional and recurrent networks for sequence modeling. arXiv preprint arXiv:1803.01271 (2018)

25. Hopfield, J.J.: Neural networks and physical systems with emergent collective computational abilities. Proc. Natl. Acad. Sci. **79**(8), 2554–2558 (1982)

26. Chollet, F.: Xception: Deep learning with depthwise separable convolutions," in Proceedings of the IEEE Conference on Computer Vision and Pattern Recognition, pp. 1251–1258 (2017)

27. Gulati, A.: Conformer: Convolution-augmented transformer for speech recognition, vol. 10 2020, pp. 5036–5040 (2020)
28. Park, D.S.: SpecAugment: A simple data augmentation method for automatic speech recognition. In: Interspeech 2019. ISCA (2019)
29. Bengio, Y., Louradour, J., Collobert, R., Weston, J.: Curriculum learning, vol. 60, 01 2009, p. 6 (2009)
30. He, K., Zhang, X., Ren, S., Sun, J.: Deep residual learning for image recognition, vol. 06 2016, pp. 770–778 (2016)

# Adversarial Training Based on Meta-Learning in Unseen Domains for Speaker Verification

Jian-Tao Zhang[1], Xin Fang[2], Jin Li[1,2], Yan Song[1(✉)], and Li-Rong Dai[1]

[1] National Engineering Research Center of Speech and Language Information Processing, University of Science and Technology of China, Hefei, China
zhangjiantao@mail.ustc.edu.cn, {songy,lrdai}@ustc.edu.cn
[2] iFLYTEK Research, iFLYTEK CO. LTD., Hefei, China
{xinfang,jinli}@iflytek.com

**Abstract.** In this paper, we propose adversarial training based on meta-learning (AML) for automatic speaker verification (ASV). Existing ASV systems usually suffer from poor performance when apply to unseen data with domain shift caused by the difference between training data and testing data such as scene noise and speaking style. To solve the above issues, the model we proposed includes a backbone and an extra domain attention module, which are optimized via meta-learning to improve the generalization of speaker embedding space. We adopt domain-level adversarial training to make the generated embedding reduce the domain differentiation. Furthermore, we also propose an improved episode-level balanced sampling to simulate the domain shift in the real-world, which is an essential factor for our model to get the improvement. In terms of the domain attention module, we use the multi-layer convolution with bi-linear attention. We experimentally evaluate the proposed method on CNCeleb and VoxCeleb, and the results show that the combination of adversarial training and meta-learning effectively improves the performance in unseen domains.

**Keywords:** Speaker verification · Meta-learning · Adversarial training · Domain invariant attention module

## 1 Introduction

Verifying whether a test utterance belongs to the enroll identity is the main target of automatic speaker verification (ASV). Compare to the traditional i-vector systems [1,2], the most recent proposed methods [3,4] rely on the profusion of deep neural network(DNN).

The success of deep learning still cannot easily solve the domain shift issues caused by complex recording devices and speaking styles in real-world applications. To better solve domain shift issues, many researchers have shifted their focus to domain adaptation and domain generalization. Aligning the distribution of source and target domain by discrepancy-based methods [5,6] and adversarial learning [7,8] are main purposes of domain adaptation. In domain generalization [9], the target domain is unseen, and the model also needs to learn embedding which is robust to the target domain.

In ASV, domain shift issues can be seen as the speaker embedding space learned from the source domain cannot be well fitted into the target domain. For getting the domain invariant embedding, adversarial training [10] is considered for the model optimization. Most recently, a new strategy of meta-learning [11–13] has been used in domain generalization. Zhang et al. [11] use the meta generalized transformation with meta-learning to get better embedding space.

However, these above methods also have some limitations. Firstly, traditional methods only pursue accuracy in the known domain and ignore the generalization ability of the model. Secondly, in the task construction of meta-learning, [11] does not fully utilize the information of domains.

In this paper, in order to enhance the generalization ability of traditional methods, we try to combine adversarial training with meta-learning. In adversarial training, we add an gradient reverse layer (GRL) [14] before the domain classification. Specifically, we propose an improved episode-level balanced sampling which makes meta-train and meta-test tasks come from different domains, and the sampling probability of each domain is based on the actual data volume. By doing so, our task sampling strategy is closer to the real-world applications and can motivate the generalization of meta-learning. With the coordination of adversarial training and meta-learning, our domain invariant attention model can learn more about domain generalization knowledge.

## 2   Overview of the Proposed Network

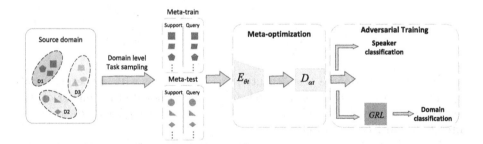

**Fig. 1.** The architecture of adversarial training based on meta-learning. Different colors represent different domains in this figure. The same symbol represents the utterance of the same speaker.

Meta-learning is an important method in machine learning [15,16]. The original intention of this method is to solve the problem of lacking valid data in downstream tasks, which can get enough knowledge from many prior tasks and reuse this knowledge in the target task. In this paper, we adopt the gradient update strategy of meta-learning to realize the generalization ability of the model to the unknown domains.

As shown in Fig. 1, we employ an improved episode-level sampling strategy to build our meta-train and meta-test tasks. Then, the embedding will be generated by the backbone $E_{\theta t}$ and the domain invariant attention module $D_{\alpha t}$. There are two parallel $FC$ layers after embedding extractor $D_{\alpha t}$. The upper branch road is to classify the speaker, and we add an gradient reverse layer (GRL) before the bottom branch road, which is used to classify the domains. Finally, we use two processes meta-train and meta-test to optimize our network parameters.

## 3  Method

### 3.1  Adversarial Training with Multi-task Learning

Previous methods are mainly based on the training strategies of meta-learning to improve the model's robustness. In this paper, we use cross-entropy loss to complete speaker classification tasks. Besides, we also use the domain label information to reconstruct the embedding space. In order to achieve the above objectives, there are two parallel $FC$ layers after embedding extractor $D_{\alpha t}$. $FC_1$ is to classify the speaker, and we add an gradient reverse layer (GRL) before the $FC_2$, which is used to classify the domains. The total loss for the model optimization is composed of speaker loss $\mathcal{L}_{c1}$ domain loss $\mathcal{L}_{c2}$ as:

$$\mathcal{L}_{total} = \mathcal{L}_{c1} + \mathcal{L}_{c2} \tag{1}$$

$\mathcal{L}_{c1}$ and $\mathcal{L}_{c2}$ denote the classification losses of speaker labels and domain labels respectively. Cross entropy will be used on the above two losses.

$$\mathcal{L}_{c1} = -\sum_{i=1}^{N} \log \frac{e^{f_1(x_i^{y_s})}}{\sum\limits_{j=1}^{S} e^{f_1(x_i^j)}} \quad \mathcal{L}_{c2} = -\sum_{i=1}^{N} \log \frac{e^{f_2(x_i^{y_d})}}{\sum\limits_{j=1}^{D} e^{f_2(x_i^j)}} \tag{2}$$

where $x_i^{y_s}$ and $x_i^{y_d}$ donate the $i$-th utterance in a batch with a speaker label $y_s$ and domain label $y_d$. $N$ is the batchsize, $S$ and $D$ donate the number of the speaker classes and domain classes separately. $f_1(\cdot)$ and $f_2(\cdot)$ donate the outputs after $FC_1$ and $FC_2$ separately.

With the effect of GRL, the $FC_2$ and embedding extractor are in a state of confrontation when distinguishing domains. By doing so, the embedding extractor can make the generated embedding reduce the domain differentiation. In coordination with the label clustering loss introduced earlier, we make all the parameters optimized simultaneously using both multi-task learning and adversarial training. Finally, our network can effectively learn domain invariant embedding based on the correct use of label information.

## 3.2   Improved Episode-Level Balanced Sampling

As we all know, in our real-world scenarios, the process between training and testing has a big difference. In order to simulate this difference, we can use episode-level sampling to build our training tasks. To put it simply, the episode-level sampling needs to select randomly $S$ identities from the seen training set. Then, we need randomly select $P$ utterances and $Q$ utterances for every identity selected before, which are named the support set and the query set separately. Now, a task is successfully created which contains a support set and a query set.

To get the domain invariant embedding, we should take full account of domain information in task construction. So we can make some improvements in episode-level sampling. In seen source domain, there will be multiple subdomains. We should select $S$ speakers in each subdomain to build tasks, not in the whole source domain. Besides, in an update iteration, we need to ensure that two tasks selected for meta-train and meta-test come from different domains to simulate the domain shift in the real world. For source domain data $D = \{D_1, D_2...D_B | B > 1\}$, which contains $B$ seen subdomains we can select proposed utterances to build our tasks set from each subdomain. The task set $T = \{T_1, T_2, ..T_B | B > 1\}$, which has $B$ sub-task sets, and each subdomain has its own sub-task set. By doing so, we can further improve the domain generalization ability of the model. Considering the difference in data volume in different subdomains, tasks from different subdomains need to be consistent with the amount of source data. In this way, the source domain data can be trained more comfortably, and balanced sampling without destroying the distribution of original data can be realized.

## 3.3   Domain-Invariant Attention Module

In this paper, we proposed the domain invariant attention module via meta-learning to learn a domain invariant embedding. The network consists of two parts, which are named backbone and domain invariant attention module. By doing so, we can reduce the training difficulty of the backbone, and deliver the ability to get domain invariant embedding to the additional attention module.

We first apply the improved episode-level balanced sampling to build our meta-learning tasks. Every task consists of a support set and a query set, we can use them to build the pair-level batches. Then, we use the backbone $E_{at}$ to get the frame-level feature of the meta-train task which is selected from the task set. The frame-level features will be sent to the domain invariant attention module, and finally output to segment-level embedding. In the meta-test, we select a task that comes from a different sub-task set compared to the task used in the meta-train.

In a meta-update, we calculate the meta-train loss $\mathcal{L}_{mtr}$ and meta-test loss $\mathcal{L}_{mte}$ by the meta-train task and meta-test task respectively. We consider global classification to calculate two losses. After optimizing by the meta-train loss, we obtain the updated parameters $\theta'_t$ of backbone,$\alpha'_t$ of domain invariant attention module.

$$\mathcal{L}_{mtr} = \mathcal{L}_{c1}\left(\mathcal{X}_{mtr}; \theta_t, \alpha_t\right) + \mathcal{L}_{c2}\left(\mathcal{X}_{mtr}; \theta_t, \alpha_t\right) \tag{3}$$

$$\theta'_t = \theta_t - \eta\nabla_{\theta t}\mathcal{L}_{mtr} \tag{4}$$

$$\alpha'_t = \alpha_t - \eta\nabla_{\alpha t}\mathcal{L}_{mtr} \tag{5}$$

where $\mathcal{X}_{mtr}$ donate the utterances of meta-train tasks, $\eta$ is the learning rate in meta-train.

Then we use the temporary update parameters to calculate the meta-test loss $\mathcal{L}_{mte}$. After the gradient backward, we update the $\alpha'_t$ of the domain invariant attention module.

$$\mathcal{L}_{mte} = \mathcal{L}_{c1}\left(\mathcal{X}_{mte}; \theta'_t, \alpha'_t\right) + \mathcal{L}_{c2}\left(\mathcal{X}_{mte}; \theta'_t, \alpha'_t\right) \tag{6}$$

$$\theta_{t+1} = \theta'_t \tag{7}$$

$$\alpha_{t+1} = \alpha_t - \mu\left(\nabla_{\alpha t'}\mathcal{L}_{mte}\right) \tag{8}$$

where $\mathcal{X}_{mte}$ donate the utterances of meta-test tasks, $\mu$ is the learning rate in meta-test.

Finally, we obtain the optimized parameter $\theta_{t+1}$ of the backbone, and the optimized parameter $\alpha_{t+1}$ of the domain invariant attention module.

## 4    Experiments and Analysis

### 4.1    Experimental Settings

**Datasets.** Our purposed method is experimentally evaluated on and CNCeleb [17]. CNCeleb is a large-scale free speaker recognition dataset that contains 2993 speakers in 11 different genres. The entire dataset is split into train data and test data. The train data contains 8 domains including entertainment, play, vlog, live broadcast, speech, advertisement, and recitation, interview from 2793 speakers, and the remaining 200 speakers containing singing, movie, drama are used for evaluation in unseen domains to achieve cross-genre evaluation. Additionally, in the Cross-Genre trial, the test samples of all pairs containing a certain genre are selected as the trial of this genre.

**Input Features.** Kaldi toolkit [18] is used to accomplish the feature extraction process. In our experiments, 41-dimensional filter bank (FBank) acoustic features are transformed from 25ms windows with a 10ms shift between frames. Energy-based voice activity detection (VAD) is used to remove silent segments. The training features are randomly truncated into short slices ranging in length from 2 to 4 s.

**Baseline Configuration.** The backbone of our model is the ResNet-18 with the frame-level output, and the baseline uses the domain invariant attention module to get the speaker embedding. The scaling factor after the L2-norm is set to 30, and batchsize is 256, with each batch consisting of 128 speakers from one domain. The networks are optimized using stochastic gradient descent (SGD), with momentum of 0.9 and weight decay of 5e-4. Additionally, the backbone of all experiments in this paper is pretrained on VoxCeleb [19,20].

## 4.2 Comparative Experiments

**Table 1.** EERs (%) of AML and ablation experiments on the Cross-Genre. AML(w/o GRL) is AML without GRL, AML(w/o ML) is AML without meta-learning.

| Method | Cross-Genre | | |
|---|---|---|---|
| | Singing | Movie | Drama |
| r-vector | 25.55 | 17.12 | 10.58 |
| baseline | 25.4 | 16.97 | 10.3 |
| AML(w/o GRL) | 23.91 | 17.22 | 10.46 |
| AML(w/o ML) | 23.66 | 16.25 | 10.00 |
| AML | **23.02** | **15.75** | **9.34** |

The results are shown in Table 1. To verify the generalization of the proposed method, our experiment is conducted on ResNet-18. Firstly, we can find that the proposed AML with ResNet-18 obtains 9.37%, 7.18%, and 9.6% relative reduction in terms of EER on singing, movie, and drama compared to the baseline with domain invariant attention module, respectively. This benefits from the adversarial training and meta-learning with a domain invariant attention module. Secondly, in the ablation results, we conduct experiments in which only one strategy of adversarial training and meta-learning with a domain invariant attention module is involved. We can find that EER on singing, movie, and drama improves 6.85%, and 4.24% on AML(w/o ML), respectively. It indicates that GRL can further motivate the robustness of the domain invariant attention module. When conducting experiments only using meta-learning with the domain invariant attention module, the performance is also better than the baseline. Inferring that meta-learning can make the domain invariant attention module learn a coincident and better embedding space for each domain. As we expected, the performance of **AML** gets further improvement, owing to achieving a better domain-invariant embedding.

Table 2 shows the comparative results on cross-genre. For cross-genre, we only use the training data of 800 speakers from 7 domains to train the entire model for fairness. By doing so, we can have the same condition with robust-MAML for the performance comparison. As the improved meta-learning method, robust-MAML can be seen that the EER is higher than that of AML. AML achieves an

**Table 2.** EERs (%) of AML and state-of-the-art methods on the Cross-Genre trials.

| Method | Metric | Cross-Genre | | |
|--------|--------|---------|-------|-----------|
| | | Singing | Movie | Interview |
| MCT [12] | cosine | 28.4 | 24.21 | 16.92 |
| Robust-MAML [12] | cosine | 27.08 | 24.21 | 16.87 |
| MGT [11] | cosine | 24.85 | 20.55 | 10.59 |
| baseline | cosine | 25.7 | 17.95 | 9.81 |
| AML | cosine | **23.96** | **15.68** | **9.38** |

EER reduction of 23.69% and 11.42% on movie and interview compared to the MGT, respectively. We believe the primary reason is that MAL uses adversarial training based on meta-learning to motivate domain invariant attention modules to get more robust features.

## 5   Conclusion

This paper proposes adversarial training based on meta-learning to improve the generalizability of speaker embedding. We present an improved episode-level balanced sampling, which can better motivate the potentiality of meta-learning. Experimental results on the CNCeleb dataset demonstrated that adversarial training based on meta-learning is more domain-invariant than the baseline and the meta-generalized transformation. In the future, we will try to train the LDA/PLDA scoring models on CNCeleb.Train. We believe that our purposed method can get better performance with the trained well back-end model. Besides, we will continue to study the training mechanism of meta-learning and tap the potential of meta-learning.

**Acknowledgements.** This work was supported by the Leading Plan of CAS (XDC08030200).

## References

1. Dehak, N., Kenny, P.J., Dehak, R., Dumouchel, P., Ouellet, P.: Front-end factor analysis for speaker verification. IEEE Trans. Audio, Speech, Lang. Process. **19**(4), 788–798 (2010)
2. Kenny, P.: Bayesian speaker verification with, heavy tailed priors. In: Proc. Odyssey 2010 (2010)
3. Snyder, D., Garcia-Romero, D., Sell, G., Povey, D., Khudanpur, S.: X-vectors: Robust dnn embeddings for speaker recognition. In: 2018 IEEE International Conference on Acoustics, Speech and Signal Processing (ICASSP), pp. 5329–5333. IEEE (2018)
4. Desplanques, B., Thienpondt, J., Demuynck, K.: Ecapa-tdnn: Emphasized channel attention, propagation and aggregation in tdnn based speaker verification. arXiv preprint arXiv:2005.07143 (2020)

5. Lin, W., Mak, M.M., Li, N., Su, D., Yu, D.: Multi-level deep neural network adaptation for speaker verification using mmd and consistency regularization. In: ICASSP 2020–2020 IEEE International Conference on Acoustics, Speech and Signal Processing (ICASSP), pp. 6839–6843. IEEE (2020)

6. Sun, B., Saenko, K.: Deep CORAL: correlation alignment for deep domain adaptation. In: Hua, G., Jégou, H. (eds.) ECCV 2016. LNCS, vol. 9915, pp. 443–450. Springer, Cham (2016). https://doi.org/10.1007/978-3-319-49409-8_35

7. Ganin, Y., et al.: Domain-adversarial training of neural networks. J. Mach. Learn. Res. **17**(1), 2030–2096 (2016)

8. Kataria, S., Villalba, J., Zelasko, P., Moro-Velázquez, L., Dehak, N.: Deep feature cyclegans: Speaker identity preserving non-parallel microphone-telephone domain adaptation for speaker verification. arXiv preprint arXiv:2104.01433 (2021)

9. Zhou, K., Yang, Y., Qiao, Y., Xiang, T.: Domain generalization with mixstyle. arXiv preprint arXiv:2104.02008 (2021)

10. Chen, Z., Wang, S., Qian, Y., Yu, K.: Channel invariant speaker embedding learning with joint multi-task and adversarial training. In ICASSP 2020–2020 IEEE International Conference on Acoustics, Speech and Signal Processing (ICASSP), pp. 6574–6578. IEEE (2020)

11. Zhang, H., Wang, L., Lee, K.A., Liu, M., Dang, J., Chen, H.: Learning domain-invariant transformation for speaker verification. In ICASSP 2022–2022 IEEE International Conference on Acoustics, Speech and Signal Processing (ICASSP), pp. 7177–7181. IEEE, (2022)

12. Kang, J., Liu, R., Li, L., Cai, Y., Wang, D., Zheng, F.T.: Domain-invariant speaker vector projection by model-agnostic meta-learning. arXiv preprint arXiv:2005.11900 (2020)

13. Qin, X., Cai, D., Li, M.: Robust multi-channel far-field speaker verification under different in-domain data availability scenarios. In: IEEE/ACM Transactions on Audio, Speech, and Language Processing, pp. 1–15, (2022)

14. Meng, Z., Zhao, Y., Li, J., Gong, Y.: Adversarial speaker verification. In: ICASSP 2019–2019 IEEE International Conference on Acoustics, Speech and Signal Processing (ICASSP), pp. 6216–6220. IEEE (2019)

15. Vilalta, R., Drissi, Y.: A perspective view and survey of meta-learning. Artif. Intell. Rev. **18**(2), 77–95 (2002)

16. Vanschoren, J. Meta-learning: A survey. arXiv preprint arXiv:1810.03548 (2018)

17. Fan, Y., et al.: Cn-celeb: a challenging chinese speaker recognition dataset. In: ICASSP 2020–2020 IEEE International Conference on Acoustics, Speech and Signal Processing (ICASSP), pp. 7604–7608. IEEE, 2020

18. Povey, D., et al.: The kaldi speech recognition toolkit. In: IEEE 2011 Workshop on Automatic Speech Recognition and Understanding, number CONF. IEEE Signal Processing Society (2011)

19. Chung, J.S., Nagrani, A., Zisserman, A.: Voxceleb2: Deep speaker recognition. arXiv preprint arXiv:1806.05622 (2018)

20. Nagrani, A., Chung, J.S., Zisserman, A.: Voxceleb: a large-scale speaker identification dataset. arXiv preprint arXiv:1706.08612 (2017)

# Multi-speaker Multi-style Speech Synthesis with Timbre and Style Disentanglement

Wei Song[✉], Yanghao Yue, Ya-jie Zhang, Zhengchen Zhang, Youzheng Wu, and Xiaodong He

JD Technology Group, Beijing, China
{songwei11,yueyanghao,zhangyajie23,zhangzhengchen1,
wuyouzheng1,hexiaodong}@jd.com

**Abstract.** Disentanglement of a speaker's timbre and style is very important for style transfer in multi-speaker multi-style text-to-speech (TTS) scenarios. With the disentanglement of timbres and styles, TTS systems could synthesize expressive speech for a given speaker with any style which has been seen in the training corpus. However, there are still some shortcomings with the current research on timbre and style disentanglement. The current method either requires single-speaker multi-style recordings, which are difficult and expensive to collect, or uses a complex network and complicated training method, which is difficult to reproduce and control the style transfer behavior. To improve the disentanglement effectiveness of timbres and styles, and to remove the reliance on single-speaker multi-style corpus, a simple but effective timbre and style disentanglement method is proposed in this paper. The FastSpeech2 network is employed as the backbone network, with explicit duration, pitch, and energy trajectory to represent the style. Each speaker's data is considered as a separate and isolated style, then a speaker embedding and a style embedding are added to the FastSpeech2 network to learn disentangled representations. Utterance level pitch and energy normalization are utilized to improve the decoupling effect. Experimental results demonstrate that the proposed model could synthesize speech with any style seen during training with high style similarity while maintaining very high speaker similarity.

**Keywords:** Speech Synthesis · Style Transfer · Disentanglement

## 1 Introduction

With the development of deep learning technology in the last decade, speech synthesis technology has evolved from traditional statistics-based speech synthesis [1] to end-to-end based [2–6] and made great advancements. The current

This work was supported by the National Key R&D Program of China under Grant No. 2020AAA0108600.

speech synthesis technology has been able to synthesize speech with high naturalness and high fidelity, and even some research [7] has been able to synthesize speech that human beings cannot distinguish between true recordings.

Although the great achievement in speech synthesis, there still exists a large improvement room for expressive speech synthesis, especially for multi-speaker multi-style Text to Speech (TTS) with cross-speaker style transfer [8–10]. Synthesizing speech with a target speaker's timbre and other speaker's style could further increase the application scenarios and expressiveness of the TTS system.

In order to synthesize more expressive speech, some researches [8,10,11] do style decoupling by using a single-speaker multi-style recording data to learn the style representation. However, it is difficult and expensive to collect these kinds of data in lots of scenarios. Some other works [9,12] try to remove this data restriction by learning a decoupled style representation, but most of them use complex network structure or rely on complicated training method, which makes it difficult to re-produce the experimental results or infeasible to control the network performance during inference.

A jointly trained reference encoder is used to learn implicit style representation in [12,13]. After the model is trained, audio with a different style or even from a different speaker could be taken as the reference audio to synthesize speech with the desired style while keeping the timbre unchanged. However, the reference-based method is unstable which usually generates unexpected style, and it's non-trivial to choose the reference audio.

Pan et al. [8] use prosody bottle-neck feature to learn a compact style representation, and paper [10] uses a separate style encoder and speaker encoder to disentangle the speaker's timbre and style, and cycle consistent loss is used to improve the disentanglement effect, a complex neural network and complicated training objectives are required to achieve good performance. Both [10] and [8] require single-speaker multi-style corpus to learning the disentanglement, which restricts the flexibility of their proposed models.

In this paper, we propose a simple but effective expressive speech synthesis network that disentangles speakers' timbres and styles, which makes it available to do multi-speaker multi-style speech synthesis. The proposed system does not require a multi-speaker multi-style corpus, each speaker's data is considered as an isolated style. The proposed network is similar to work [9], but we use FastSpeech2 [4] as the network backbone, which removes the skip/repeat pronunciation issue caused by the attention mechanism. Prosody features (duration, pitch, and energy) are used directly to improve the disentanglement effectiveness of timbre and style, furthermore, utterance level pitch and energy normalization (UttNorm) are used to prevent identity leakage from prosody features.

## 2    The Proposed Model

The proposed network utilizes FastSpeech2 [4] as the network backbone and applies utterance level pitch and energy normalization to achieve a better decoupling effect. MelGAN [14] is used as the neural vocoder to convert acoustic features to speech.

## 2.1    The Network Structure of Proposed Network

FastSpeech2 [4] network architecture is used as the network backbone for the proposed model, which consists of a phoneme encoder to learn syntactic and semantic features, a mel-spectrogram [3] decoder to generate frame-level acoustic features, and a variance adaptor to learn style-related features. The network structure is shown in Fig. 1 (a).

A style embedding is introduced to the network to learn a style-dependent variance adaptor, the network structure of variance adapter is illustrated in Fig. 1 (b) and Fig. 1 (c). To decouple timbre and style, the speaker embedding is moved to the input of the decoder to make the decoder timbre dependent.

One important purpose of the proposed model is to remove the dependency of the single-speaker multi-style corpus, so each speaker's corpus is considered as a unique style and we could learn style representation from other speakers' corpus. Actually, during the training procedure, the speaker id and style id is identical, then it is very important to prevent style embedding from leaking into the backbone network, which means style embedding should only be used for style feature prediction and never be exposed to the backbone network. So in this proposed network, style embedding is only used in the variance adaptor to ensure that the whole network learns the speaker's timbre by speaker embedding instead of style embedding, and style is only affected by style embedding.

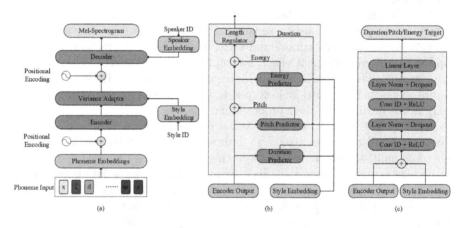

**Fig. 1.** An illustration of the proposed network architecture. (a) Network structure of the acoustic model. (b) The variance adaptor structure with style embedding as conditional input. (c) The structure of variance predictor.

## 2.2    Utterance Level Feature Normalization

The speaking style is represented in many aspects, such as the duration of each syllable, the fundamental frequency (F0), and the trajectory of pitch and energy.

However, these scalars could still contain speaker identity and impact the decoupling effect. For example, female speakers usually have higher F0 [15] than male speakers, and excited speakers express higher energy value, so the pitch and energy features contain speaker timbre information to some extent, then it's essential to normalize the style features to disentangle a speaker's timbre and style.

In this paper, instead of speaker level normalization (SpkNorm), utterance level pitch and energy normalization (UttNorm) are used to remove speaker identity from the style features for better timbre and style disentanglement. The style difference in each utterance will lead to a statistics difference between speaker level and utterance level statistics, this difference could cause identity leakage and affect the decoupling effect, especially for the unprofessional, noisy recordings. UttNorm could eliminate this statistic difference and improve the timbre and style disentanglement effectiveness.

## 3 Experimental Setup

Three open-sourced Chinese mandarin corpora and three internal Chinese mandarin corpora with distinctive styles are used to train the proposed model. The open-sourced corpus includes CSMSC[1], which is recorded by a female speaker, and the MST-Originbeat[2] [16], which consists of recordings from a female speaker and a male speaker. Details of the training data are listed in Table 1.

**Table 1.** The description of the dataset used in the experiment. There are three open-sourced corpus (with a * in the speaker name) and three internal corpora.

| Speaker | #Utterance | Style | Gender |
| --- | --- | --- | --- |
| CSMSC* | 10000 | normal | female |
| Originbeat-S1* | 5000 | normal | female |
| Originbeat-S2* | 5000 | normal | male |
| C1 | 10000 | children story | female |
| F1 | 450 | news-broadcasting | female |
| M1 | 35000 | story-telling | male |

The training data waves are converted to 16 kHz, 16bit depth, and then scaled to 6dB in our experiments. The extracted phoneme labels and processed speech waves are aligned by the MFA [17] tool to detect the phoneme boundary. The 80-band mel-scale spectrogram is extracted as the training target with a 12 ms hop size and 48 ms window size. Pitch is extracted by using the PyWORLD[3]

---

[1] https://www.data-baker.com/open_source.html.

[2] http://challenge.ai.iqiyi.com/detail?raceId=5fb2688224954e0b48431fe0.

[3] https://github.com/JeremyCCHsu/Python-Wrapper-for-World-Vocoder.

toolkit. Both pitch and energy trajectories are normalized in utterance level to remove speaker identity.

The proposed model is trained by Adam [18] optimizer with a batch size of 32 and noam [19] learning rate schedule. The learning rate is warmed up to a maximum value of 1e-3 in the first 4000 steps and then exponentially decayed. The model is trained by 400,000 steps and the network is regularized by weight decay with a weight of 1e-6.

## 4    Experimental Results

This chapter shows the experimental results of the proposed timbre and style disentanglement network. We encourage the readers to listen to the synthesized speeches on our demo page[4].

### 4.1    Subjective Evaluation

To evaluate the proposed method, Mean Opinion Score (MOS) evaluation is conducted to evaluate the speaker similarity and style similarity for the synthesized speech of different speakers and different styles. The target speakers are from the 3 open-sourced corpora, and our internal news-broadcasting (F1), children story (C1), and story-telling (M1) styles are used as the target style. Twenty utterances are synthesized for each speaker and style combination, listened to by 15 testers. During each MOS evaluation, the speaker similarity or style similarity is the only point that testers need to focus on.

**Table 2.** The 5 points MOS results with a confidence interval of 95%. 5 means the style or the timbre is exactly the same as the reference, and 1 means totally different.

| Speaker | Speaker Similarity | | | Style Similarity | | |
|---|---|---|---|---|---|---|
| | C1 | F1 | M1 | C1 | F1 | M1 |
| CSMSC | $4.50 \pm 0.05$ | $4.47 \pm 0.06$ | $4.49 \pm 0.04$ | $4.31 \pm 0.03$ | $4.62 \pm 0.03$ | $4.11 \pm 0.06$ |
| Originbeat-S1 | $4.36 \pm 0.08$ | $4.66 \pm 0.02$ | $4.29 \pm 0.06$ | $4.13 \pm 0.03$ | $4.18 \pm 0.05$ | $4.08 \pm 0.08$ |
| Originbeat-S2 | $4.36 \pm 0.07$ | $4.59 \pm 0.04$ | $4.23 \pm 0.07$ | $4.23 \pm 0.04$ | $4.30 \pm 0.03$ | $4.10 \pm 0.08$ |

MOS results of speaker similarity and style similarity are shown in Table 2, we could see that for a given target speaker, the speaker similarity of speeches in different style are very high and consistent, and the style similarity is also very high with the target style, which indicates that the proposed model achieves excellent disentanglement effect for speaker's timbre and style.

---

[4] https://weixsong.github.io/demos/MultiSpeakerMultiStyle/.

## 4.2    Ablation Study of Utterance Level Feature Normalization

To verify the effectiveness of UttNorm, another female corpus(112 sentences) crawled from a podcast, even with some background noise is chosen to demonstrate the impact.

In our experiments, due to the large variety of this noisy corpus from the oral podcast, a speaker similarity MOS evaluation and a preference evaluation are conducted to evaluate the performance of different normalization methods. We use speaker F1 as the target speaker and the style from this noisy corpus as the target style, then 20 sentences in the podcast domain are synthesized and listened to by 15 testers.

According to the MOS results for speaker similarity, when UttNorm is used, the speaker similarity increases from $3.82 \pm 0.05$(SpkNorm) to $3.91 \pm 0.06$(UttNorm), which showed that UttNorm could facilitate the decoupling effect of timbre and style for data with high variance, especially found data, noisy data or podcast data.

From the AB preference evaluation results, the proportions of preference are 0.17(SpkNorm), 0.35(No Preference), 0.48(UttNorm), and the $p$-value is less than 0.001. It could be found that the listeners strongly prefer the results from UttNorm, which proves the effectiveness of utterance normalization.

## 4.3    Demonstration of the Proposed Model

Synthesized speeches with a given speaker and different styles are shown in Fig. 2 (a), and speeches for a given style and different speakers are shown in Fig. 2 (b). The fundamental frequency is consistent for the three different styles in Fig. 2 (a), and all the speakers' pitch trajectories follow basically the same curve but with different fundamental frequency in Fig. 2 (b), this explains the decoupling effectiveness to some extent.

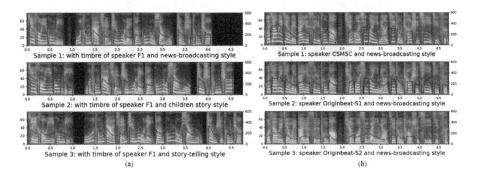

**Fig. 2.** A mel-spectrogram illustration of the proposed multi-speaker multi-style TTS model. (a) Speaker F1 is selected as the target speaker, the target style is the original news-broadcasting style, children story style, and story-telling style. (b) News-broadcasting style is selected as target style, the speakers are CSMSC, Originbeat-S1, and Originbeat-S2 respectively, from top to bottom.

## 4.4    Style Transition Illustration

As both a speaker ID and a style ID should be sent to the network to synthesize the target speaker's speech with a given style, we could also use the speaker ID to generate a style embedding that represents the source style, and the embedding from the style ID represents the target style. By combining the style embeddings from source and target with different weights, we could generate speech with different target style intensities.

To demonstrate the continuity of the learned style embedding representation, a style transition example is given in Fig. 3. From the pitch trajectory in each synthesized speech we could find that the fundamental frequency is identical for different target style weight, indicating that the proposed model keep the timbre unchanged when synthesizing different style speech and achieves outstanding timbre and style disentanglement effect.

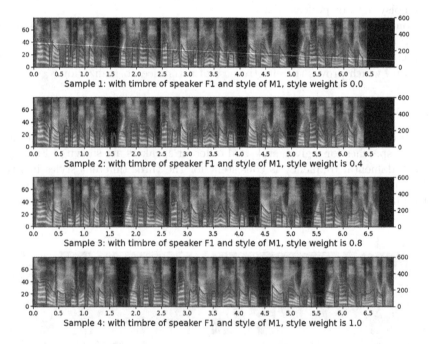

**Fig. 3.** A style transition illustration. Speaker F1 is selected as the target speaker and style M1 is selected as the target style, this figure shows the gradual style transition from style F1 to style M1 from top to bottom, with different target style weights.

# 5 Conclusions

A simple but effective speaker's timbre and style disentanglement network is proposed in this paper, which eliminates the reliance on a single-speaker multi-style corpus. The proposed network learns a style-dependent variance adaptor and a speaker-dependent mel-spectrogram prediction decoder. The style-related features are predicted by the variance adaptor with the guidance of style embedding, while the timbre is learned by the mel-spectrogram decoder with the control of speaker embedding. Utterance level feature normalization is proposed to prevent speaker information leakage from the style feature. Experimental results showed that the proposed model achieves good timbre and style disentanglement effect, for a given speaker the proposed model could synthesize speech with any style seen during training, even when the target style corpus only contains a few hundred training utterances. Furthermore, the proposed model learns a continuous style representation, which could generate speech that gradually transits from source style to target style.

# References

1. Taylor, P.: Text-to-speech synthesis. Cambridge University Press (2009)
2. Sotelo, J., et al.: Char2Wav: End-to-end speech synthesis. In: Proceedings ICLR, Toulon (2017)
3. Wang, Y., et al.: Tacotron: Towards end-to-end speech synthesis. In: Proceedings Interspeech, Stockholm (2017)
4. Ren, Y., Hu, C., Qin, T., Zhao, S., Zhao, Z., Liu, T.-Y.: FastSpeech 2: Fast and High-ality End-to-End Text to Speech. arXiv preprint arXiv:2006.04558(2020)
5. Ping, W., et al.: Deep voice 3: Scaling text-to-speech with convolutional sequence learning. arXiv:1710.07654 (2017)
6. Tan, X., Qin, T., Soong, F., Liu, T.-Y.: A Survey on Neural Speech Synthesis, arXiv preprint arXiv:2106.15561 (2021)
7. Tan, X., et al.: NaturalSpeech: End-to-End Text to Speech Synthesis with Human-Level Quality. arXiv preprint arXiv:2205.04421 (2022)
8. Pan, S., He, L.: Cross-speaker style transfer with prosody bottleneck in neural speech synthesis. arXiv preprint arXiv:2107.12562 (2021)
9. Xie, Q., etb al.: Multi-speaker multi-style text-to-speech synthesis with single-speaker single-style training data scenarios. arXiv preprint arXiv:2112.12743 (2021)
10. An, X., Soong, F.K., Xie, L.: Disentangling style and speaker attributes for tts style transfer. IEEE/ACM TASLP **30**, 646–658 (2022)
11. Li, T., Yang, S., Xue, L., Xie, L.: Controllable emotion transfer for end-to-end speech synthesis. In: Proceedings ISCSLP 2021. IEEE, 2021, pp. 1–5 (2021)
12. Wang, Y., et al.: Style tokens: Unsupervised style modeling, control and transfer in end-to-end speech synthesis, In: International Conference on Machine Learning. PMLR, 2018, pp. 5180–5189 (2018)
13. Zhang, Y.-J., Pan, S., He, L., Ling, Z.-H.: Learning latent representations for style control and transfer in end-to-end speech synthesis. In: Proceedings ICASSP. IEEE, 2019, pp. 6945–6949 (2019)
14. Kumar, K., et al.: MelGAN: Generative adversarial networks for conditional waveform synthesis. In: Proceedings NIPS, Vancouver (2019)

15. Zen, H., Tokuda, K., Black, A,W,: Statistical parametric speech synthesis. Speech Commun. **51**(11), 1039–1064 (2009)
16. Xie, Q., et al.: The multi-speaker multi-style voice cloning challenge 2021. In: Proceedings ICASSP. IEEE, (2021)
17. McAuliffe, M., Socolof, M., Mihuc, S., Wagner, M., Sonderegger, M.: Montreal Forced Aligner: Trainable Text-Speech Alignment Using Kaldi. In: Interspeech, 2017, pp. 498–502 (2017)
18. Kingma, D.P., Ba, J.: Adam: A method for stochastic optimization. In: Proceedings ICLR, San Diego (2015)
19. Vaswani, A., et al.: Attention is all you need. In: Proceedings of NIPS, Long Beach (2017)

# Multiple Confidence Gates for Joint Training of SE and ASR

Tianrui Wang[1](✉), Weibin Zhu[1], Yingying Gao[2], Junlan Feng[2], Dongjie Zhu[1], Surong Zhu[1], and Shilei Zhang[2]

[1] Institute of Information Science, Beijing Jiaotong University, Beijing, China
{20120318,wbzhu,21120323,20126520}@bjtu.edu.cn
[2] China Mobile Research Institute, Beijing, China
{gaoyingying,fengjunlan,zhangshilei}@chinamobile.com

**Abstract.** Joint training of speech enhancement model (SE) and speech recognition model (ASR) is a common solution for robust ASR in noisy environments. SE focuses on improving the auditory quality of speech, but the enhanced feature distribution is changed, which is uncertain and detrimental to the ASR. To tackle this challenge, an approach with multiple confidence gates for jointly training of SE and ASR is proposed. A speech confidence gates prediction module is designed to replace the former SE module in joint training. The noisy speech is filtered by gates to obtain features that are easier to be fitting by the ASR network. The experimental results show that the proposed method has better performance than the traditional robust speech recognition system on test sets of clean speech, synthesized noisy speech, and real noisy speech.

**Keywords:** Robust ASR · Joint Training · Deep learning

## 1 Introduction

The performance of ASR in a noisy environment will greatly deteriorate, and it has been a persistent and challenging task to improve the adaptability of ASR in a noisy environment [1].

From the perspective of features, some researchers have proposed feature construction strategies that are more robust to noise, in order to reduce the difference in features caused by noise and thereby improve the robustness of ASR [2–4]. But these methods are difficult to work with a low signal-to-noise ratio (SNR). From the perspective of model design, a direct approach is to set a SE in front of the ASR [5]. Although the enhanced speech is greatly improved currently in human hearing [6–9], the enhanced feature distribution is changed also, which may be helpful for the hearing but not always necessarily beneficial for ASR. In order to make the ASR adaptable to the enhanced feature, the ASR can be retrained with the data processed by the enhanced model [10]. However, the performance of this method is severely affected by the SE effect. Especially in the case of low SNR, the SE may corrupt the speech structure and bring

L. Zhenhua et al. (Eds.): NCMMSC 2022, CCIS 1765, pp. 141–148, 2023.
https://doi.org/10.1007/978-981-99-2401-1_13

somehow additional noise. In order to reduce the influence of SE and achieve better recognition accuracy, researchers proposed the joint training strategy [11]. Let ASR constrain SE and train the model with the recognition accuracy of ASR as the main goal. However, the models are difficult to converge during joint training, and the final recognition performance improvement is limited due to the incompatibility between two goals, for SE it's speech quality, while for ASR it's recognition accuracy.

During joint training, the ASR network needs to continuously fit the feature distribution, but the feature distribution keeps changing under the action of SE, which makes it difficult for the ASR to converge. We believe that the ASR network has strong fitting and noise-carrying capabilities. We need neither to do too much processing on the feature distribution nor to predict the value of clean speech. We only need to filter out the feature points that don't contain speech. This method can avoid the problem that ASR cannot converge during joint training, and also give full play to the noise-carrying ability of ASR itself. So, the multiple confidence gates for joint training of SE and ASR is proposed. A speech confidence gates prediction module is designed to replace the speech enhancement module in joint training. Each gate is a confidence spectrum with the same size as the feature spectrum (the probability that each feature point contains speech). The noisy speech is filtered by each confidence gate respectively. Then the gated results are combined to obtain the input feature for ASR. The experimental results show that the proposed method can effectively improve the recognition accuracy of the joint training model.

## 2   Our Method

The overall diagram of the proposed system is shown in Fig. 1. It is mainly comprised of two parts, namely the multiple confidence gates enhancement module (MCG) and ASR. Firstly, We convert the waveform to logarithmic Fbank (log-Fbank) $X \in \mathbb{R}^{T \times Q}$ as the input to the model by the short-time fast Fourier transform (STFT) and auditory spectrum mapping [12], where $T$ and $Q$ denote the number of frames and the dimension of the Bark spectrum respectively. Secondly, The MCG predicts multiple confidence gates based on the noisy log-Fbank, and the noisy features are filtered by each gate respectively. Then the filtered results are combined into inputs of ASR by a convolution block (CB). Different confidence gates correspond to the selections of speech feature points with different energy thresholds, and each process will be described later.

### 2.1   Multiple Confidence Gates Enhancement Module

The multiple confidence gates enhancement module is used to predict the confidence gates to filter the noisy spectrum. The main body of MCG is the convolutional recurrent network (CRN) [7], it is an encoder-decoder architecture for speech enhancement. Both the encoder and decoder are comprised of convolution blocks (CB). And each CB is comprised of 2D convolution, batch normalization (BN) [13], and parametric rectified linear unit (PReLU) [14]. Between the

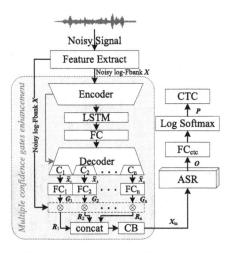

**Fig. 1.** Architecture of the our method. (Color Figure Online)

encoder and decoder, long short-term memory (LSTM) layer [15] is inserted to model the temporal dependencies. Additionally, skip connections are utilized to concatenate the output of each encoder layer to the input of the corresponding decoder layer (blue line in Fig. 1).

Since we want to determine the confidence that each feature point contains speech, we change the output channel number of the last decoder to $(C_1 + C_2 + \cdots + C_n)$, where $C_1, C_2, \cdots, C_n$ are the number of channels of the input features for corresponding fully connected layers $(FC_1, FC_2, \cdots, FC_n)$. Each FC can output a confidence gate $G \in \mathbb{R}^{T \times Q}$ for every feature point,

$$G_n = \mathrm{sigmoid}(\tilde{X}_n \cdot W_n + b_n) \tag{1}$$

where $\tilde{X}_n \in \mathbb{R}^{T \times Q \times C_n}$ represents the $C_n$ channels of the decoder output. $W_n \in \mathbb{R}^{C_n \times 1}$ and $b_n$ are the parameters of $FC_n$. sigmoid [16] is used to convert the result into confidence (the probability that a feature point contains speech).

The label of confidence gate $G$ is designed based on energy spectra. Different thresholds are used to control different filtering degrees. The greater the energy threshold, the fewer the number of speech points in the label, and the greater the speech energy of the corresponding spetra, as shown in Fig. 2. We first compute the mean $\mu \in \mathbb{R}^{Q \times 1}$ of the log-Fbank of the clean speech set $X = [\dot{X}_1, \cdots, \dot{X}_d] \in \mathbb{R}^{D \times T \times Q}$ and standard deviation $\sigma \in \mathbb{R}^{Q \times 1}$ of log-Fbank means,

$$\mu = \sum_{i=1}^{D} \left[ \left( \sum_{t=1}^{T} \mathbb{X}_{i,t} \right) / T \right] / D \tag{2}$$

$$\sigma = \sqrt{\sum_{i=1}^{D} \left[ \left( \sum_{t=1}^{T} \mathbb{X}_{i,t} \right) / T - \mu \right]^2 / D} \tag{3}$$

where $\dot{X}$ represents log-Fbank of the clean speech. $D$ represents the clip number of audio in clean speech set $\mathbb{X}$. Then the thresholds $\kappa = (\mu + \varepsilon \cdot \sigma) \in \mathbb{R}^{Q \times 1}$ for bins are controlled according to different offset values $\varepsilon$, and the confidence gate label $\dot{G}$ is 1 if the feature point value is larger than $\kappa$, 0 otherwise, as shown in Fig. 2.

$$\dot{G}_{t,q} = \begin{cases} 1\,, \dot{X}_{t,q} \geq \kappa \\ 0\,, \dot{X}_{t,q} < \kappa \end{cases} \tag{4}$$

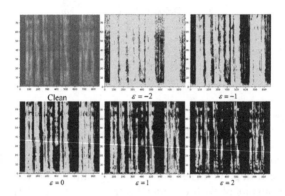

**Fig. 2.** $\dot{G}$ obtained by different $\varepsilon$. The greater $\varepsilon$, the greater the speech energy of the corresponding choosed points.

After (1), we can get $n$ confidence gates $[G_1, \cdots, G_n]$ corresponding to different $\varepsilon$. Then noisy log-Fbank are filtered by the confidence gates, resulting in different filtered results $[R_1, \cdots, R_n]$,

$$R_n = G_n \otimes X \tag{5}$$

where $\otimes$ is the element-wise multiplication. And these filtered results are then concatenated in channel dimension and input to a CB to obtain the input $X_{\text{in}}$ of ASR.

## 2.2    Automatic Speech Recognition

Since we focuses more on the exploration for acoustic information processing, the Conformer acoustic encoder [17] with CTC is employed as ASR module. Conformer first processes the enhanced input $X_{\text{in}} \in \mathbb{R}^{T \times Q}$ with a convolution subsampling layer. Then the downsampled feature is processed by several conformer blocks. Each block is comprised of four modules stacked together, i.e., a half-step residual feed-forward module, a self-attention module, a convolution module, and a second FFN is followed by a layer norm [18]. Finally, a fully connected layer maps the features to the word probabilities with 4233-class and then computes the CTC loss.

## 2.3 Loss Function

The loss function of our framework is comprised of four components.

$$L = L_G + L_R + L_O + L_{\text{CTC}} \tag{6}$$

In the MCG module, we measured the predicted confidence gate as,

$$L_G = \sum_{i=1}^{n} ||\boldsymbol{G}_n - \dot{\boldsymbol{G}}_n||_1 \tag{7}$$

where $||\cdot||_1$ represents the 1 norm. In addition, to strengthen the filtering ability of the module on noise, we also compute filtered results from the clean speech through the same processing, which are denoted as $\left[\dot{\boldsymbol{R}}_1, \cdots, \dot{\boldsymbol{R}}_n\right]$, and calculate the difference between them and the filtered results of noisy speech as,

$$L_R = \sum_{i=1}^{n} ||\boldsymbol{R}_n - \dot{\boldsymbol{R}}_n||_1 \tag{8}$$

After the ASR, In order to reduce the noise-related changes brought by SE to ASR, we also measured the difference between the $\dot{\boldsymbol{O}}$ computed from clean speech and that from noisy speech as,

$$L_O = ||\boldsymbol{O} - \dot{\boldsymbol{O}}||_1 \tag{9}$$

Note that the gradients of the processing of all clean speech are discarded. And $L_{\text{CTC}}$ is the connectionist temporal classification [19] for ASR.

## 3 Experiments

### 3.1 Dataset

We use AISHELL1 [20] as the clean speech dataset, which includes a total of 178 h of speech from 400 speakers, and it is divided into training, development, and testing sets. The noise data used in training, development, and test set are from DNS challenge (INTERSPEECH 2020) [21], ESC50 [22], and Musan [23], respectively. During the mixing process, a random noise cut is extracted and then mixed with the clean speech under an SNR selected from −5 dB to 20 dB. For the training and development sets, half of the data in AISHELL are noised. For the test set, the test set of AISHELL is copied into two copies, one with noise and one without noise. In addition, we randomly selected 2600 speech clips from the noisy speech recorded in real scenarios as an additional test set, and call it APY[1].

---

[1] https://www.datatang.com/dataset/info/speech/191.

## 3.2    Training Setup and Baseline

We trained all models on our dataset with the same strategy. The initial learning rate is set to 0.0002, which will decay 0.5 when the validation loss plateaued for 5 epochs. The training is stopped if the validation loss plateaued for 20 epochs. And the optimizer is Adam [24]. We set a Conformer, a traditional joint training systems, and a separately training (ST) system as the baselines.

**Conformer:** The dimension of log-Fbank is 80. The number of conformer blocks is 12. Linear units is 2048. The convolution kernel is 15. Attention heads is 4. Output size is 256. And we replaced the CMVN with BN to help Conformer integrate with the SE module.

**DCRN+Conformer:** DCRN is an encoder-decoder model for speech enhancement in the complex domain [7]. The 32ms Hanning window with 25% overlap and 512-point STFT are used. The channel number of encoder and decoder is $\{32, 64, 128, 256, 256, 256\}$. Kernel size and stride are (5,2) and (2,1). One 128-units LSTM is adopted. And a 1024-units FC layer is after the LSTM.

**DCRN+Conformer(ST):** Separately training is to train a DCRN first, then the conformer is trained based on the processed data by the DCRN. The parameter configuration is the same as that of the joint training.

**MCG+Conformer:** The channel numbers of MCG are $\{32, 48, 64, 80, 96\}$. Kernel size is (3,3), strides are $\{(1,1), (1,1), (2,1), (2,1), (1,1), (1,1)\}$. One 128-units LSTM is adopted. And a 1920-units FC layer is after the LSTM. The channel number of the last decoder is $10n$ and $C_1 = \cdots = C_n = 10$. The best results are obtained when $n = 3$ and $\varepsilon = [-1, 1, 2]$.

## 3.3    Experimental Results and Discussion

We trained the models as described in Sect. 3.2, and counted the substitution error (S), deletion error (D), insertion error (I), and character error rate (CER) of different systems on the clean, noisy, and APY test set, shown as Table 1.

**Table 1.** Recognition results of the models on test set

| Model | Clean | | | | Noisy | | | | APY | | | |
|---|---|---|---|---|---|---|---|---|---|---|---|---|
| | S | D | I | CER | S | D | I | CER | S | D | I | CER |
| Conformer | 1.58 | 0.05 | 0.04 | 11.58 | 2.81 | 0.15 | 0.08 | 20.98 | 2.73 | 0.21 | 0.18 | 44.97 |
| +DCRN | 1.43 | 0.05 | 0.03 | 10.33 | 2.75 | 0.20 | 0.07 | 20.49 | 2.98 | 0.17 | 0.22 | 48.82 |
| +DCRN(ST) | 1.44 | 0.05 | 0.03 | 10.45 | 2.74 | 0.23 | 0.06 | 20.84 | 3.15 | 0.21 | 0.18 | 50.53 |
| +MCG | **1.27** | **0.04** | **0.03** | **9.34** | **2.25** | **0.14** | **0.05** | **16.88** | **2.39** | **0.17** | **0.12** | **39.22** |

It can be seen from the results, our method achieved the lowest CER on all test sets, moreover, the insertion and substitution errors of our model are minimal on the noisy test set, which demonstrates that the multiple confidence gates scheme can filter out the non-speech parts without destroying the speech structure, thereby greatly improving the recognition performance. The performance improvement on APY also indicates that the change of feature by the proposed front-end module is less affected by noise and more friendly to ASR.

**Table 2.** Recognition results models trained by different confidence gates on test set

| n | $\varepsilon$ | Clean CER | Noisy CER | APY CER |
|---|---|---|---|---|
| 1 | $[0]$ | 9.457 | 17.134 | 39.361 |
| 2 | $[-1, 1]$ | 9.669 | 17.626 | 41.211 |
| 3 | $[-1, 1, 2]$ | **9.343** | **16.882** | **39.223** |
| 4 | $[-2, -1, 1, 2]$ | 9.498 | 17.440 | 40.513 |

In addition, we have done some experiments on the design of the confidence gates. We set different thresholds to control the filtering ability of the gates. It can be seen that the best performance of the model is achieved with $n = 3$ and $\varepsilon = [-1, 1, 2]$. As can be seen from Fig. 2, as $\varepsilon$ increases, the higher the energy of selected feature points, the stronger the filtering ability of the gate. So designing multiple gates with different filtering abilities can improve the fitting and generalization ability of model to achieve better results.

## 4   Conclusions

In this paper, we propose the multiple confidence gates enhancement for joint training of SE and ASR. A speech confidence gates prediction module is designed to replace the former speech enhancement model. And the noisy speech is filtered by each confidence gate respectively. Then the gated results are combined to obtain the input for ASR. With the help of the proposed method, the incompatibility between SE and ASR during joint training is mitigated. And the experimental results show that the proposed method can filter out the non-speech parts without destroying the speech structure, and it's more friendly to ASR during joint training.

## References

1. Pedro, J.M.: Speech recognition in noisy environments. Ph.D. thesis, Carnegie Mellon University (1996)
2. Xue, F., Brigitte, R., Scott, A., James, R.G.: An environmental feature representation for robust speech recognition and for environment identification. In: INTER-SPEECH, pp. 3078–3082 (2017)
3. Neethu, M.J., Dino, O., Zoran, C., Peter, B., Steve R.: Deep scattering power spectrum features for robust speech recognition. In: INTERSPEECH, pp. 1673–1677 (2020)

4. Mirco, R., et al.: Multi-task self-supervised learning for robust speech recognition. In: ICASSP. IEEE, pp. 6989–6993 (2020)
5. Keisuke K., Tsubasa O., Marc D., Tomohiro N.: Improving noise robust automatic speech recognition with single-channel time-domain enhancement network. In: ICASSP. IEEE. pp. 7009–7013 (2020)
6. Braun, S., Tashev, I.: Data augmentation and loss normalization for deep noise suppression. In: Karpov, A., Potapova, R. (eds.) SPECOM 2020. LNCS (LNAI), vol. 12335, pp. 79–86. Springer, Cham (2020). https://doi.org/10.1007/978-3-030-60276-5_8
7. Ke, T., DeLiang, W.: Complex spectral mapping with a convolutional recurrent network for monaural speech enhancement. In: ICASSP. IEEE, pp. 6865–6869 (2019)
8. Yanxin, H.: Dccrn: Deep complex convolution recurrent network for phase-aware speech enhancement. In: ArXiv preprint arXiv:2008.00264 (2020)
9. Tianrui, W., Weibin, Z.: A deep learning loss function based on auditory power compression for speech enhancement. In: ArXiv preprint arXiv:2108.11877 (2021)
10. Archiki, P., Preethi, J., Rajbabu, V.: An investigation of end-to-end models for robust speech recognition. In: ICASSP, IEEE. pp. 6893–6897 (2021)
11. Masakiyo, F., Hisashi, K.: Comparative evaluations of various factored deep convolutional rnn architectures for noise robust speech recognition. In: ICASSP IEEE, pp. 4829–4833 (2018)
12. Gilbert, S., Truong, N.: Wavelets and filter banks (1996)
13. Sergey, I., Christian, S.: Batch normalization: Accelerating deep network training by reducing internal covariate shift. In: The 32nd International Conference on Machine Learning, vol. 1, pp. 448–456 (2015)
14. Kaiming, H., Xiangyu, Z., Shaoqing, R., Jian., S.: Delving deep into rectifiers: Surpassing humanlevel performance on imagenet classification. In: ICCV, pp. 1026–1034 (2015)
15. Klaus, G., Rupesh, K.S., Jan, K., Bas, R.S., Jurgen, S.: Lstm: A search space odyssey. IEEE Trans. Neural Netw. **28**(10), 2222–2232 (2017)
16. George, C.: Approximation by superpositions of a sigmoidal function. In: Mathematics of Control, Signals, and Systems, vol. 2, no. 4, pp. 303–314 (1989)
17. Anmol G., et al.: Conformer: Convolution-augmented transformer for speech recognition. In: ArXiv preprint arXiv:2005.08100 (2020)
18. Jimmy Lei, B., Jamie Ryan, K., Geoffrey, E.H.: Layer normalization. In: ArXiv preprint arXiv:1607.06450 (2016)
19. Alex, G.: Connectionist temporal classification, pp. 61–93 (2012)
20. Hui, B., Jiayu, D., Xingyu, N., Bengu, W., Hao, Z.: Aishell-1: An open-source mandarin speech corpus and a speech recognition baseline. In: OCOCOSDA. IEEE, pp. 1–5 (2017)
21. Chandan, K.A.R., et al.: The interspeech 2020 deep noise suppression challenge: Datasets, subjective speech quality and testing framework. In: ArXiv preprint arXiv:2001.08662 (2020)
22. Karol J. P.: ESC: Dataset for Environmental Sound Classification. In: The 23rd Annual ACM Conference on Multimedia. pp. 1015–1018, ACM Press (2015)
23. David, S., Guoguo, C., Daniel, P.: Musan: A music., speech., and noise corpus. In: ArXiv preprint arXiv:1510.08484 (2015)
24. Diederik, P.K., Jimmy Lei, B.: Adam: A method for stochastic optimization. In: ICLR (2015)

# Detecting Escalation Level from Speech with Transfer Learning and Acoustic-Linguistic Information Fusion

Ziang Zhou[1] , Yanze Xu[1] , and Ming Li[1,2](✉)

[1] Data Science Research Center, Duke Kunshan University, Kunshan, China
{ziang.zhou372,ming.li369}@dukekunshan.edu.cn
[2] School of Computer Science, Wuhan University, Wuhan, China

**Abstract.** Textual escalation detection has been widely applied to e-commerce companies' customer service systems to pre-alert and prevent potential conflicts. Similarly, acoustic-based escalation detection systems are also helpful in enhancing passengers' safety and maintaining public order in public areas such as airports and train stations, where many impersonal conversations frequently occur. To this end, we introduce a multimodal system based on acoustic-linguistic features to detect escalation levels from human speech. Voice Activity Detection (VAD) and Label Smoothing are adopted to enhance the performance of this task further. Given the difficulty and high cost of data collection in open scenarios, the datasets we used in this task are subject to severe low resource constraints. To address this problem, we introduce transfer learning using a multi-corpus framework involving emotion detection datasets such as RAVDESS and CREMA-D to integrate emotion features into escalation signals representation learning. On the development set, our proposed system achieves 81.5% unweighted average recall (UAR), which significantly outperforms the baseline of 72.2%.

**Keywords:** escalation detection · transfer learning · emotion recognition · multimodal conflict detection

## 1 Introduction

Escalation level detection system has been applied in a wide range of applications, including human-computer interaction and computer-based human-to-human conversation [36]. For instance, there are e-commerce companies [23] that have been equipped with textual conversational escalation detectors. Once an increasing escalation level of the customers is detected, the special agents will take over and settle their dissatisfaction, effectively preventing the conflict from worsening and protecting the employees' feelings. In public areas like transportation centers, and information desks, where many impersonal interactions occur, it is also essential to detect the potential risk of escalations from conversation to guarantee public security. Therefore, audio escalation level analysis is instrumental and crucial.

© The Author(s), under exclusive license to Springer Nature Singapore Pte Ltd. 2023
L. Zhenhua et al. (Eds.): NCMMSC 2022, CCIS 1765, pp. 149–161, 2023.
https://doi.org/10.1007/978-981-99-2401-1_14

We adopted two escalation datasets, Aggression in Trains (TR) [22] and Stress at Service Desks (SD) [21] from a previous escalation challenge [36]. These two datasets provide conversation audio recorded on the train and at the information desk respectively. About four hundred training audios from the SD dataset, with an average length of 5 s, are used for training. Five hundred audios from the TR dataset are used for testing. Given datasets with limited scales, learning effective escalation signals from scratch would be challenging. Thus, supervised domain adaptation [2] became a better option, for we can adapt a more general feature distribution backed by sufficient data to adapt to the escalation signal domain using limited resources. Since emotion is an obvious indicator in potential conversational conflicts, we have good reasons to assume that the encoding ability of emotion features would be a good starting point to support the modeling of escalation signals from conversations. Through pretraining on large-scale emotion recognition datasets, our model will be more capable of capturing emotion features and knowledge to support fine-tuning using the escalation datasets. Recent research [30] on small sample set classification tasks also showed promising results on pattern recognition via transfer learning.

## 2    Related Works

### 2.1    Conflict Escalation Detection

Several conflict escalation research has been done in recent years, focusing on the count of overlaps and interruptions in speeches. In [13], the number of overlaps is recorded in the hand-labeled dataset and used in conflict prediction. And [8,16] uses a support vector machine (SVM) to detect overlap based on acoustic and prosodic features. Kim et al. in [17] analyzed the level of conversation escalation based on a corpus of French TV debates. They proposed an automatic overlap detection approach to extract features and obtained 62.3% unweighted accuracy on the corpus. Effective as they seem, these methods are considered impractical in this Escalation Sub-task. First, the length of audio files in [8] ranges from 3 to 30 min, and the length of conversation audio in [17] is 30 s. While in our escalation detection task, the average length of the corpus is 5 s. Most of the time, an audio piece only contains a single person's voice. Thus, focusing on overlap detection seems to be ineffective. Besides, we did not spot a significant difference in overlap frequency among different escalation classes based on conversation script analysis. Second, with a total training corpus duration of fewer than 30 min, the model built on overlap counts will easily suffer from a high bias and low variance [4,28].

### 2.2    Transfer Learning

In [12], Gideon et al. demonstrate that emotion recognition tasks can benefit from advanced representations learned from paralinguistic tasks. This suggests that emotion representation and paralinguistic representation are correlated in nature. Also, in [39], supervised transfer learning has brought improvements to music classification tasks as a pre-training step. Thus it occurs to us that utilizing transfer learning to gain emotion feature encodability from large-scale emotion recognition datasets might as well benefit the escalation detection task. Research [5] on discrete class emotion recognition mainly focuses on emotions, including happiness, anger, sadness, frustration and neutral.

### 2.3  Textual Embeddings

Emotions are expressive in multiple modalities. As shown in [6, 33], multimodal determination has become increasingly important in emotion recognition. In the TR [22] and SD [21] datasets, manual transcriptions for the conversations are also provided besides the audio signals. Given various lengths of transcriptions, we look for textual embeddings that agree in size with each other. In [34, 35], Reimers et al. proposed Sentence-BERT (SBERT) to extract sentimentally meaningful textual embeddings in sentence level. Using conversation transcriptions as input, we encoded them into length-invariant textual embeddings, utilizing the pre-trained multilingual model.

## 3  Datasets and Methods

An overview of our solution pipeline is shown in Fig. 1. We apply librosa toolkit [27] to extract Mel Frequency Cepstral Coefficient (MFCC), which is then fed to the residual network backbone [14] to pretrain the emotion encoder. The embedding extractor is pre-trained on an aggregated dataset of four emotion recognition datasets, learning emotion representations from a broader source.

### 3.1  Datasets

**Escalation Datasets.** For this escalation level detection task, we employed two datasets: Dataset of Aggression in Trains (TR) [22] and Stress at Service Desk Dataset (SD) [21]. The TR dataset monitors the misbehaviors in trains and train stations, and the SD dataset consists of problematic conversations that emerged at the service desk [36]. The escalation level has been classified into three stages: low, mid, and high. The higher escalation level suggests that the conflict will likely grow more severe. Moreover, the Dutch datasets have an average of 5 s for each conversation clip. The SD dataset is used for training, and the TR dataset is used for testing. More details regarding these datasets can be found in the overview of the previous challenge [36].

**Emotion Recognition Datasets.** Previous work [12] has highlighted the correlation between Emotion Recognition Tasks and Paralinguistics tasks. Hence we assume that the escalation level detection and emotion recognition tasks share certain distributions in their feature space. Thus, we aggregated four well-known audio emotion datasets for joint sentimental analysis: **RAVDESS** [26] is

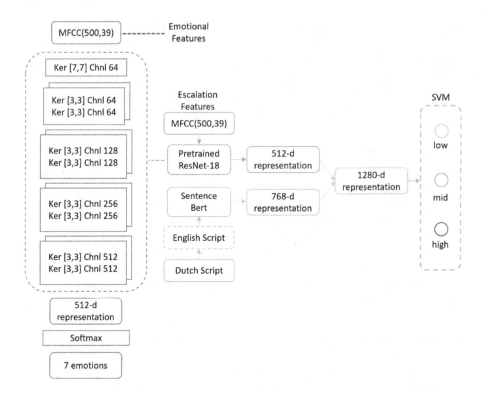

**Fig. 1.** Pipeline of the Escalation Detection System

a gender-balanced multimodal dataset. Over 7000 pieces of audio are carefully and repetitively labeled, containing emotions like calm, happy, sad, angry, fearful, surprise, and disgust. **CREMA-D** [7] is a high quality visual-vocal dataset, containing 7442 clips from 91 actors. **SAVEE** [11] is a male-only audio dataset with same emotion categories with RAVDESS Dataset. **TESS** [9] is a female-only audio dataset collected from two actresses whose mother languages are both English. TESS contains 2800 audios covering the same emotion categories mentioned above.

With four audio emotion recognition datasets incorporated together, we have 2167 samples for each of Angry, Happy and Sad emotions, 1795 samples for Neutral, 2047 samples for Fearful, 1863 samples for Disgusted and 593 samples for Surprised emotions.

## 3.2   Methods

**Voice Activity Detection.** Voice activity detection is a process of identifying the presence of human speech in an audio clip [19]. The SD dataset [21] is collected at a service desk and therefore contains background noises. In case the background noise undermines the paralinguistic representations, we implement the WebRTC-VAD [1] tool to label non-speech segments from the audio before feeding the whole pieces for feature extraction.

**Transfer Learning.** Transfer learning has proved effective in boosting performance on low-resource classification tasks [42]. Under our previous assumption, emotion features are essential indicators in escalation-level detection; hence we expect our model to include the capability to involve emotion patterns in the representation learning process. The emotion datasets mentioned in 3.1 are combined to train the ResNet-18 backbone as the emotion recognition model. Note that the emotion classifier and escalation level classifier share the same configuration for a reason. Although linear probing is much cheaper computationally, fine-tuning all parameters of the source model achieves state-of-the-art performance more often [10]. Therefore, we adopt full fine-tuning where all layers of the pre-trained model are updated during the fine-tuning stage on the escalation datasets.

**Features.** Automatic emotion and escalation detection has been a challenging task for the fact that emotion can be expressed in multiple modalities [38]. In multimodal emotion recognition, visual, audio and textual features are the most commonly studied channels. In our task, raw Dutch audio and manually transcribed texts are provided. Thus acoustic and linguistic features can be fused to determine the escalation predictions jointly. Although more recent clustering-based feature extraction approaches have proven instrumental in human speech emotion recognition, e.g., adaptive normalization [15] and pitch-conditioned text-independent feature construction [40], they may further bias our training process due to the extremely low resource limitation. Since MFCC has been widely used in speech emotion recognition tasks [18,20,24], we adopt MFCC as our acoustic feature in this task.

To label the silent intervals, we first apply the WebRTC-VAD [1] to filter out the low-energy segments. Next, MFCCs are calculated from the filtered audio fragments. After that, the MFCCs are applied to fine-tune the emotion classifier, which was previously pre-trained on the aggregated large-scale emotion datasets. On top of the ResNet backbone, we also adopted the Global Average Pooling (GAP) layer structure [25], granting us the compatibility to process the input of variant length during the evaluation phase. According to Tang et al. [37], simply replacing fully connected layers with linear SVMs can improve classification performance on multiple image classification tasks. Therefore, our work does not construct an end-to-end detection system. Instead, we employ Support Vector Machine (SVM) [25] to conduct the backend classification task.

For the textual embeddings extraction, we adopt the pre-trained multilingual model `distiluse-base-multilingual-cased-v1` from Universal Sentence Encoder (USE) [41] from Sentence-BERT(SBERT) [34,35] to extract the sentence-level embeddings. We also compared the Unweighted Average Recall (UAR) metric by extracting Dutch embeddings directly and by extracting embeddings from Dutch-to-English translation. The experiment result shows that the sentence embeddings from the English translation outperformed the original Dutch transcriptions; thus, we adopted the former for textual embedding extraction.

## 4    Experimental Results

For the Escalation Sub-challenge, we aim to build a multimodal model to determine whether the escalation level in a given conversation is low, medium or high. UAR has been a reliable metric in evaluating emotion recognition tasks under data imbalance constraints. So we followed the metric choice of UAR by previous work on similar settings [31,32]. This section will introduce our experiment setup, results and several implementation details.

### 4.1    Feature Configuration

In the audio preprocessing stage, we first applied the open-source tool WebRTC-VAD [1] to filter out the silent segments in the audio from the temporal domain. The noise reduction mode of WebRTC-VAD is set to 2. Next, we extract MFCCs from the filtered audio segments. The window length of each frame is set to 0.025 s, the window step is initialized to 0.01 s, and the window function is hamming function. The number of mel filters is set to 256. Also, the frequency range is 50 Hz to 8,000 Hz. The pre-emphasize parameter is set to 0.97. The representation dimension is set to 512.

### 4.2    Model Setup

As for the emotion classification task, both the architecture and the configuration of the representation extractor are the same with the escalation model, except that the former is followed by a fully connected layer that maps a 128 dimension representation to a seven dimension softmax probability vector and the latter is followed by a linear SVM classifier of three levels. Weighted Cross-Entropy Loss is known as capable of offsetting the negative impact of imbalanced data distribution [3] and is set as the loss function for that reason. The optimizer is Stochastic Gradient Descent (SGD), with the learning rate set to 0.001, weight decay set to 1e-4, and momentum set to 0.8. The maximum training epochs is 50, with an early stop of 5 non-improving epochs. In the fine-tuning stage, the system configurations remain unchanged, except that the training epochs are extended to 300, and no momentum is applied to the optimizer to reduce overfitting.

The dimension of textual embeddings extracted from the multilingual pre-trained model `distiluse-base-multilingual-cased-v1` is 768-d [35]. In the fusion stage of our experiment, textual embeddings will be concatenated with the 512-d acoustic representations, forming 1280-d embeddings at the utterance level.

## 4.3   Results

Prior to fine-tuning the emotion recognition model on the escalation datasets, we first train the ResNet-18 architecture on emotion recognition datasets to learn emotion representations from audio. The highest UAR achieved by our emotion recognition model is 65.01%. The model is selected as the pre-trained model to be fine-tuned on the Escalation dataset.

To learn better escalation signals in the fine-tuning process, we introduced three factors that may impact model performance positively. Besides fine-tuning the pretrained emotion recognition model, VAD and acoustic-linguistic information fusion are also tested to boost the escalation level detection ability further. We start by analyzing whether voice activity detection will leverage the performance of the development set. Then, we fine-tune the emotion recognition model pre-trained on the four audio emotion datasets to analyze any notable improvement. Finally, we examine whether the fusion between textual embeddings extracted from SBERT [34,35] and acoustic embeddings can improve the performance.

To evaluate the effect of VAD on the prediction result, we did several controlled experiments on the development set. Table 1 demonstrates the effect of VAD on various metrics. First, we calculated features from unprocessed audio and fed them into the embedding extractor. With acoustic embeddings alone, we scored 0.675 on the UAR metric using the Support Vector Machines (SVM) as the backend classifier. With all conditions and procedures remaining the same, we added VAD to the audio pre-processing stage, filtering out non-speech voice segments. The result on the UAR metric has increased from 0.675 to 0.710. This shows that VAD is critical to the escalation representation learning process.

We believe that the escalation detection tasks, to some degree, share certain advanced representations with emotion recognition tasks. [12] Thus, we also experimented with fine-tuning parameters on the escalation dataset with the model pre-trained on the emotion datasets. Table 2 shows the experiment results after implementing transfer learning to our system. We have witnessed a positive impact of VAD on our experiment results on the development set. Thus the base experiment has been implemented with VAD applied. We can see that, after applying the pre-trained model to the MFCC+VAD system, with acoustic embeddings alone, the score reached 0.810 on the metric UAR, which turns out to be a significant improvement. This also proves that emotion features can benefit paralinguistic tasks by transfer learning. Additionally, the MFCC+VAD+PR system may have already been stable enough that the textual embedding fusion brings no noticeable improvement.

Besides the experiments recorded above, we also implemented various trials involving different features, networks, and techniques. As listed in Table 3, we recorded other meaningful experiments with convincing performance on the devel set that could be applied to final model fusion. Other standard acoustic features like the Log filterbank are also under experiment. Label Smoothing technique [29] is applied on the MFCC+VAD+PR model but brings a slightly negative impact. According to the experiment results, Voice Activity Detection has again

**Table 1.** Effects of Voice Activity Detection (VAD) **TE**: Textual Embeddings fused.

| Model Name | Precision | UAR | F1-Score |
|---|---|---|---|
| MFCC | 0.640 | 0.675 | 0.647 |
| MFCC+VAD | 0.675 | 0.710 | 0.688 |
| MFCC+TE | 0.652 | 0.690 | 0.664 |
| MFCC+VAD+TE | 0.676 | 0.721 | **0.691** |
| Baseline Fusion [36] | – | **0.722** | – |

**Table 2.** Effects of fine-tuning **PR**: Pre-trained Emotion Recognition Model applied.

| Model Name | Precision | UAR | F1-Score |
|---|---|---|---|
| MFCC+VAD | 0.675 | 0.710 | 0.688 |
| MFCC+VAD+PR | 0.807 | **0.810** | 0.788 |
| MFCC+VAD+PR+TE | 0.807 | **0.810** | 0.788 |
| Baseline Fusion [36] | – | 0.722 | – |

**Table 3.** Extra Experiments. **LS**: Label Smoothing.

| Model Name | Precision | UAR | F1-Score |
|---|---|---|---|
| Logfbank | 0.670 | 0.743 | 0.684 |
| Logfbank+VAD | 0.711 | 0.778 | 0.733 |
| MFCC+VAD+PR+LS | 0.781 | **0.781** | **0.761** |
| MFCC+VAD+ResNet-9 | 0.727 | 0.749 | 0.725 |
| Baseline Fusion [36] | – | 0.722 | – |

been proven effective in enhancing model performance on the development set. ResNet-9 without being pre-trained is also implemented, whose classification result is 74.9% UAR on the devel set.

To further improve our model performance, we proposed model fusion on three models of the best performance on the development set. The fusion is conducted in two ways, early fusion and late fusion. We also proposed two approaches to deal with the embeddings in the early fusion stage. The first approach is concatenating the embeddings, and the second is simply taking the mean value of the embeddings. Both scenarios employ SVM as the classifier. As for the late fusion, we proposed a voting mechanism among the three models' decisions. The fusion result is shown in Table 4. By implementing late fusion, we got the best system performed on the devel set with a UAR score of 81.5%.

**Table 4.** Model Fusion. Selected models: MFCC+VAD+PR, MFCC+VAD+PR+LS, Logfbank+VAD

| Fusion Approach | Precision | UAR | F1-Score |
|---|---|---|---|
| Concatenate | 0.783 | 0.800 | 0.779 |
| Mean | 0.789 | 0.805 | 0.789 |
| Voting | 0.810 | **0.815** | **0.803** |

## 5 Discussion

Our proposed best fusion model exceeded the devel set baseline by 12.8%. It is worth mentioning that the WebRTC-VAD system is not able to tease out every non-speech segment. Instead, its value cast more light on removing the blank or noisy segments at the beginning and end of an audio clip. This is reasonable since the unsounded segments in a conversation are also meaningful information to determine the escalation level and emotion. Dialogues would be more likely labeled as high escalation level if the speaker is rushing through the conversation and vice versa. Thus we agreed that a more complicated Neural-Network-based voice activity detector may be unnecessary in this task.

The significant improvement of our system on the devel set has again proved that emotion recognition features and paralinguistic features share certain advanced representations. Just as we mentioned in the related work part, they benefit from each other in the transfer learning tasks. However, due to the small scale of the training set, the overfitting problem is highly concerned. Thus we chose ResNet-18 to train the model on a combined emotion dataset, containing 12,000+ labeled emotion clips. This architecture has also been proved to be effective in the Escalation detection task.

For this task, we adopted Sentence-BERT [34] as the textual embeddings extractor. We utilized the pre-trained multilingual BERT model which is capable of handling Dutch, German, English, etc. We chose to translate the raw Dutch text to English text before feeding them into the embedding extractor, for we did a comparative experiment, which showed that the English textual embeddings alone significantly outperformed the Dutch textual embeddings. The UAR on the devel set achieved by English Embeddings alone is around 45%, whose detection ability is very likely to be limited by the dataset scale and occasional errors in translation. Had we have richer textual data, the linguistic embeddings should be of more help.

An unsuccessful attempt is adding denoising into the preprocessing attempt. Denoising should be part of the preprocessing stage since most of the collected audios contain background noise from public areas. According to [23], they first denoise the police body-worn audio before feature extraction, which turns out rewarding for them in detecting conflicts from the audios. However, our attempt does not improve the performance. Our denoised audio is agreed to be clearer in human perception and contains weaker background noises. However, the performance on the devel set is significantly degraded. Our assumption is that, unlike the police body-warn audio, which mostly contains criminality-related scenarios, the conversation audios in TR and SD datasets happen with richer contextual environments. The speech enhancement system might also affect the signal of speech which might be the reason for the performance degradation.

## 6    Conclusions

In this paper, we proposed a multimodal solution to tackle the task of escalation level detection under extremely low resource contraints. We applied Voice Activity Detection to pre-process the escalation datasets. We also pre-trained an emotion recognition model with ResNet backbone and fine-tune the parameters with the escalation dataset. We also validated that the learning process of escalation signals can benefit from emotion representations learning. By integrating linguistic information in the classification process, the model can become more stable and robust. The single best model can achieve 81.0% UAR, compared to 72.2% UAR baseline. By doing the late fusion the models after fusion are able to achieve the 81.5% UAR. Future efforts will be focusing on addressing the over-fitting problem.

**Acknowledgements.** This research is funded in part by the Synear and Wang-Cai donation lab at Duke Kunshan University. Many thanks for the computational resource provided by the Advanced Computing East China Sub-Center.

## References

1. Webrtc-vad (2017). https://webrtc.org/
2. Abdelwahab, M., Busso, C.: Supervised domain adaptation for emotion recognition from speech. In: 2015 IEEE International Conference on Acoustics, Speech and Signal Processing (ICASSP), pp. 5058–5062. IEEE (2015)
3. Aurelio, Y.S., de Almeida, G.M., de Castro, C.L., Braga, A.P.: Learning from imbalanced data sets with weighted cross-entropy function. Neural Process. Lett. **50**(2), 1937–1949 (2019)
4. Brain, D., Webb, G.I.: On the effect of data set size on bias and variance in classification learning. In: Proceedings of the Fourth Australian Knowledge Acquisition Workshop, University of New South Wales, pp. 117–128 (1999)
5. Busso, C., et al.: Iemocap: interactive emotional dyadic motion capture database. Lang. Resour. Eval. **42**(4), 335–359 (2008)
6. Busso, C., et al.: Analysis of emotion recognition using facial expressions, speech and multimodal information. In: Proceedings of the 6th International Conference on Multimodal Interfaces, pp. 205–211 (2004)
7. Cao, H., Cooper, D.G., Keutmann, M.K., Gur, R.C., Nenkova, A., Verma, R.: Crema-d: Crowd-sourced emotional multimodal actors dataset. IEEE Trans. Affect. Comput. **5**(4), 377–390 (2014)
8. Caraty, M.-J., Montacié, C.: Detecting speech interruptions for automatic conflict detection. In: D'Errico, F., Poggi, I., Vinciarelli, A., Vincze, L. (eds.) Conflict and Multimodal Communication. CSS, pp. 377–401. Springer, Cham (2015). https://doi.org/10.1007/978-3-319-14081-0_18

9. Dupuis, K., Pichora-Fuller, M.K.: Toronto emotional speech set (tess)-younger talker_happy (2010)

10. Evci, U., Dumoulin, V., Larochelle, H., Mozer, M.C.: Head2toe: Utilizing intermediate representations for better transfer learning. In: International Conference on Machine Learning, pp. 6009–6033. PMLR (2022)

11. Fayek, H.M., Lech, M., Cavedon, L.: Towards real-time speech emotion recognition using deep neural networks. In: 2015 9th International Conference on Signal Processing and Communication Systems (ICSPCS), pp. 1–5. IEEE (2015)

12. Gideon, J., Khorram, S., Aldeneh, Z., Dimitriadis, D., Provost, E.M.: Progressive Neural Networks for Transfer Learning in Emotion Recognition. In: Proceedings Interspeech 2017, pp. 1098–1102 (2017). https://doi.org/10.21437/Interspeech. 2017-1637

13. Grèzes, F., Richards, J., Rosenberg, A.: Let me finish: automatic conflict detection using speaker overlap. In: Proceedings Interspeech 2013, pp. 200–204 (2013). https://doi.org/10.21437/Interspeech. 2013-67

14. He, K., Zhang, X., Ren, S., Sun, J.: Deep residual learning for image recognition. In: Proceedings of the IEEE Conference on Computer Vision and Pattern Recognition, pp. 770–778 (2016)

15. Huang, C., Song, B., Zhao, L.: Emotional speech feature normalization and recognition based on speaker-sensitive feature clustering. Int. J. Speech Technol. 19(4), 805–816 (2016). https://doi.org/10.1007/s10772-016-9371-3

16. Kim, S., Valente, F., Vinciarelli, A.: Annotation and detection of conflict escalation in Political debates. In: Proceedings Interspeech 2013, pp. 1409–1413 (2013). https://doi.org/10.21437/Interspeech. 2013-369

17. Kim, S., Yella, S.H., Valente, F.: Automatic detection of conflict escalation in spoken conversations, pp. 1167–1170 (2012). https://doi.org/10.21437/Interspeech. 2012-121

18. Kishore, K.K., Satish, P.K.: Emotion recognition in speech using mfcc and wavelet features. In: 2013 3rd IEEE International Advance Computing Conference (IACC), pp. 842–847. IEEE (2013)

19. Ko, J.H., Fromm, J., Philipose, M., Tashev, I., Zarar, S.: Limiting numerical precision of neural networks to achieve real-time voice activity detection. In: 2018 IEEE International Conference on Acoustics, Speech and Signal Processing (ICASSP), pp. 2236–2240. IEEE (2018)

20. Lalitha, S., Geyasruti, D., Narayanan, R., M, S.: Emotion detection using mfcc and cepstrum features. Procedia Computer Science 70, 29–35 (2015). https://doi. org/10.1016/j.procs.2015.10.020, https://www.sciencedirect.com/science/article/ pii/S1877050915031841, proceedings of the 4th International Conference on Eco-friendly Computing and Communication Systems

21. Lefter, I., Burghouts, G.J., Rothkrantz, L.J.: An audio-visual dataset of human-human interactions in stressful situations. J. Multimodal User Interfaces 8(1), 29–41 (2014)

22. Lefter, I., Rothkrantz, L.J., Burghouts, G.J.: A comparative study on automatic audio-visual fusion for aggression detection using meta-information. Pattern Recogn. Lett. 34(15), 1953–1963 (2013)

23. Letcher, A., Trišović, J., Cademartori, C., Chen, X., Xu, J.: Automatic conflict detection in police body-worn audio. In: 2018 IEEE International Conference on Acoustics, Speech and Signal Processing (ICASSP), pp. 2636–2640. IEEE (2018)

24. Likitha, M.S., Gupta, S.R.R., Hasitha, K., Raju, A.U.: Speech based human emotion recognition using mfcc. In: 2017 International Conference on Wireless Communications, Signal Processing and Networking (WiSPNET), pp. 2257–2260 (2017). https://doi.org/10.1109/WiSPNET.2017.8300161

25. Lin, M., Chen, Q., Yan, S.: Network in network. arXiv preprint arXiv:1312.4400 (2013)

26. Livingstone, S.R., Russo, F.A.: The ryerson audio-visual database of emotional speech and song (ravdess): a dynamic, multimodal set of facial and vocal expressions in north american english. PLoS ONE **13**(5), e0196391 (2018)

27. McFee, B., et al.: librosa: Audio and music signal analysis in python. In: Proceedings of the 14th Python in Science Conference. vol. 8, pp. 18–25. Citeseer (2015)

28. Mehta, P., et al.: A high-bias, low-variance introduction to machine learning for physicists. Phys. Rep. **810**, 1–124 (2019)

29. Müller, R., Kornblith, S., Hinton, G.E.: When does label smoothing help? In: Wallach, H., Larochelle, H., Beygelzimer, A., d' Alché-Buc, F., Fox, E., Garnett, R. (eds.) Advances in Neural Information Processing Systems. vol. 32. Curran Associates, Inc. (2019). https://proceedings.neurips.cc/paper/2019/file/f1748d6b0fd9d439f71450117eba2725-Paper.pdf

30. Ng, H.W., Nguyen, V.D., Vonikakis, V., Winkler, S.: Deep learning for emotion recognition on small datasets using transfer learning. In: Proceedings of the 2015 ACM on International Conference on Multimodal Interaction, pp. 443–449 (2015)

31. Peng, M., Wu, Z., Zhang, Z., Chen, T.: From macro to micro expression recognition: Deep learning on small datasets using transfer learning. In: 2018 13th IEEE International Conference on Automatic Face & Gesture Recognition (FG 2018), pp. 657–661. IEEE (2018)

32. Polzehl, T., Sundaram, S., Ketabdar, H., Wagner, M., Metze, F.: Emotion classification in children's speech using fusion of acoustic and linguistic features. In: Proceedings Interspeech 2009, pp. 340–343 (2009). https://doi.org/10.21437/Interspeech. 2009–110

33. Poria, S., Hazarika, D., Majumder, N., Naik, G., Cambria, E., Mihalcea, R.: Meld: A multimodal multi-party dataset for emotion recognition in conversations. In: Proceedings of the 57th Annual Meeting of the Association for Computational Linguistics, pp. 527–536 (2019)

34. Reimers, N., Gurevych, I.: Sentence-bert: Sentence embeddings using siamese bert-networks. In: Proceedings of the 2019 Conference on Empirical Methods in Natural Language Processing and the 9th International Joint Conference on Natural Language Processing (EMNLP-IJCNLP), pp. 3982–3992 (2019)

35. Reimers, N., Gurevych, I.: Making monolingual sentence embeddings multilingual using knowledge distillation. In: Proceedings of the 2020 Conference on Empirical Methods in Natural Language Processing (EMNLP), pp. 4512–4525 (2020)

36. Schuller, B.W., et al.: The INTERSPEECH 2021 Computational Paralinguistics Challenge: COVID-19 Cough, COVID-19 Speech, Escalation & Primates. In: Proceedings INTERSPEECH 2021, 22nd Annual Conference of the International Speech Communication Association. ISCA, Brno, Czechia (September 2021), to appear

37. Tang, Y.: Deep learning using linear support vector machines (2013). https://doi.org/10.48550/ARXIV.1306.0239, https://arxiv.org/abs/1306.0239

38. Tzirakis, P., Trigeorgis, G., Nicolaou, M.A., Schuller, B.W., Zafeiriou, S.: End-to-end multimodal emotion recognition using deep neural networks. IEEE J. Selected Topics Signal Process. **11**(8), 1301–1309 (2017). https://doi.org/10.1109/JSTSP.2017.2764438

39. van den Oord, A., Dieleman, S., Schrauwen, B.: Transfer learning by supervised pre-training for audio-based music classification. In: Conference of the International Society for Music Information Retrieval, Proceedings, p. 6 (2014)

40. Wu, C., Huang, C., Chen, H.: Text-independent speech emotion recognition using frequency adaptive features. Multimedia Tools Appl. **77**(18), 24353–24363 (2018). https://doi.org/10.1007/s11042-018-5742-x

41. Yang, Y., et al.: Multilingual universal sentence encoder for semantic retrieval. In: Proceedings of the 58th Annual Meeting of the Association for Computational Linguistics: System Demonstrations, pp. 87–94 (2020)

42. Zhao, W.: Research on the deep learning of the small sample data based on transfer learning. In: AIP Conference Proceedings, vol. 1864, p. 020018. AIP Publishing LLC (2017)

# Pre-training Techniques for Improving Text-to-Speech Synthesis by Automatic Speech Recognition Based Data Enhancement

Yazhu Liu[ID], Shaofei Xue[(✉)][ID], and Jian Tang[ID]

AIspeech Ltd., Suzhou, China
{yazhu.liu,shaofei.xue,jian.tang}@aispeech.com

**Abstract.** As the development of deep learning, neural network (NN) based text-to-speech (TTS) that adopts deep neural networks as the model backbone for speech synthesis, has now become the mainstream technology for TTS. Compared to the previous TTS systems based on concatenative synthesis and statistical parametric synthesis, the NN based speech synthesis shows conspicuous advantages. It needs less requirement on human pre-processing and feature development, and brings high-quality voice in terms of both intelligibility and naturalness. However, robust NN based speech synthesis model typically requires a sizable set of high-quality data for training, which is expensive to collect especially in low-resource scenarios. It is worth investigating how to take advantage of low-quality material such as automatic speech recognition (ASR) data which can be easily obtained compared with high-quality TTS material. In this paper, we propose a pre-training technique framework to improve the performance of low-resource speech synthesis. The idea is to extend the training material of TTS model by using ASR based data augmentation method. Specifically, we first build a framewise phoneme classification network on the ASR dataset and extract the semi-supervised <linguistic features, audio> paired data from large-scale speech corpora. We then pre-train the NN based TTS acoustic model by using the semi-supervised <linguistic features, audio> pairs. Finally, we fine-tune the model with a small amount of available paired data. Experimental results show that our proposed framework enables the TTS model to generate more intelligible and natural speech with the same amount of paired training data.

**Keywords:** Pre-training techniques · neural network · text-to-speech · automatic speech recognition

## 1 Introduction

Recent advances in neural network (NN) based text-to-speech (TTS) have significantly improved the naturalness and quality of synthesized speech. We are now able to generate high-quality human-like speech from given text with less

L. Zhenhua et al. (Eds.): NCMMSC 2022, CCIS 1765, pp. 162–172, 2023.
https://doi.org/10.1007/978-981-99-2401-1_15

requirement on human pre-processing and feature development [1–5]. However, such models typically require tens of hour transcribed dataset consisting of high-quality text and audio training pairs, which are expensive and time consuming to collect. Requiring large amounts of data limits the overall naturalness and applicability especially in low-resource scenarios.

A series of extended technologies have been developed to improve the data efficiency for NN based TTS training. Most of these existing methods can be grouped into three categories: dual transformation, transfer learning and self-supervised/semi-supervised training. Firstly, dual transformation mainly focuses on the dual nature of TTS and automatic speech recognition (ASR). TTS and ASR are two dual tasks and can be leveraged together to improve each other. Speech chain technique is presented in [6] to construct a sequence-to-sequence model for both ASR and TTS tasks as well as a loop connection between these two processes. The authors in [7] develop a TTS and ASR system named LRSpeech which use the back transformation between TTS and ASR to iteratively boost the accuracy of each other under the extremely low-resource setting. In [8], it proposes an almost unsupervised learning method that only leverages few hundreds of paired data and extra unpaired data for TTS and ASR by using dual learning. Secondly, although paired text and speech data are scarce in low-resource scenarios, it is abundant in rich-resource scenarios. Transfer learning approaches try to implement adaptation methods and retain the satisfactory intelligibility and naturalness. Several works attempt to help the mapping between text and speech in low-resource languages with pre-training the TTS models on rich-resource languages [9–12]. In order to alleviate the difference of phoneme sets between rich and low-resource languages. The work in [13] proposes to map the embeddings between the phoneme sets from different languages. In [14], international phonetic alphabet (IPA) is adopted to support arbitrary texts in multiple languages. Besides that, voice conversion (VC) [15, 16] is also an effective way to improve the data efficiency in low-resource TTS training. Recent work in [17] brings significant improvements to naturalness by combining multi-speaker modelling with data augmentation for the low-resource speaker. This approach uses a VC model to transform speech from one speaker to sound like speech from another, while preserving the content and prosody of the source speaker. Finally, self-supervised/semi-supervised training strategies are leveraged to enhance the language understanding or speech generation capabilities of TTS model. For example, paper [18] aims to lower TTS systems' reliance on high quality data by providing them the textual knowledge, which is extracted from BERT [19] language models during training. They enrich the textual information through feeding the linguistic features that extracted by BERT from the same input text to the decoder as well along with the original encoder representations. In [20], the researchers propose a semi-supervised training framework to allow Tacotron to utilize textual and acoustic knowledge contained in large, publicly available text and speech corpora. It first embeds each word in the input text into word vectors and condition the Tacotron encoder on them. Then an unpaired speech corpus is used to pre-train the Tacotron decoder in the acoustic domain. Finally, the model is fine-tuned using available paired data. An unsupervised pre-training mechanism that uses Vector-Quantization Variational-Autoencoder (VQ-VAE) [21] to extract the unsupervised linguistic units from the untranscribed speech is investigated in [22]. More recently, an unsupervised TTS system based on an alignment module that outputs pseudo-text and another synthesis module that uses pseudo-text for training and real text for inference, is presented in [23].

The motivation of this work is to develop novel techniques to alleviate the data demand for training NN based TTS. We propose a semi-supervised pre-training technique framework to improve the performance of speech synthesis by extending the training material of TTS model with ASR based data augmentation. Specifically, we first build a frame-wise phoneme classification network on ASR dataset and extract the semi-supervised <linguistic features, audio> paired data from large-scale speech corpora. Then, we pre-train the NN based TTS acoustic model by using the semi-supervised <linguistic features, audio> pairs. Finally, we fine-tune the model with a small amount of available paired data.

It should be noticed that similar semi-supervised pre-training work has been related in [20]. However, our work is different in several ways, constituting the main contributions of our work. Firstly, the semi-supervised <linguistic features, audio> paired data for pre-training TTS model are extracted from a frame-wise phoneme classification network, which is built from the beginning based on the ASR dataset. It makes us possible to pre-train the entire TTS acoustic model, while the encoder and decoder are separately pre-trained in [20]. Secondly, the acoustic model of TTS system implemented in our work is different. We choose to use AdaSpeech [5] which involves the adaptive custom voice technique by inserting speaker embedding as the conditional information. Finally, we investigate and analyze the effectiveness of building low-resource language TTS systems with the help of semi-supervised pre-training on the rich-resource language.

The rest of this paper is organized as follows: In Sect. 2, we briefly review the architecture of TTS model used in this work. In Sect. 3, our proposed novel techniques to improve the performance of low-resource TTS are described. Section 4 shows our experimental setups and detailed results on Mandarin and Chinese Dialects tasks. Several conclusions are further drawn in Sect. 5.

## 2   TTS Model

As the development of deep learning, NN based TTS that adopts deep neural networks as the model backbone for speech synthesis, has now become the mainstream technology for TTS. Compared to the previous TTS systems based on concatenative synthesis and statistical parametric synthesis, the NN based speech synthesis shows conspicuous advantages. It needs less requirement on human pre-processing and feature development, and brings high-quality voice in terms of both intelligibility and naturalness. A NN based TTS system often consists of three basic components: a text analysis module, an acoustic model (abbr. TTS-AM), and a vocoder. The text analysis module converts a text sequence into the linguistic features, and then TTS-AM transforms linguistic features to the acoustic features, finally the vocoder synthesizes the waveform based on the acoustic features.

## 2.1   Text Analysis Module

In the TTS system, the text analysis module has an important influence on the intelligibility and naturalness of synthesized speech. The typical text analysis module in a Chinese TTS system consists of text normalization (TN), Chinese word segmentation (CWS), part-of-speech (POS) tagging, grapheme-to-phoneme (G2P) conversion, and prosody prediction. It extracts various linguistic features from the raw text, aiming to provide enough information for training the TTS-AM.

## 2.2   Acoustic Model

In this work, we choose to use AdaSpeech, which is a non-autoregressive model based on the transformer architecture. The basic model backbone consists of a phoneme encoder, a spectrogram decoder, an acoustic condition modeling that captures the diverse acoustic conditions of speech in speaker level, utterance level and phoneme level. And a variance adaptor which provides variance information including duration, pitch and energy into the phoneme hidden sequence. The decoder generates spectrogram features in parallel from the predicted duration and other information.

## 2.3   Vocoder

The vocoder in our work is based on LPCNet [24–26]. It introduces conventional digital signal processing into neural networks, and uses linear prediction coefficients to calculate the next waveform point while leveraging a lightweight RNN to compute the residual. This makes it possible to match the quality of state-of-the art neural synthesis systems with fewer neurons, significantly reducing the complexity. The LPCNet is a good compromise between quality and inference speed for a TTS system. As the LPCNet uses bark-frequency cepstrum as input, we modify the AdaSpeech to generate bark-frequency cepstrum as output. No external speaker information, such as speaker embedding, is referred in the building of LPCNet model.

## 3   The Proposed Approach

In this section, the proposed semi-supervised pre-training framework on TTS modeling is detailed. The Illustration of our general framework is shown in Fig. 1. We first describe the structure of frame-wise phoneme classification model and the alignment module that greedily proposes a pairing relationship between speech utterances and phoneme transcripts. After that, the pre-training and fine-tuning procedures of our method are presented.

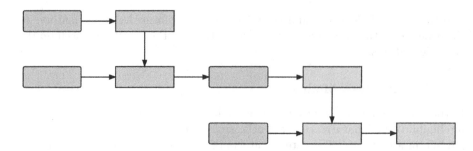

**Fig. 1.** Illustration of the proposed semi-supervised pre-training framework on TTS modeling.

### 3.1 Frame-Wise Phoneme Classification

**DFSMN Model.** DFSMN is an improved FSMN architecture by introducing the skip connections and the memory strides [27]. The DFSMN component consists of four parts: a ReLU layer, a linear projection layer, a memory block and a skip connection from the bottom memory block, except for the first one that without the skip connection from the bottom layer. By adding the skip connections between the memory blocks of DFSMN components, the output of the bottom layer memory block can directly flow to the upper layer. During back-propagation, the gradients of higher layer can also be assigned directly to lower layer that help to overcome the gradient vanishing problem. Since the information of adjacent frames in speech signals always have strong redundancy due to the overlap. The strides for look-back and look-ahead are used to help the DFSMN layer remove the redundancy in adjacent acoustic frames. DFSMN is able to model the long-term dependency in sequential signals while without using recurrent feedback. In practice, DFSMN models usually contain DFSMN layers around ten to twenty. We follow the model topology in [28] and implement a DFSMN with ten DFSMN layers followed by two fully-connected ReLU layers, a linear layer and a softmax output layer. To avoid the mismatch in G2P conversion, we share the same phoneme set between phoneme classification and TTS tasks. We adopt IPA as described in [14] to support arbitrary texts in Mandarin and multiple Chinese Dialects evaluations in our work.

**Alignment Module.** After training DFSMN based phoneme classification model, the semi-supervised paired data for pre-training TTS model has to be prepared. Pseudo phoneme transcript of the training set is first generated by greedy decoding over the output of DFSMN. Instead of extracting the phoneme duration with soft attention mechanism as described in [20,22], the alignment between pseudo phoneme transcript and speech sequence is derived from a forced alignment procedure computed by Kaldi [29] with a phonetic decision tree. This improves the alignment accuracy and reduces the information gap between the model input and output.

## 3.2 Semi-supervised Pre-training

In the baseline AdaSpeech, the model should simultaneously learn the textual representations, acoustic representations, and the alignment between them. The encoder takes a source phoneme text as input and produces sequential representations of it. The decoder is then conditioned on the phoneme representations to generate corresponding acoustic representations, which are then converted to waveforms. [20] proposes two types of pre-training methods to utilize the external textual and the acoustic information. For textual representations, they pre-train encoder by the external word-vectors. For acoustic representations, they pre-train the decoder by untranscribed speech. Although [20] shows that the proposed semi-supervised pre-training helps the model synthesizes more intelligible speech, it finds that pre-training the encoder and decoder separately at the same time does not bring further improvement than only pre-training the decoder. However, there is a mismatch between pre-training only the decoder and fine-tuning the whole model. To avoid potential error introduced by this mismatch and further improve the data efficiency by using only speech, we instead use the semi-supervised paired data generated by the frame-wise phoneme classification model as described in Sect. 3.1. It helps to alleviate the mismatch problem and makes pre-training the entire model possible.

## 3.3 AdaSpeech Fine-Tuning

The AdaSpeech in pre-trained model is applied as a multi-speaker TTS-AM, which means we do not use the adaptive custom voice technique as described in [5]. After that, the AdaSpeech is fine-tuned with some high-quality paired speech data from the target speaker. In this procedure, the inputs of the model are phoneme sequences derived from the normalized text.

# 4 Experiments

In this section, we evaluate the performance of the proposed approach on two type of TTS tasks including single-speaker Mandarin dataset and multi-speaker Chinese Dialects dataset. For both two experiments, we use a 3000-hour Mandarin dataset which consists of 1000-hour low-quality transcribed ASR data (1000 h-TD) and 2000-hour low-quality untranscribed data (2000 h-UTD) for pre-training. The data are collected from many domains, such as voice search, conversation, video and the sample rate of the data is 16 kHz.

In the ASR setup, waveform signal is analyzed using a 25-ms Hamming window with a 10-ms fixed frame rate. 40-dimensional filterbank features are used for training DFSMN phoneme classification models. The features are pre-processed with the global mean and variance normalization (MVN) algorithm. We use

11 frames (5-1-5) of filter-banks as the input features of neural networks. The DFSMNs model stacked with 1 Relu Layer (2048 hidden nodes), 10 DFSMN layers (2048 memory block size, 512 projection size, 10*[2048-512]), 4 ReLU layers (2*2048-1024-512) and 1 Softmax output layer.

In the TTS setup, 20-dimensional features, which consist of 18 Bark-scale cepstral coefficients and 2 pitch parameters (period, correlation), are extracted from 16k audio using a 25-ms Hamming window with a 10-ms fixed frame rate. The TTS-AM of AdaSpeech consists of 6 feed-forward Transformer blocks for the phoneme encoder and the decoder. The hidden dimension of phoneme embedding, speaker embedding and self-attention are all set to 384. The number of attention heads is 4 in the phoneme encoder and the decoder. The pre-train and fine-tune models are separately trained in a distributed manner using stochastic gradient descent (SGD) optimization on 16 GPUs and 4 GPUs.

### 4.1   Single-Speaker Mandarin Task

**Experimental Setup.** In the single-speaker Mandarin task, we evaluate our method with the Chinese Standard Mandarin Speech Corpus (CSMSC). CSMSC has 10,000 recorded sentences read by a female speaker, totaling 12 h of natural speech with phoneme-level text grid annotations and text transcriptions. The corpus is randomly partitioned into non-overlapping training, development and test sets with 9000, 800 and 200 sentences respectively. We conduct several experimental setups to investigate the influence of semi-supervised pre-training. All parameters of the TTS-AM are directly updated during the fine-tuning stage. For better comparing the efficiency of pre-training on TTS-AM, we use the same LPCNet which is trained on full 12 h CSMSC. The performance of the overall quality samples is evaluated using the mean opinion score (MOS). Listeners are asked to rate the overall naturalness and prosodic appropriateness of samples on a scale from 1 and 5. Then these synthesized samples are mixed with real speech samples and presented to listeners independently in random order. 15 raters who are native Mandarin speakers are included in the subjective test.

**Performance of Phoneme Classification Models** The phoneme error rate (PER) performance of using different amounts of low-quality transcribed data to build phoneme classification models is shown in Table 1. We evaluate with three test sets, including the CSMSC development set (Test-c), a 5 h dataset that randomly sampled from the 1000 h-TD (Test-i) and exists in each training data, a 4 h dataset that randomly sampled from the 2000 h-UTD which never exists in the training sets (Test-o). It can be observed that increasing the amount of training data yields a large improvement on the PER. To better evaluate the relationship between PER and pre-training efficiency, we use the recognized phoneme transcripts of all training sets for alignment to generate the semi-supervised paired data.

**Table 1.** PER% performance of phoneme classification models on three evaluation tasks.

| Training Data Size | Test-c | Test-i | Test-o |
|---|---|---|---|
| 100 h | 9.3 | 18.8 | 26.9 |
| 1000 h | 6.6 | 12.2 | 15.4 |

**Results on Different Phoneme Classification Models** In this section, the results of implementing different phoneme classification models to generate semi-supervised data for TTS-AM pre-training are presented. The mean MOS scores on CSMSC test set using two different DFSMN models are gradually explored. DFSMN-1000 h indicates that we use the 1000 h-TD to train the DFSMN model. DFSMN-100 h means the DFSMN model is built with 100 h subset of the 1000 h-TD. When generating the semi-supervised data for pre-training TTS-AM, we choose to use the same 100 h subset data. The results shown in Table 2 confirm that our proposed semi-supervised pre-training method brings conspicuous improvement on the MOS especially when we utilize only 15 min paired data. Besides that, the MOS on DFSMN-1000 h pre-trained model is slightly better then the DFSMN-100 h pre-trained model. It indicates that achieving higher accuracy semi-supervised paired data is also a feasible way for improving the intelligibility and naturalness of synthesized speech. In the next experiment, we choose DFSMN-1000 h model to generate all semi-supervised data.

**Table 2.** The mean MOS on CSMSC of using different phoneme classification models to generate TTS-AM pre-training data.

| Fine-tuning Data Size | Without Pre-training | Pre-trained Model | | Ground Truth |
|---|---|---|---|---|
| | | DFSMN-100 h | DFSMN-1000 h | |
| 15 min | 2.80 | 3.55 | **3.70** | 4.71 |
| 2 h | 3.99 | **4.09** | 4.03 | |
| 10 h | 4.11 | 4.05 | **4.16** | |

**Results on Different Amounts of Pre-training Data.** In this experiment, we compare the results of using different amounts of data for pre-training. We utilize Size-$N$ for labelling the data size used in TTS-AM pre-training. Thus, Size-100 h indicates that we use the same 100 h subset data as in above experiment. Size-1000 h stands that we use the whole 1000 h-TD for TTS-AM pre-training. Size-3000 h means we expand the dataset for pre-training TTS-AM by including the 2000 h-UTD. As shown in Table 3, several conclusions can be drawn from the results. Firstly, the results suggest that expanding the pre-training data size directly helps the speech synthesis performance. The MOS of fine-tuning Size-3000 h TTS-AM with 15 min paired data is similar with directly training on 2 h paired data. Secondly, it seems that the difference between using pre-training model and without pre-trained model is small when enough paired TTS data

**Table 3.** The mean MOS on CSMSC of using different amounts of pre-training data.

| Fine-tuning Data Size | Without Pre-training | Pre-train Data Size | | | Ground Truth |
|---|---|---|---|---|---|
| | | Size-100 h | Size-1000 h | Size-3000 h | |
| 15 min | 2.80 | 3.70 | 3.83 | **3.93** | 4.71 |
| 2 h | 3.99 | 4.03 | **4.09** | 4.01 | |
| 10 h | 4.11 | **4.16** | 4.10 | 4.08 | |

has been involved in the building process. For example, when fine-tuning on 10 h TTS data, we observe no conspicuous improvement on the MOS evaluation.

### 4.2    Multi-speaker Chinese Dialects Task

**Experimental Setup.** In the multi-speaker Chinese Dialects task, we evaluate our method with the two Chinese Dialect speech corpuses including Shanghainese and Cantonese. The Shanghainese corpus consists of 3 female speakers, each has 1 h recorded speech. The Cantonese corpus has 2 female speakers, each also has 1 h recorded speech. The pre-trained TTS-AM model used in this task is trained with 3000 h Mandarin dataset. For better comparison, the fine-tuning procedure is separately utilized for each speaker. And we build the speaker-dependent LPCNet to convert acoustic features into wave. We also use the MOS score to evaluate the performance of overall quality samples and share the same rating rules as in above experiments. We find 15 native Shanghainese speakers and 15 native Cantonese speakers to implement the subjective test.

**Results on Low-Resource Languages** Table 4 and Table 5 show the TTS performance of our proposed method on low-resource Shanghainese and Cantonese Corpuses. We investigate the MOS on fine-tuning with 15 min and 1 h datasets. It is obvious that the pre-training on rich-resource Mandarin benefits the building of low-resource Chinese Dialects TTS-AMs. For example, when conducting experiment on 15 min dataset, the mean MOS score of Shanghainese increases from 2.97 to 3.30 and the mean MOS score of Cantonese increases

**Table 4.** The mean MOS on Shanghainese.

| Fine-tuning Data Size | Without Pre-training | Pre-trained Model | Ground Truth |
|---|---|---|---|
| 15 min | 2.97 | **3.30** | 4.30 |
| 1 h | 3.39 | **3.51** | |

**Table 5.** The mean MOS on Cantonese.

| Fine-tuning Data Size | Without Pre-training | Pre-trained Model | Ground Truth |
|---|---|---|---|
| 15 min | 2.66 | **3.09** | 4.53 |
| 1 h | 3.54 | **3.99** | |

from 2.66 to 3.09. With the help of proposed technique, we can generate more intelligible and natural speech with the same amount of low-resource data.

## 5 Conclusions

In this paper, a novel semi-supervised pre-training technique framework that extends the training material of TTS model by using ASR based data augmentation method is proposed to improve the performance of speech synthesis. We first build a frame-wise phoneme classification network on the ASR dataset and extract the semi-supervised <linguistic features, audio> paired data from large-scale speech corpora. After that, the semi-supervised <linguistic features, audio> pairs is used to pre-train the NN based TTS acoustic model. Finally, we fine-tune the model with a small amount of available paired data. Experimental results show that our proposed framework can benefits the building of low-resource TTS system by implementing semi-supervised pre-training technique. It enables the TTS model to generate more intelligible and natural speech with the same amount of paired training data.

## References

1. Wang, Y., et al.: Tacotron: towards end-to-end speech synthesis. arXiv preprint arXiv:1703.10135 (2017)
2. Shen, J., et al.: Natural TTS synthesis by conditioning WaveNet on MEL spectrogram predictions. In: 2018 IEEE International Conference on Acoustics, Speech and Signal Processing (ICASSP), pp. 4779–4783 (2018)
3. Ren, Y., et al.: FastSpeech: fast, robust and controllable text to speech. In: Advances in Neural Information Processing Systems, vol. 32 (2019)
4. Ren, Y., et al.: FastSpeech 2: fast and high-quality end-to-end text to speech. arXiv preprint arXiv:2006.04558 (2020)
5. Chen, M., et al.: AdaSpeech: adaptive text to speech for custom voice. arXiv preprint arXiv:2103.00993 (2021)
6. Tjandra, A., Sakti, S., Nakamura, S.: Listening while speaking: speech chain by deep learning. In: IEEE Automatic Speech Recognition and Understanding Workshop (ASRU), pp. 301–308 (2017)
7. Xu, J., et al.: LRSpeech: extremely low-resource speech synthesis and recognition. In: Proceedings of the 26th ACM SIGKDD International Conference on Knowledge Discovery and Data Mining, pp. 2802–2812 (2020)
8. Ren, Y., Tan, X., Qin, T., Zhao, S., Zhao, Z., Liu, T.-Y.: Almost unsupervised text to speech and automatic speech recognition. In: International Conference on Machine Learning, pp. 5410–5419. PMLR (2019)
9. Azizah, K., Adriani, M., Jatmiko, W.: Hierarchical transfer learning for multilingual, multi-speaker, and style transfer DNN-based TTS on low-resource languages. IEEE Access 8, 179 798–179 812 (2020)
10. de Korte, M., Kim, J., Klabbers, E.: Efficient neural speech synthesis for low-resource languages through multilingual modeling. arXiv preprint arXiv:2008.09659 (2020)

11. Zhang, W., Yang, H., Bu, X., Wang, L.: Deep learning for Mandarin-Tibetan cross-lingual speech synthesis. IEEE Access **7**, 167 884–167 894 (2019)
12. Nekvinda, T., Dušek, O.: One model, many languages: meta-learning for multilingual text-to-speech. arXiv preprint arXiv:2008.00768 (2020)
13. Tu, T., Chen, Y.-J., Yeh, C.-C., Lee, H.-Y.: End-to-end text-to-speech for low-resource languages by cross-lingual transfer learning. arXiv preprint arXiv:1904.06508 (2019)
14. Hemati, H., Borth, D.: Using IPA-based Tacotron for data efficient cross-lingual speaker adaptation and pronunciation enhancement. arXiv preprint arXiv:2011.06392 (2020)
15. Mohammadi, S.H., Kain, A.: An overview of voice conversion systems. Speech Commun. **88**, 65–82 (2017)
16. Karlapati, S., Moinet, A., Joly, A., Klimkov, V., Sáez-Trigueros, D., Drugman, T.: CopyCat: many-to-many fine-grained prosody transfer for neural text-to-speech. arXiv preprint arXiv:2004.14617 (2020)
17. Huybrechts, G., Merritt, T., Comini, G., Perz, B., Shah, R., Lorenzo-Trueba, J.: Low-resource expressive text-to-speech using data augmentation. In: ICASSP 2021–2021 IEEE International Conference on Acoustics, Speech and Signal Processing (ICASSP), pp. 6593–6597 (2021)
18. Fang, W., Chung, Y.-A., Glass, J.: Towards transfer learning for end-to-end speech synthesis from deep pre-trained language models. arXiv preprint arXiv:1906.07307 (2019)
19. Devlin, J., Cheng, M.-W., Kenton, L., Toutanova, K.: BERT: pre-training of deep bidirectional transformers for language understanding. In: Proceedings of NAACL-HLT, pp. 4171–4186 (2019)
20. Chung, Y.-A., Wang, Y., Hsu, W.-N., Zhang, Y., Skerry-Ryan, R.: Semi-supervised training for improving data efficiency in end-to-end speech synthesis. In: ICASSP 2019–2019 IEEE International Conference on Acoustics, Speech and Signal Processing (ICASSP), pp. 6940–6944 (2019)
21. Van Den Oord, A., Vinyals, O., et al.: Neural discrete representation learning. In: Advances in Neural Information Processing Systems, vol. 30 (2017)
22. Zhang, H., Lin, Y.: Unsupervised learning for sequence-to-sequence text-to-speech for low-resource languages. arXiv preprint arXiv:2008.04549 (2020)
23. Ni, J., et al.: Unsupervised text-to-speech synthesis by unsupervised automatic speech recognition. arXiv preprint arXiv:2203.15796 (2022)
24. Valin, J.-M., Skoglund, J.: LPCNet: improving neural speech synthesis through linear prediction. In: 2019 IEEE International Conference on Acoustics, Speech and Signal Processing (ICASSP), pp. 5891–5895 (2019)
25. Skoglund, J., Valin, J.-M.: Improving Opus low bit rate quality with neural speech synthesis. arXiv preprint arXiv:1905.04628 (2019)
26. Valin, J.-M., Skoglund, J.: A real-time wideband neural vocoder at 1.6 kb/s using LPCNet. arXiv preprint arXiv:1903.12087 (2019)
27. Zhang, S., Lei, M., Yan, Z., Dai, L.: Deep-FSMN for large vocabulary continuous speech recognition. In: IEEE International Conference on Acoustics, Speech and Signal Processing (ICASSP), pp. 5869–5873 (2018)
28. Zhang, S., Lei, M., Yan, Z.: Automatic spelling correction with transformer for CTC-based end-to-end speech recognition. arXiv preprint arXiv:1904.10045 (2019)
29. Povey, D., et al.: The Kaldi speech recognition toolkit. In: IEEE 2011 Workshop on Automatic Speech Recognition and Understanding (2011)

# A Time-Frequency Attention Mechanism with Subsidiary Information for Effective Speech Emotion Recognition

Yu-Xuan Xi[1], Yan Song[1(✉)], Li-Rong Dai[1], and Lin Liu[2]

[1] National Engineering Laboratory for Speech and Language Information Processing, University of Science and Technology of China, Hefei, China
xyxah96@mail.ustc.edu.cn, {songy,lrdai}@ustc.edu.cn
[2] iFLYTEK Research, iFLYTEK CO., LTD., Hefei, China
linliu@iflytek.com

**Abstract.** Speech emotion recognition (SER) is the task of automatically identifying human emotions from the analysis of utterances. In practical applications, the task is often affected by subsidiary information, such as speaker or phoneme information. Traditional domain adaptation approaches are often applied to remove unwanted domain-specific knowledge, but often unavoidably contribute to the loss of useful categorical information. In this paper, we proposed a time-frequency attention mechanism based on multi-task learning (MTL). This uses its own content information to obtain self attention in time and channel dimensions, and obtain weight knowledge in the frequency dimension through domain information extracted from MTL. We conduct extensive evaluations on the IEMOCAP benchmark to assess the effectiveness of the proposed representation. Results demonstrate a recognition performance of 73.24% weighted accuracy (WA) and 73.18% unweighted accuracy (UA) over four emotions, outperforming the baseline by about 4%.

**Keywords:** speech emotion recognition · convolutional neural network · multi-task learning · attention mechanism

## 1 Introduction

Speech emotion recognition (SER) involves the automatic identification of human emotions from the analysis of spoken utterances. The field has drawn increasing research interest in recent years, largely due to the rapid growth of speech-based human-computer interaction applications, such as intelligent service robotics, automated call centres, remote education and so on.

Traditional SER techniques typically follow a conventional pattern recognition pipeline. This mainly focuses on robust and discriminative feature extraction, an effective classifier, and often a combination of both. More recently, SER methods based on deep neural networks (DNN), convolutional neural networks (CNN) [12,20], and recurrent neural networks (RNN) [11] have been widely proposed and evaluated.

© The Author(s), under exclusive license to Springer Nature Singapore Pte Ltd. 2023
L. Zhenhua et al. (Eds.): NCMMSC 2022, CCIS 1765, pp. 173–184, 2023.
https://doi.org/10.1007/978-981-99-2401-1_16

Despite good progress, several issues remain with current SER research. Speech emotion is generally psychological in nature, and as such may be affected by the factors like speaker characteristics, utterance content, and language variations. Many researchers use text data in conjunction with speech data information – and it is known that incorporating text information greatly improves SER accuracy [6,18,19]. Therefore, several domain-adaptation and multi-task learning (MTL) methods have been proposed to utilize emotion information from multiple corpora, and to address data scarcity issues [1–4,17,24]. In [7], Gao et al. try to propose a domain adversarial training method to cope with the non-affective information during feature extraction. Gat et al. [8] propose a gradient-based adversary learning framework which normalize speaker characteristics from the feature representation. However, in the elimination of domain specific information, part of the wanted emotional information is inevitably also lost.

In order to make use of subsidiary information, while retaining emotional information as much as possible, we considered the use of an attention module. Squeeze-and-Excitation Networks [10] design an attention module in the channel dimension. Similar methods can also be applied to other dimensions [15,22,25]. In [27], Zou et al. propose SER system using multi-level acoustic information with a co-attention module. Meanwhile we were inspired from the image processing domain by work from Hou *et al.* [9] who extended the SE module to H and W dimensions to better exploit longer or wider parts of an image. In addition we note that Fan *et al.* [5] proposed a frequency attention module for SER. Furthermore, they used gender information to calculate attention weights, although only apply this at the end of the feature extractor.

In this paper, we design two attention modules to exploit different information in the dimensions of frequency and time, effectively extending [5] into more dimensions. In addition, we make use of gender information within the network.

We then evaluate the effectiveness of the proposed framework through extensive experiments on IEMOCAP. The results demonstrate recognition performance of 73.24% weighted accuracy (WA) and 73.18% unweighted accuracy (UA) over four emotions, outperforming the state-of-the-art method by about 5%.

The rest of paper is organized as follows. We describe the architecture of the proposed framework in Sect. 2 and expound the methods used in Sect. 3. A detailed discussion of our experiments and performance evaluation is reported in Sect. 4 before we conclude our work in Sect. 5.

## 2 Overview of the Speech Emotion Recognition Architecture

Modern CNN/DNN based SER methods typically input the spectrum of the speech signal, but we note that this differs in a fundamental way from a traditional image recognition task in which the two dimensions of an image are often functionally equivalent. By contrast, the semantic information encoded in

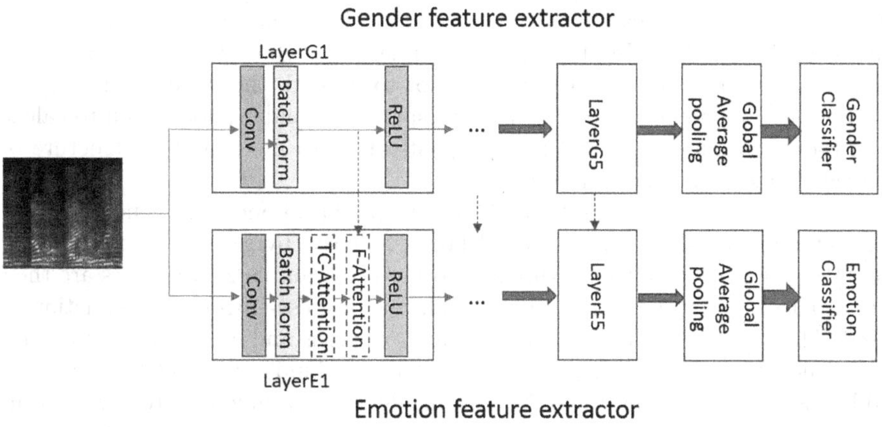

**Fig. 1.** The above part is gender classification network, and the below part is emotion recognition network. The two parts are connected by attention module. In training, the loss functions of gender and emotion are calculated respectively, and then train the whole network together.

the two spectrogram dimensions are highly dissimilar. SER incorporates a lot of subsidiary information (or noise), and this has different effects on the time and frequency axes of the spectrogram. For example, in the time dimension, individual speakers might have different ways or speeds of speaking as emotion changes, but it is difficult for us to find rules to exploit this because their speech content also changes, and this has a great impact on our task. Therefore, in the time dimension, it makes sense to pay more attention to the phonetic content, that is, the phone information.

In the frequency dimension, by contrast, individual speakers will have utterances characterised by dissimilar frequency distributions, with the gender of the speaker often being an important factor. For example, the predominant frequency band occupied by male voices tends to be lower than that of female voices [14]. When we operate on the frequency dimension, it makes more sense to take account of the gender identity of the speaker.

In order to use this kind of information in the frequency domain, we designed a network structure similar to multi-task learning. Two branches in the network are used to extract gender information, and emotion information respectively.

Given input utterance data, we load this into two parallel networks, one of which is used as an emotion feature extractor, and uses emotion label training, the other is used as gender feature extractor and uses gender label training. In the former network, we designed two different attention modules, one uses its own features in time and channel dimension, and the other uses a gender feature in frequency dimension contributed by the first network. Based on the previous analysis, we have adopted different methods to calculate the attention weight of these two attention modules. In terms of time and channel dimensions, our focus on the current feature is basically related to the content of that location.

Therefore, we use the feature itself to calculate the attention weight and get the self-attention module. In the frequency dimension, which frequencies are concerned about the current features has a lot to do with the speaker information of the speech. Therefore, we use the corresponding gender information to calculate the attention weight and get the joint-attention module. The structure of the whole model is shown in Fig. 1.

We employ a VGG5 model for our network backbone. The gender feature extractor contains five convolutional layers followed by global average pooling (GAP) to downsample the feature size to $1 \times 1$. FC and Softmax layers are then utilized to predict the final gender label, and a cross-entropy loss function is used. The structure of the emotion feature extractor and classifier are basically the same as in the gender part, except that two kinds of attention layers are added between batch-norm and ReLU. The detailed parameters are described in Sect. 4.2.

## 3    Methods

### 3.1    TC Self-attention Module

Strip pooling [9] is an improvement of the generic SE module. It is intended to identify the uneven length and width of image regions. For the SER task, we can use a similar structure. For the feature map $x \in \mathbb{R}^{C \times T \times F}$, we know that the content of speech information will change in the T dimension. For different phonemes, we want to have different channel attention weights for this part. We thus first apply an average pooling in the F dimension, and then use a $1 \times 1$ convolution layer to reduce the feature to D dimensions.

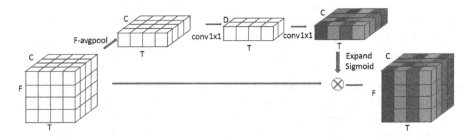

**Fig. 2.** Structure of TC self-Attention module

$$x_{i,j}^{FP} = \frac{1}{F} \sum_{0 < k \leq F} x_{i,j,k}$$
$$x^{D} = Conv_{C \to D}^{1 \times 1}(x^{FP})$$

(1)

This amounts to classifying the feature data at each time point, so that each time point can derive a D-dimensional feature vector. The feature vector changes with time and aims to encapsulate the phoneme information at that time point. Then we use a 1×1 convolution layer to change it back to the C dimension, and use a sigmoid to map the value between 0 and 1. Hence we derive a channel dimension attention weight at each time point according to the current phoneme information.

$$x^{att} = Sigmoid(Conv_{D \to C}^{1 \times 1}(x^D))$$
$$x^{out} = x \times Expand(x^{att})$$

(2)

The structure of the whole TC self-attention pooling is shown in Fig. 2

## 3.2   F Domain-Attention Module

In the frequency dimension, the performance of different gender speakers is very different. Therefore, we hope to use the speaker's gender information to add attention system to the frequency dimension.

In order to obtain gender information, we add a frequency feature extractor and a classifier, and the parameter setting of the feature extractor's convolution layer is completely consistent with that of the emotional feature extractor, so that the emotional features and gender features on the same layer can correspond in the frequency dimension.

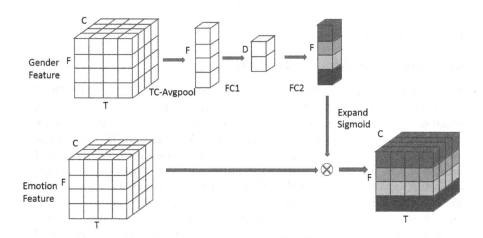

**Fig. 3.** Structure of the F domain-Attention module.

During experimental evaluation, we find that the accuracy of the gender classification exceeds 95%, so we are confident that its features carry useful gender information.

The influence of gender on frequency dimension is very different from that of the phoneme on the time dimension, so the attention layer structure we designed is also different. The degree of attention paid to any frequency location should be determined by the gender information of the whole sentence, so we should use gender information across the whole (single speaker) utterance length to obtain the attention weight for the frequency dimension.

For gender feature $x^{gen} \in \mathbb{R}^{C \times T \times F}$, we first perform average pooling in the C and T dimensions, then two fully connected layers are used to reduce it to D dimensions and restore it back to the F dimension. Again we use a sigmoid to map the value between 0 and 1. In this way, the attention weight in F dimensions is obtained through gender information.

$$
\begin{aligned}
x_k^{TCP} &= \frac{1}{C \times T} \sum_{\substack{0 < i \leq C \\ 0 < j \leq T}} x_{i,j,k}^{gen} \\
x^F &= FC2(ReLU(FC1(x^{TCP}))) \\
x^{att} &= Sigmoid(x^F)
\end{aligned}
\tag{3}
$$

By multiplying the weight and emotional features, we can focus on different frequency bands for different gender speech segments as follows:

$$
x^{out} = x^{emo} \times Expand(x^{att}) \tag{4}
$$

The structure of the whole F domain-attention pooling framework is illustrated in Fig. 3.

## 4   Experiments

### 4.1   Dataset and Acoustic Features

We use the Interactive Emotional Dyadic Motion Capture database (IEMOCAP) for all experiments. IEMOCAP contains approximately 12 h of audiovisual data recorded by 10 skilled actors. The entire database is divided into 5 sections, each containing one male and one female actor. According to the recording scenarios, it can be further subdivided into an improvised speech section, and a scripted speech section. Each utterance in the dataset is annotated by multiple annotators into 8 emotion labels. Following previous works, we choose 4 emotion types for our experiments (namely neutral, happy, angry and sad) from the improvised speech for study – since scripted data may contain undesired contextual information. Adopting the methodology of previous works, we performed a 5-fold cross-validation using a leave-one-out strategy. In each training process, 8 speakers are used as training data, one of the remaining speakers is used as verification data and the final speaker as the test data.

Magnitude spectrograms are utilized as input features, extracted over 40 ms Hamming windows with a 10 ms shift between windows and an FFT size of 1600 points. Then 0–4 kHz spectrograms are utilized since human vocal expression is thought to be mainly located in this frequency range [14]. The speech utterances are cut into 2 s portions with 1 s overlap, and zero-padding applied for utterances

**Table 1.** Detailed network parameters

| Gender network | Emotion network |
|---|---|
| Conv(16@3*3)+BN | Conv(16@3*3)+BN |
| – | TC-A(16,2) + F-A(400, 40) |
| ReLU+Maxpool(2,2) | ReLU+Maxpool(2,2) |
| Conv(32@3*3)+BN | Conv(32@3*3)+BN |
| – | TC-A(32,4) + F-A(200,20) |
| ReLU+Maxpool(2,2) | ReLU+Maxpool(2,2) |
| Conv(48@3*3)+BN | Conv(48@3*3)+BN |
| – | TC-A(48,8) + F-A(100,20) |
| ReLU+Maxpool(2,2) | ReLU+Maxpool(2,2) |
| Conv(64@3*3)+BN | Conv(64@3*3)+BN |
| – | TC-A(64,16) + F-A(50,16) |
| ReLU | ReLU |
| Conv(80@3*3)+BN | Conv(80@3*3)+BN |
| – | TC-A(80,16) + F-A(50,16) |
| ReLU+Maxpool(2,2) | ReLU+Maxpool(2,2) |
| Global Avgpooling | Global Avgpooling |
| FC(80,2) | FC(80,4) |

shorter than 2 s. Thus the input spectrograms have a size of $400 \times 200$. For each spectrogram, we then apply mu-law expansion.

The SER performance is evaluated using the standard metrics: Weighted Accuracy (WA), which is the classification accuracy of all utterances, Unweighted Accuracy (UA), which averages the accuracy of each individual emotion class, and F1-score.

### 4.2 System Description

As noted, we adopt a VGG5 model for our experiments, with detailed parameters of the network listed in Table 1. Conv(C@K) represents the convolution layer with output channel C and kernel size K. TC-A(C, D) means our TC self-attention module where C is the input channel, D is the middle layer channel. F-A(F, D) means our F domain-attention module where F is input frequency, D is middle layer frequency. Maxpool(2,2) expresses maxpooling, and we use kernel size K and stride S. FC($C_{in}$,$C_{out}$) denotes a fully connected layer with input channel $C_{in}$ and output channel $C_{out}$.

**Baseline.** There is no attention module in the baseline network, so it does not need to use gender knowledge, and only use emotion labels to train the emotion network.

The CNN training makes use of the PyTorch deep learning framework. The optimization method is standard Stochastic Gradient Descent (SGD) with a mini-batch size of 128. We use a Nesterov momentum of 0.9 and a weight decay of 0.0001. The CNNs are trained over 40 epochs with initial learning rate of 0.05, reducing by a factor of 10 at the 21, 31 and 41 epochs.

**TC Self-attention Module.** In this variant, only the TC self-attention module is added to the network, and thus gender information is not needed. We add the TC self attention module after layer five. In fact, we did many experiments to study in which layers it is best for the module to be located. Some of those results are listed in Table 2.

**Table 2.** Accuracy achieved with the TC self-attention module inserted after different layers (as numbered).

| Methods | WA | UA | F1-score |
|---|---|---|---|
| Baseline | 68.53% | 62.09% | 59.03% |
| TC-A module12 | 69.42% | 65.13% | 61.22% |
| TC-A module45 | 69.43% | **66.24%** | **62.10%** |
| TC-A module234 | **70.01%** | 65.37% | 61.58% |
| TC-A module12345 | 69.64% | 64.07% | 61.17% |
| SE module234 | 69.43% | 62.38% | 60.87% |

It can be seen from the table that the TC-A module is generally beneficial for SER. After much experimentation, we conclude that adding two or three TC-A modules to the network has a better effect, which may be due to the consequence of the improved matching of network parameters and task data. We also experimented with the traditional SE module, and found that the proposed TC-A module performs better.

**F Domain-Attention Module.** In the F-A module, we need to incorporate gender information when training the network. We thus combine gender and emotional losses:

$$Loss = Loss_{emo} + \lambda Loss_{gen} \tag{5}$$

In the experiment, the value of $\lambda$ is set to 1. As in Sect. 4.2, we conducted many experiments to explore the usage location of F-A modules. Some of those results are listed in Table 3. From this we can see that inserting the F-A module after the final two convolution layers achieves the best performance.

**Table 3.** Accuracy of F domain-attention module inserted after different convolutional layers (as numbered).

| Methods | WA | UA | F1-score |
|---|---|---|---|
| F-A module12 | 69.42% | 65.13% | 62.32% |
| F-A module45 | **71.04%** | **67.87%** | **63.20%** |
| F-A module5 | 70.46% | 66.41% | 63.09% |

**Time-Frequency Attention Mechanism.** We next integrated the two kinds of attention modules and repeated our experimental evaluation. The results, summarised in Table 4, indicate that combining both TC-A and F-A (at the 4th and 5th layers respectively) achieves the best performance.

**Table 4.** The best results achieved by using two attention modules separately and together.

| Methods | WA | UA | F1-score |
|---|---|---|---|
| TC-A module | 70.01% | 66.24% | 62.10% |
| F-A module | 71.04% | 67.87% | 63.20% |
| TF-A module | **73.24%** | **73.18%** | **65.39%** |

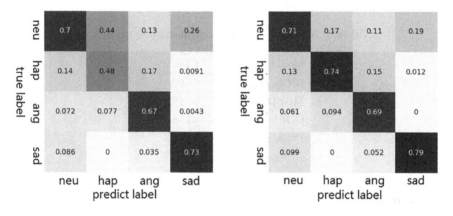

**Fig. 4.** Confusion matrices for the baseline method (left) and our TF-A module (right).

We also present confusion matrices for the baseline and combined TF-A module in Fig. 4. This shows that the proposed method can improve the accuracy of all kinds of emotion categories, especially happy and angry, which are characterised by less data – thus the improvement in UA is greater than in WA. The two emotional expressions of happy and angry are relatively intense, which may explain why the proposed attention mechanism is able to perform so well.

**Comparison to State-of-the-Art Systems** We compare the proposed TF-A model to the state-of-the-art published results in Table 5. Due to the existence of many test methods which use IEMOCAP in different ways, we only listed results which adopted a similar evaluation methodology as in our tests. Some experiments were evaluated using WA and UA, namely LSTM-ELM [20], CNN-Att [12], DED [26], NAS [23], co-att [27]. Among these models, the proposed time-frequency attention mechanism outperforms other, more complex, systems.

In addition, we also found some experimental results using f1 score for comparison with our results shown in Table 6.

**Table 5.** Accuracy comparison with existing models.

| Model | WA | UA |
|---|---|---|
| LSTM-ELM [20] | 62.85% | 63.89% |
| CNN-Att [12] | 70.18% | 66.38 % |
| DED [26] | 69.0% | 70.1% |
| NAS [23] | 70.54% | 56.94% |
| co-att [27] | 69.80% | 71.05% |
| Proposed TF-A | 73.24% | 73.18% |

**Table 6.** Accuracy comparison with existing models.

| Model | F1-score |
|---|---|
| ICON [21] | 63.0% |
| MCED [13] | 60.1% |
| COPYPASTE [16] | 63.78 |
| Proposed TF-A | 65.39% |

## 5  Conclusion

In this paper, we proposed a time-frequency attention mechanism for the SER task. In order to better utilize both phone information and gender information in speech, we designed two different attention modules for time and frequency. We also used a structure similar to multi task learning, and built an additional network branch to obtain the domain information – which comprises only a simple gender indication. In experiments using the IEMOCAP dataset, we demonstrate good performance. In our future research, we aim to design more complex systems, which can use speaker and other domain information, or in turn use emotional information to help speaker classification.

## References

1. Atmaja, B.T., Akagi, M.: Multitask learning and multistage fusion for dimensional audiovisual emotion recognition. In: ICASSP 2020, pp. 4482–4486 (2020)
2. Deng, J., Xu, X., Zhang, Z., Frühholz, S.: Semisupervised autoencoders for speech emotion recognition. IEEE/ACM Trans. Audio Speech Lang. Process. **26**, 31–43 (2018)

3. Deng, J., Xu, X., Zhang, Z., Frühholz, S., Schuller, B.: Universum autoencoder-based domain adaptation for speech emotion recognition. IEEE Signal Process. Lett. **24**, 500–530 (2017)

4. Deng, J., Zhang, Z., Eyben, F., Schuller, B.: Autoencoder-based unsupervised domain adaptation for speech emotion recognition. IEEE Signal Process. Lett. **21**, 1068–1072 (2014)

5. Fan, W., Xu, X., Xing, X., Huang, D.: Adaptive domain-aware representation learning for speech emotion recognition. In: Proceedings of Interspeech 2020, pp. 4089–4093 (2020). https://doi.org/10.21437/Interspeech.2020-2572

6. Feng, H., Ueno, S., Kawahara, T.: End-to-end speech emotion recognition combined with acoustic-to-word ASR model. In: Proceedings of Interspeech 2020, pp. 501–505 (2020). https://doi.org/10.21437/Interspeech.2020-1180

7. Gao, Y., Okada, S., Wang, L., Liu, J., Dang, J.: Domain-invariant feature learning for cross corpus speech emotion recognition. In: ICASSP 2022–2022 IEEE International Conference on Acoustics, Speech and Signal Processing (ICASSP), pp. 6427–6431 (2022). https://doi.org/10.1109/ICASSP43922.2022.9747129

8. Gat, I., Aronowitz, H., Zhu, W., Morais, E., Hoory, R.: Speaker normalization for self-supervised speech emotion recognition. In: ICASSP 2022–2022 IEEE International Conference on Acoustics, Speech and Signal Processing (ICASSP), pp. 7342–7346 (2022). https://doi.org/10.1109/ICASSP43922.2022.9747460

9. Hou, Q., Zhang, L., Cheng, M.M., Feng, J.: Strip pooling: rethinking spatial pooling for scene parsing. In: Proceedings of the IEEE/CVF Conference on Computer Vision and Pattern Recognition, pp. 4003–4012 (2020)

10. Hu, J., Shen, L., Albanie, S., Sun, G., Wu, E.: Squeeze-and-excitation networks. In: Computer Vision and Pattern Recognition (CVPR) (2018)

11. Lee, J., Tashev, I.: High-level feature representation using recurrent neural network for speech emotion recognition. In: Proceedings of Interspeech, pp. 1537–1540 (2015)

12. Li, P., Song, Y., McLoughlin, I., Guo, W., Dai, L.: An attention pooling based representation learning method for speech emotion recognition. In: Proceedings of Interspeech (2018)

13. Li, R., Zhao, J., Jin, Q.: Speech emotion recognition via multi-level cross-modal distillation. In: Proceedings of Interspeech 2021, pp. 4488–4492 (2021). https://doi.org/10.21437/Interspeech.2021-785

14. McLoughlin, I.V.: Speech and Audio Processing: a MATLAB-Based Approach. Cambridge University Press, Cambridge (2016)

15. Miao, X., McLoughlin, I., Yan, Y.: A new time-frequency attention tensor network for language identification. Circ. Syst. Signal Process. **39**(5), 2744–2758 (2020)

16. Pappagari, R., Villalba, J., Żelasko, P., Moro-Velazquez, L., Dehak, N.: CopyPaste: an augmentation method for speech emotion recognition. In: ICASSP, pp. 6324–6328 (2021). https://doi.org/10.1109/ICASSP39728.2021.9415077

17. Parthasarathy, S., Busso, C.: Jointly predicting arousal, valence and dominance with multi-task learning. In: Proceedings of Interspeech 2017, pp. 1103–1107 (2017). https://doi.org/10.21437/Interspeech.2017-1494

18. Pepino, L., Riera, P., Ferrer, L., Gravano, A.: Fusion approaches for emotion recognition from speech using acoustic and text-based features. In: ICASSP 2020, pp. 4482–4486 (2020)

19. Priyasad, D., Fernando, T., Denman, S., Sridharan, S., Fookes, C.: Attention driven fusion for multi-modal emotion recognition. In: ICASSP 2020, pp. 6484–6488 (2020)

20. Satt, A., Rozenberg, S., Hoory, R.: Efficient emotion recognition from speech using deep learning on spectrograms. In: Proceedings of Interspeech, pp. 1089–1093 (2017)

21. Sebastian, J., Pierucci, P.: Fusion techniques for utterance-level emotion recognition combining speech and transcripts. In: Proceedings of Interspeech 2019, pp. 51–55 (2019). https://doi.org/10.21437/Interspeech.2019-3201

22. Woo, S., Park, J., Lee, J.-Y., Kweon, I.S.: CBAM: convolutional block attention module. In: Ferrari, V., Hebert, M., Sminchisescu, C., Weiss, Y. (eds.) ECCV 2018. LNCS, vol. 11211, pp. 3–19. Springer, Cham (2018). https://doi.org/10.1007/978-3-030-01234-2_1

23. Wu, X., Hu, S., Wu, Z., Liu, X., Meng, H.: Neural architecture search for speech emotion recognition. In: ICASSP 2022–2022 IEEE International Conference on Acoustics, Speech and Signal Processing (ICASSP), pp. 6902–6906 (2022). https://doi.org/10.1109/ICASSP43922.2022.9746155

24. Xi, Y., Li, P., Song, Y., Jiang, Y., Dai, L.: Speaker to emotion: domain adaptation for speech emotion recognition with residual adapters. In: Asia-Pacific Signal and Information Processing Association (APSIPA) (2019)

25. Xia, W., Koishida, K.: Sound event detection in multichannel audio using convolutional time-frequency-channel squeeze and excitation. In: Proceedings of Interspeech, pp. 3629–3633 (2019)

26. Yeh, S.L., Lin, Y.S., Lee, C.C.: A dialogical emotion decoder for speech emotion recognition in spoken dialog. In: ICASSP 2020, pp. 6479–6483 (2020)

27. Zou, H., Si, Y., Chen, C., Rajan, D., Chng, E.S.: Speech emotion recognition with co-attention based multi-level acoustic information. In: ICASSP 2022–2022 IEEE International Conference on Acoustics, Speech and Signal Processing (ICASSP), pp. 7367–7371 (2022). https://doi.org/10.1109/ICASSP43922.2022.9747095

# Interplay Between Prosody and Syntax-Semantics: Evidence from the Prosodic Features of Mandarin Tag Questions

Jinting Yan[1], Yuxiao Yang[2(✉)], and Fei Chen[3]

[1] School of Education, Cangzhou Normal University, Cangzhou, China
[2] Foreign Studies College, Hunan Normal University, Changsha 410081, China
davidyyx@qq.com
[3] School of Foreign Languages, Hunan University, Changsha, China

**Abstract.** The investigation of Mandarin prosody has been focusing on the suprasegmental characteristics of lexical tones in previous studies, the understanding of Mandarin sentence prosody, on the other hand, is still limited. To bridge this gap, this study probed the prosodic features of Mandarin tag questions in comparison with those from the declarative counterparts. The aim was to verify the hypothesis that the statement parts in the tag questions would be prosodically comparable with the declarative counterparts, whereas the tag *"dui bu dui?"* *(analogous to English question tags e.g. "is it, doesn't it?")*, being an interrogative marker, would show focal characteristics. 20 adult Mandarin speakers recorded 12 Mandarin sentences (6 tag questions + 6 declarative counterparts); the pitch fluctuation scale, duration ratio, and intensity ratio of which were extracted using Mini Speech Lab (Zhu & Shi, 2020) to compare the focal differences between the two types of Mandarin sentences. The statement parts in the tag questions exhibited focal characteristics similar to Mandarin general questions, while the tag *"dui bu dui?"* showed the characteristics of post-focal compression. These findings were at odds with our hypothesis. Results of the current study suggested that the focal positions of Mandarin questions might not be consistently associated with the interrogative markers as previous study would suggest, but are contingent upon the syntax-semantics of the corresponding utterances.

**Keywords:** Acoustics of prosody · Mandarin tag questions · Mandarin declarative sentences · Syntax-semantics

## 1 Introduction

Prosody is concerned with suprasegmental properties such as tone, intonation, rhythm, and stress, which are used to express different pragmatic, linguistic, emotional, and

This research was supported by the National Social Science Foundation of China for post funded projects (20FYYB042) and Humanities and Social Science Project of Ministry of Education of China (22YJC740093).

idiosyncratic functions (Cutler & Isard 1980). Mandarin is a tonal language, and the acoustic features of its four lexical tones had drawn much attention from the standpoints of either L1 or L2 acquisition. Nonetheless, prosody at sentence level also plays a critical role in Mandarin as it serves as a landmark differentiating various utterance functions, e.g. interrogative, exclamatory and declarative etc. The overlap between lexical tone and sentence prosody contributes to the complexity of Mandarin suprasegmental characteristics, yet it is crucial to pay special attention to sentence prosody in relation to its syntactic-semantic function (Hsu & Xu 2020; Lin 2020; Shen 1989; Wang et al., 2018). To date, research focusing on Mandarin sentence prosody is still limited compared to that on tones, and extant literature is primarily concerned with the prosody of declarative statements or general/wh- questions (Liu et al. 2006; Hsu & Xu 2020; Shen 1989). The empirical investigation of other types of Mandarin sentences is by far scarce and needs to be carried out to enrich the holistic understanding of Mandarin prosody and its interaction with syntax-semantics.

To bridge this gap, the current study probes the prosodic features of a rarely discussed type of Mandarin tag questions (henceforth TQs) consisting of a statement and an interrogative phrase dui bu dui? (yes or no?), which bear some syntactic resemblance with English tag questions, e.g., "It is going to stop (statement), isn't it (interrogation)?". Shao (1996: p. 123) delineated three basic characteristics of dui bu dui Mandarin TQs: 1) the tag dui bu dui? cannot be used without the preceding statement; 2) the interrogative mood is constituted solely by the tag dui bu dui?; 3) the answers to TQs have to be either yes or no. As such, the source of information stems from the statement, and the interrogative mood is expressed via the tag "dui bu dui?". It had been argued that the interrogative mood of Mandarin TQs could be reduced inasmuch as questioners might already have an anticipatory positive answer for the content being inquired (Yan 2017: p. 77), yet this argument was primarily drawn from impressionistic speculations, and empirical evidence needs to be replenished to verify if the deduction holds true.

It had been well documented that Mandarin prosody is closely related to sentence foci (Chen 2006; Cooper et al., 1985; Liu & Xu 2005; Hsu & Xu 2020; Xu 1999). Foci are the contents that speakers intend to make prominent throughout the utterance, which can be realized via syntactic structure or prosody etc. A consensus had been reached that the pitch height on the focal point of a sentence could rise, causing an expansion of the pitch range on the corresponding syllable, and a sharp decrease of pitch could be observed on the following syllable, rendering a compression of pitch range. In the meantime, the duration of the focal syllable could be lengthened compared to other syllables (Chen & Gussenhoven 2008; Chen et al. 2014; Xu et al. 2012). Yet less is known regarding the intensity representation of sentence foci. Notably, the findings mentioned herein were primarily concluded from the natural/logical foci in declarative sentences (henceforth DS), but much needs to be explored in light of questions.

Mandarin questions could be generally divided into two types: 1) general questions without interrogative markers (syntactically unmarked yes/no questions); 2) questions with interrogative markers (wh-words, Verb-not-Verb, particle ma, A or B, yes or no tag). The first type of Mandarin questions are syntactically identical with DSs, and their interrogative mood is expressed mainly through the change of prosody, that is to say, when no logical focus is involved, the entire question is believed to be focalized, resulting in an overall higher pitch and a wider pitch range in comparison with the syntactically identical DS; in duration, word-final syllables are longer in yes/no questions, and the longest duration occurs at the sentence-final syllable; for intensity, both yes/no questions and DSs display an overall descending of energy, yet a stark surge of intensity could be observed at the final syllable of the yes/no questions (Kochanski & Shih 2003; Shen 1989). Regarding the second type of Mandarin questions, extant studies on wh-questions and "Verb-not-Verb" questions suggested that the interrogative markers in these questions could be focalized, such that the pitch, duration, and intensity of the wh-words and "Verb-not-Verb" phrases could be amplified (Hsu & Xu 2020, Shen 1989; Yan et al. 2014). Most recently, Hsu & Xu (2020) found that the sentence-final particle ma, being the interrogative marker of yes/no questions in Mandarin, could also exhibit some acoustic prominence, even in post-focal positions. The authors ascribed this phenomenon as being triggered by the syntactic-semantic function of sentence-final particles in Mandarin, which could override the effect of post-focal compression. Meanwhile, researchers found that the prosody of pre-focal parts in questions could parallel with that in DSs (Yan et al. 2016). That being said, little is known regarding the Mandarin TQs, and the prosodic features of the two components of TQs need to be empirically investigated.

To this end, the current study seeks to unveil the prosodic characteristics of Mandarin TQs via a production experiment. The acoustic analysis will be administered following Shi & Wang (2014) via three acoustic parameters including fluctuation scale (pitch), duration ratio, and intensity ratio, which had been verified effective and widely adopted in the analysis of Mandarin prosody. Drawing upon previous findings on Mandarin interrogative questions, it was assumed that the tag "dui bu dui?", being an interrogative marker signaling sentence type, would be focalized (more prominent in the three parameters), while the statement part of TQs would prosodically parallel with DSs, being the information source in the pre-focal position.

## 2    Method

### 2.1    Participants

Twenty native Mandarin speakers were recruited in the experiment, including 8 males and 12 females (age range: 19–30 years). The participants were teachers and students specializing in broadcasting from Cangzhou Normal University, Hebei province. All participants thereby communicate in standard Mandarin in daily life, and their Mandarin proficiency was above "advanced level in the secondary class (over 87 out of 100)". The participants were all confirmed as having no speaking or hearing disorders and were rewarded monetarily for their participation.

### 2.2    Stimuli

The testing materials are illustrated in Table 1, which encompassed two types of sentences, Mandarin TQ and DS. TQs consisted of six statements adapted from Shen (1985) and the tag "dui bu dui?"; DSs contained only the statements. As shown in Table 1, each statement comprised three prosodic words constituted by 10 syllables (W1: S1–S3; W2:

S4–S6; W3: S7–S10). The six statements were designed to eliminate possible bias from lexical tones: statement 1 contained only level tone "ˉ", statement 2—rising tone "ˊ", statement 3—dipping tone "ˇ" and statement 4—falling tone "ˋ". Statements 5 and 6 were incorporated to cope with tone sandhi in Mandarin (a dipping tone altering to a rising tone when preceding another dipping tone). Statement 5 and 6 are in complement to each other to assure that a dipping tone in its citation form appears at each syllabic position. The statements were constant across TQs and DSs and will be compared to unveil possible prosodic differences.

**Table 1.** Testing materials transcribed in Pinyin with English translations. TQs = statements + tag; DSs = statements; W = prosodic word; S = syllable.

| | Components of TQs and DSs | | | | | | | | | | | | |
|---|---|---|---|---|---|---|---|---|---|---|---|---|---|
| | statements | | | | | | | | | | tag | | |
| | W1 | | | W2 | | | W3 | | | | W4 | | |
| | S1 | S2 | S3 | S4 | S5 | S6 | S7 | S8 | S9 | S10 | S11 | S12 | S13 |
| 1. | Zhāng | Zhōng | bīng | xīng | qī | tiān | xiū | shōu | yīn | jī | , | duì | bu | duì ? |
| | Zhang Zhongbing is going to repair the ratio on Sunday | | | | | | | | | | , | isn't he | | ? |
| 2. | Wú | Guó | huá | Chóng | Yáng | jié | huí | Yáng | chéng | hú | , | duì | bu | duì ? |
| | Wu Guohua goes back to Yangcheng Lake on Double Ninth Festival | | | | | | | | | | , | doesn't he | | ? |
| 3. | Lǐ | Xiǎo | bǎo | wǔ | diǎn | zhěng | xiě | yǎn | jiǎng | gǎo | , | duì | bu | duì ? |
| | Li Xiaobao will write the speech at five o' clock sharp | | | | | | | | | | , | won't he | | ? |
| 4. | Zhào | Shù | qìng | bì | yè | hòu | dào | jiào | yù | bù | , | duì | bu | duì ? |
| | Zhao shuqing went to the Ministry of Education after graduation | | | | | | | | | | , | didn't he | | ? |
| 5. | Lǐ | Jīn | bǎo | wǔ | shí | zhěng | jiāo | jiǎng | huà | gǎo | , | duì | bu | duì ? |
| | Li Jinbao will hand in the speech at five o' clock sharp | | | | | | | | | | , | won' he | | ? |
| 6. | Lǐ | Xiǎo | Gāng | wǔ | diǎn | bàn | xiě | bān | jiǎng | cí | , | duì | bu | duì ? |
| | Li Xiao Gang will write the award speech at half past five | | | | | | | | | | , | won't he | | ? |

### 2.3 Procedure

The recordings of the testing materials were administered in a phonetic laboratory in Cangzhou Normal University using a cardioid condenser microphone (Takstar PCM-5520, Huizhou) placed 15 cm away from the participants. A desktop computer (Microsoft Surface Pro 7) was connected to the microphone to monitor the ongoing recording progress. Praat (Boersma & Weenink, 2019) was used to collect mono-sound track at a sampling rate of 11, 025 Hz with 16-bit digitization. Prior to the formal recording process, all participants were given 10 min to get familiarized with the test materials; they were asked to read the testing sentences twice at a natural pace, yielding in total 240 sentences (20 speakers × 6 sentences × 2 times) for the data analysis.

## 2.4   Acoustic Analysis

To minimize the effect of individual variabilities and bias from various testing materials, the acoustic properties in this study were examined through three normalized parameters: 1) fluctuation scale, 2) duration ratio and 3) intensity ratio over the two types of testing sentences. Fluctuation scale refers to the pitch variation across syllables calculated as normalized percentages relative to the speaker's total utterances; the upper limit is 100% and the lower limit is 0%. The normalization was proceeded via the equation $Qx = (Kx - Kmin) / (Kmax - Kmin)$. Qx is the fluctuation of the target syllable; Kx refers to the highest/lowest pitch values of the target syllable transferred in semitones; Kmin is the lowest pitch value (semitone) of the speaker's total utterances, and Kmax is the highest. Pitch values were measured and transferred using Mini Speech Lab (Zhu & Shi, 2020) as depicted in the left panel in Fig. 1.

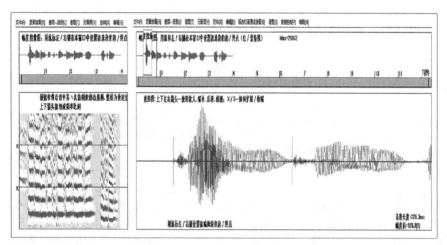

**Fig. 1.** Measurement of pitch (left panel) and aggregated amplitude (right panel) panel using Mini Speech Lab.

Duration ratio was calculated to quantify the temporal variation across syllables relative to the speaker's utterances. It was calculated via the equation $Dx = (Sx + Gx) /S\#$, in which Dx is the duration ratio of the target syllable; Sx represents the duration of the target syllable; Gx is the pause following the target syllable; S# refers to the average syllabic duration (including pauses) from the speaker's utterance. A syllable is considered to be lengthened when its duration ratio exceeds 1.

Regarding intensity ratio, the calculation of which is in line with that of duration: the aggregated amplitude of the target syllable divided by the mean aggregated amplitude of syllables from the speaker's utterance. Aggregated amplitude is an intensity measurement automatically calculated via Mini Speech Lab (see the right panel in Fig. 1), meaning the product of the average sample amplitude and the selected duration (in seconds), which is believed to be more fine-grained and informative than raw intensity in dB (Liang & Shi 2008). Likewise, a syllable is considered enhanced when its intensity ratio exceeds 1.

## 3   Results

### 3.1   Fluctuation Scale

For the pitch comparison between the two types of testing sentences, Fig. 2 illustrates the average fluctuation scales of TQs and the DSs. The percentages marked on the top and the bottom of prosodic words represent their highest and lowest fluctuation values. The gap between the highest and the lowest fluctuation values is the pitch range of the corresponding prosodic words.

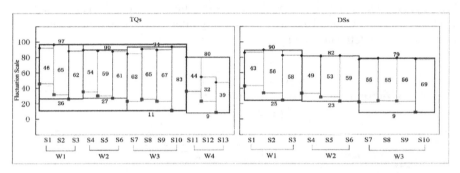

**Fig. 2.** Comparison of fluctuation scales between TQs and DSs. The thin columns indicate the fluctuation of single syllables; thicker columns indicate that of prosodic words; the thickest column frames the statement part from TQs.

As depicted in Fig. 2, the statements from TQs and DSs exhibit observable differences in fluctuation scale, in that the upper level was 7% higher in TQs than DSs (97% vs. 90% on S2), and the lower level was 2% higher in TQs than DSs (11% vs. 9% on S9). The overall pitch of the TQs statements was higher than DSs as early as in the beginning of the whole sentence, and the disparity became the largest at the end of the statements. Moreover, the pitch range of the three prosodic words in TQs statements was, to varying degrees, expanded, which was 6% larger than DSs in W1 (71% vs. 65%), 4% larger in W2 (63% vs. 59%) and 13% larger in W3 (83% vs. 70%). As such, the largest difference was observed with W3 between TQs statements and DSs. Concerning the tag "dui bu dui?", its overall pitch was relatively low compared with the statements in TQs and exhibited a descending trajectory with an upper level at 80% in the first syllable, and the lower level at 9% in the last syllable. The pitch range of each syllable in the tag (S11: 44%, S12: 32%, and S13: 39%) was also smaller relative to the statements.

For statistical analysis, independent-samples T-tests were conducted to compare the fluctuation scales of the three prosodic words between TQs statements and DSs. Results showed that TQs statements were significantly higher than DSs across all three prosodic words in terms of the upper level of fluctuation scale [W1: $t(29) = 3.51, p < 0.01$; W2: $t(38) = 2.88, p < 0.01$; W3: $t(29) = 5.25, p < 0.001$]. Regarding the pitch range of the three prosodic words, T-tests yielded no significant difference between TQs statements and DSs in either W1 [$t(38) = 1.89, p = 0.07$] or W2 [$t(38) = 0.91, p = 0.37$], yet the pitch range of W3 was significantly larger in TQs statements than DSs [$t(38) = 3.53, p$

< 0.01]. For the comparison between tag and statements within TQs, results of T-tests demonstrated that the upper level of the fluctuation scale at the first syllable in the tag was significantly lower than the last syllable in the statements [t (30) = 4.07, p < 0.001], so was the pitch range [t (38) = 10.15, p = < 0.001], indicating that the tag might have been compressed, highlighting the preceding statements.

## 3.2  Duration Ratio

As regards duration, Fig. 3 demonstrates the comparison between TQs and DSs in duration ratios. In contrast with DSs, more syllables in the statements in TQs are observed with lengthening: S2 (1.01), S3 (1.39), S6 (1.25), S7 (1.21), S8 (1.06), and S10 (1.31), all of which had duration ratios exceeding 1. In the meantime, only four syllables in the DSs were lengthened [S3 (1.24); S6 (1.19); S7 (1.09); S10 (1.12)]. Additionally, all syllables in TQs statements were larger in duration ratio in comparison with DSs. Furthermore, the duration ratio of the last syllable in TQs statements (S10: 1.31) was 0.34 higher than its preceding syllable (S9: 0.97) and 0.06 higher than the last syllable of W2 (S6: 1.25), and the incremental degree between the last two syllables in TQs statements was markedly higher than that in DS (0.34 vs. 0.22). As for the tag, the duration ratios of its three syllables were 0.72, 0.43, and 0.91 respectively, showing no signs of lengthening; the mid-syllable bu was particularly short in duration followed by a relatively long syllable at the end of the sentence.

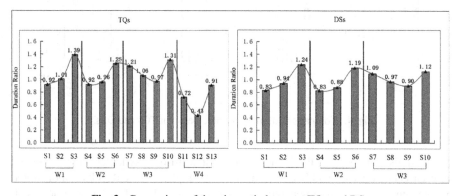

**Fig. 3.** Comparison of duration ratio between TQs and DSs.

For statistical analysis, the duration ratios of the word-final syllables were computed in independent samples T-tests to compare the differences between the TQs statements and DSs. Results revealed that the duration ratios of the final syllables in W1 and W3 from TQs were significantly higher than those from DSs [W1: t (36) = 3.8, p < 0.01; W2: t (36) = 4.26, p < 0.001]; no significant difference was found in the final syllable in W2 [t (36) = 1.54, p = 0.13]. As such, the overall temporal characteristics of TQs share some commonalities with general questions in Mandarin in that the word-final syllables are prolonged to a degree greater than that in DSs. Regarding the tag, results from T-tests revealed that its first syllable was significantly shorter than the last syllable in statements [t (27) = 14.61, p < 0.001], aligning with the compression pattern from fluctuation scales.

## 3.3  Intensity Ratio

Concerning intensity, Fig. 4 displays the intensity ratios of syllables and prosodic words from TQs and DSs. As depicted, more syllables were observed with over 1 intensity ratios in TQs statements [S1 (1.31), S2 (1.15), S3 (1.59), S5 (1.01), S6 (1.18), S7 (1.14), S8 (1.01)] than DSs [S1 (1.11), S2 (1.13), S3 (1.45), S6 (1.10), S7 (1.03)], and each syllable in TQs statements was higher than DSs in this parameter. Notably, even though the overall intensity of both types of sentences exhibited a declining trajectory, the sentence-final syllable (S10) in TQs exhibited a conspicuous rise relative to S9 (0.95 vs. 0.84), yet the same pattern was not observed in DSs (0.79 vs. 0.78). The intensity ratios of the three syllables in the tag, on the other hand, were relatively low compared to the preceding statements; the intensity of the mid-syllable bu constituted the lowest point throughout the whole sentence, in keeping with the duration patterns observed earlier.

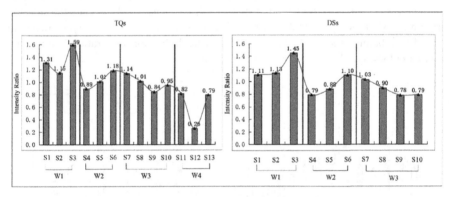

**Fig. 4.** Comparison of intensity ratio between TQs and DSs.

For statistical analysis, the intensity ratios of word-final syllables were compared between TQs statements and DSs via independent samples T-tests. Results demonstrated that the intensity ratios in TQs statements were unanimously higher than DSs across all three words [W1: t (37) = 2.13, p < 0.05; W2: t (28) = 2.32, p < 0.05; W3: t (34) = 3.92, p < 0.001]. As for the comparison between the tag and the statements within TQs, the results of T-tests showed that the first syllable in the tag was significantly lower than the last syllable in statements [t (38) = 2.76, p < 0.01], reinforcing the speculation that the tag was prosodically compressed.

## 4  Discussion

This study attempts to explore the interplay between prosody and syntax-semantics via the acoustic investigation of the under-studied Mandarin TQs in comparison with

DSs. On the basis of previous literature, it was assumed that the statement part of TQs, being the source of information, would prosodically parallel with DSs; the tag "dui bu dui?", on the other hand, could display some acoustic prominence, being focalized as an interrogative marker. The results of the production test, however, run counter to our expectation in that TQs statements manifested remarkable differences from DSs; the pitch height and range along with the duration and intensity of the word-final syllables were significantly higher in TQs statements than DSs, and the disparities reached the greatest at the sentence-final syllables.

The statement part of Mandarin TQs is generally considered as of declarative function, and the interrogative mood is expressed solely through the tag "dui bu dui?" (Shao, 1996: p. 123). The empirical findings from this study, however, suggested that the statement part in TQs could largely diverge from DSs and in effect pattern closer to the syntactically unmarked yes/no questions in Mandarin given their similar patterns of overall prosodic rise especially at sentence-final syllables. This implies that the statement part might be focalized in TQs. Nevertheless, it is worth noticing that the magnitude of the acoustic rise in TQs statements observed in the current study could be smaller than what had been found with syntactically unmarked yes/no questions from previous studies (Kochanski & Shih (2003) and Shen (1989). This might be due to the fact that the tag "dui bu dui?" in TQs helps fulfill the interrogative purpose which could only be delivered via prosodic means in syntactically unmarked yes/no questions. An alternative account is that the degree of interrogation in TQs is milder relative to yes/no questions, since TQs are asked with a bias of positive answers as introduced earlier, while yes/no questions do not involve such inclination, imposing stronger interrogation.

As regards the interrogative marker "dui bu dui?", its prosodic feature revealed in the current study was largely at odds with what can be expected from a sentence focus. On the contrary, the first syllable of the tag dui was significantly lower than the last syllables in the statements of TQs in all three acoustic parameters, constituting a post-focal compression. That is to say, the tag "dui bu dui?" in Mandarin TQs could serve as an interrogative marker without focus, which is somewhat surprising provided the vast amount of previous evidence on the focal features of interrogative phrases in Mandarin questions. In the meantime, the three syllables in the tag descended dramatically in pitch, analogous to a declarative phrase. The mid-syllable bu was produced exceptionally short and low in intensity, followed by a longer syllable dui at the end, serving as a plausible turn-taking strategy of seeking for rapid confirmation from the interlocutor.

As such, the present study unveiled some intriguing phenomena that the statement part of Mandarin TQs could exhibit the prosodic features of syntactically unmarked yes/no questions (focalized), in contrast with the syntactically identical DSs; the tag "dui bu dui?", on the other hand, was compressed in pitch, duration, and intensity. These findings ran counter to our expectation that the statement part of TQs could parallel with DSs in prosody considering their similar function, while the tag would be focalized as an interrogative marker. These observations might be accounted for by the interaction between prosody and syntax-semantics. The prominent prosodic features of interrogative markers revealed by previous studies were primarily concerned with wh-phrases, Verb-not-Verb and sentence-final particles, among which wh-phrases serve not only as the interrogative marker (syntax), but also contain much semantic information e.g. shen me—what, na li—where, shui—who etc.; the Verb-not-Verb phrases are also semantically rich as verbs are generally considered as highly informative across utterances. Yet, the tag "dui bu dui?" in Mandarin TQs are semantically impoverished, which only fulfills the syntactic function of signaling interrogative mood, thus shall not be focalized, but serve as a focus marker highlighting the preceding statement for the complete conveyance of utterance meaning. It also could be why the tag was produced rapidly with low pitch and intensity, since a mere syntactic function might not require three well-articulated syllables. Therefore, it could be deduced that only when an interrogative marker performs both syntactic and semantic functions, can it receive sentence focus, and thus become prosodically prominent in terms of the relevant acoustic properties. However, one may argue that the sentence-final particles "ma" and "ne", being interrogative markers without semantic functions, were observed with some acoustic prominence (at least not compressed if not focalized) as reported by Hsu & Xu (2020). This might be due to the fact that one group of participants in this study were Taiwan Mandarin speakers who have a propensity of stressing sentence-final particles (Kuang & Kuo 2011). Another more plausible account is that the sentence-final particles in Hsu & Xu (2020), namely, ma, ne, and ba, not only decided the syntactic type of the sentences, but also dictated the semantic information of the preceding wh-indeterminates, i.e. shen me, na li and shui function as "what, where and who" when preceding ma, "something, somewhere and someone" when preceding ne and ba. Hence, the particles ma and ne in Hsu and Xu (2020), at some level, could be regarded as interrogative markers carrying both syntactic and semantic functions, resulting in relative acoustic prominence.

To this point, the syntax-semantics accounts of prosody in Mandarin TQs accommodate much observation from the current and previous studies. The viewpoint that not all interrogative markers in Mandarin could receive sentence foci needs to be attached with great importance, since they could be prosodically compressed, and serve only as focus markers highlighting other parts of the utterances when performing only syntactic functions. This argument needs to be further corroborated. Future studies need to be carried out probing the prosodic features of statement (unambiguous) + ma questions in Mandarin; if the proposal posited in the current study holds true, then it could be expected that the sentence-final ma, performing only syntactic function in this context, would be in keeping with dui bu dui in light of non-prominence in prosody. Besides, Mandarin TQs using other types of tags other than dui bu dui are also worth investigating to enrich our understanding of the interaction between prosody and syntax-semantics.

# References

Chen, Y.: Durational adjustment under corrective focus in standard Chinese. J. Phon. **34**, 176–201 (2006)

Chen, Y., Gussenhoven, C.: Emphasis and tonal implementation in standard Chinese. J. Phon. **36**(4), 724–746 (2008)

Chen, Y., Xu, Y., Guion-Anderson, S.: Prosodic realization of focus in bilingual production of southern min and mandarin. Phonetica **71**, 249–270 (2014)

Cooper, W.E., Eady, S.J., Mueller, P.R.: Acoustical aspects of contrastive stress in question-answer contexts. J. Acoust. Soc. Am. **77**, 2142–2156 (1985)

Cutler, A., Isard, S.D.: The production of prosody. In: Language Production, vol. 1, pp. 245–269. Academic Press, London (1980)

Hsu, Y.Y., Xu, A.: Interaction of prosody and syntax-semantics in Mandarin wh -indeterminates. J. Acoust. Soc. Am. **148**(2), EL119–EL124 (2020)

Kochanski, G.P., Shih, C.: Prosody modeling with soft template. Speech Commun. **39**, 311–352 (2003)

Kuang, J., Kuo, G.: Comparison of prosodic properties of intonation in Beijing Mandarin and Taiwan Mandarin. J. Acoust. Soc. Am. **129**(4), 2680 (2011)

Liang, L., Shi, F.: Acoustic studies on lexical stress of bi-syllabic words in standard Chinese. In: Phonetic Conference of China PCC 2008, Beijing (2008)

Lin, Y., Ding, H., Zhang, Y.: Prosody dominates over semantics in emotion word processing: evidence from cross-channel and cross-modal stroop effects. J. Speech Lang. Hear. Res. **63**(3), 896–912 (2020)

Liu, F., Surendran, D., Xu, Y.: Classification of statement and question intonations in Mandarin. In: Speech Prosody Dresden, 2–5 May 2006

Liu, F., Xu, Y.: Parallel encoding of focus and interrogative meaning in Mandarin intonation. Phonetica **62**, 70–87 (2005)

Mini Speech Lab 2.0 [computer program] Homepage. https://www.freeber.cn/Index/show/catid/49/id/207.html. Accessed 6 2020

Praat: Doing phonetics by computer (version 6.0.49) Homepage [computer program]. http://www.praat.org/. Accessed 2 Mar 2019

Shao, J.: Study of Interrogative Sentences in Modern Chinese. East China Normal University Press, Shanghai (1985)

Shen, J.: Pitch range of tone and intonation in Beijing dialect. In: Lin, T., Wang, L. (eds.) Working Papers in Experimental Phonetics, pp. 73–130. Peking University Press, Beijing (1985)

Shen, X.: The Prosody of Mandarin Chinese. University of California Press, Berkeley (1989)

Shi, F., Wang, P.: Boundary tone and focus tone. J. Chin. Linguist. **42**(1), 93–108 (2014)

Wang, B., Xu, Y., Ding, Q.: Interactive prosodic marking of focus, boundary and newness in Mandarin. Phonetica **75**(1), 24–56 (2018)

Xu, Y.: Effects of tone and focus on the formation and alignment of F0 contours. J. Phon. **27**, 55–105 (1999)

Xu, Y., Chen, S., Wang, B.: Prosodic focus with and without post-focus compression: A typological divide within the same language family? Linguist. Rev. **29**(1), 131–147 (2012)

Yan, Y.: Study of Tag Questions in Modern Chinese. Shanghai People's Publishing House, Shanghai (2017)

Yan, J., Wang, P., Shi, F.: Acoustic analysis on mark multiplexing in interrogative sentences: for the positive and negative interrogative sentences. Lang. Teach. Linguist. Stud. **5**, 97–102 (2014)

Yan, J., Wang, P., Shi, F.: An acoustic analysis of specific interrogation of Mandarin. J. Sino-Tibetan Linguist. **9**, 121–129 (2016)

# Improving Fine-Grained Emotion Control and Transfer with Gated Emotion Representations in Speech Synthesis

Jianhao Ye[✉], Tianwei He, Hongbin Zhou, Kaimeng Ren, Wendi He, and Heng Lu

Ximalaya Inc., Beijing, China
{jianhao.ye,tianwei.he,hongbin.zhou,irving.ren,
cloris.he,bear.lu}@ximalaya.com

**Abstract.** Fine-grained emotion strength control and prediction is recently studied in text-to-speech to adjust local emotion intensity in an utterance. Due to the lack of fine-grained emotion strength labelling data, emotion or style strength extractor is usually learned at the whole utterance scale through a ranking function. However, such utterance-based extractor is then used to provide fine-grained emotion strength labels, conditioning on which a fine-grained emotional speech synthesis model is separately trained. To bridge the granularity gap between emotion strength extraction and emotional synthesis speech generation, a simple yet effective component called Emotion Gate is designed to learn fine-grained emotion strengths in an end-to-end way, which are then used to create scaled emotion representations that serve as a condition of emotional speech synthesis. Furthermore, beside predicting from a jointly trained emotion strength predictor, our proposed method also allows to manually assign and control the fine-grained emotion strengths during inference. In experiment part, the proposed method is evaluated in both non-transferred emotional speech synthesis and cross-speaker transferred scenarios. Both objective and subjective evaluations show the effectiveness and superiority of the proposed method over the state-of-the-art baseline systems. The audio samples in our experiments can found in the demo page: https://kingstorm.github.io/emotiongate/.

**Keywords:** text-to-speech · emotion strength · end-to-end · style transfer · emotional speech synthesis

## 1 Introduction

Recent years have witnessed a growing trend of using text-to-speech (TTS) systems to produce audiobooks, in which the synthesized speech needs to be more expressive and emotional compared to regular synthesized speech [1,2]. The emotional utterances can either be synthesized by end-to-end TTS models [3–6] trained with an emotional database or models with cross-speaker emotion transfer [2,7,8] for speakers whose databases do not contain emotional data. Beside

L. Zhenhua et al. (Eds.): NCMMSC 2022, CCIS 1765, pp. 196–207, 2023.
https://doi.org/10.1007/978-981-99-2401-1_18

emotional speech synthesis, the capability of controlling emotion expressiveness in the synthesized speech is another important aspect of TTS audiobooks creation. Nevertheless, most of the above methods do not provide a direct way to control emotion expressiveness at a fine-grained level. Although the speech synthesis with reference-based methods [2,7] can be guided with given reference audios, it is usually hard to select a suitable reference for expected emotion expressiveness.

One of the attempts to address the emotion controlling issue mentioned above is introducing emotion strength as an attribute to control emotion expressiveness. To model emotion strengths, [9] proposes to use relative attributes [10] and train a ranking function to label the emotion strength of each utterance. Then the emotion strengths are used as conditions in the TTS model training and can be controlled during inference. More recently, models with fine-grained emotion strengths control are built in [11,12] using a similar method by directly employing a ranking function trained at utterance level to predict fine-grained emotion strengths. The experimental results in [11] show the model built with phoneme-level emotion strengths has stronger ability of emotion expressiveness control compared to Global Style Tokens (GST) [2]. Moreover, [12] employs the same methods and find modelling at syllable-level is advantageous to that at phoneme-level for Mandarin speech synthesis.

Although models built with ranking-function-based emotion strengths provide a fine-grained control of emotion expressiveness, there are still some remaining issues to be addressed. First, The ranking function is trained on utterance level, yet the input features at inference phase for emotion strengths extraction are at fine-grained level. Second, the ranking function for emotion strength extraction and the TTS model for emotional speech reconstruction are independently learned, which makes the whole system not end-to-end optimized. Last but not least, previous methods based on the ranking function have rarely been evaluated in cross-speaker emotion transfer scenarios.

In this paper, inspired by [3,13] that achieves style control by scaling up and down the style embedding, a simple yet effective component called Emotion Gate (EG) is proposed to address the above issues by constructing gated emotion representations. The proposed EG consists of an emotion strength extractor which extracts fine-grained emotion strengths from speech segments and a gating mechanism which change the Norm of emotion embedding by scaling them using the extracted emotion strengths. The two parts work together to produce gated emotion representations which condition a TTS model to train and generate emotional speech. Since the resulting gated emotion representations directly connect the EG with the TTS model, the EG can be jointly trained with the whole model in an end-to-end paradigm. Therefore, without two-stage processing, the proposed method naturally bridges the gap between utterance-level training and fine-grained inference as in ranking function based method.

Similar to [11], the EG also allows the fine-grained emotion strengths to be manually assigned or predicted by a jointly trained emotion strength predictor during inference. We conduct experiments by building Mandarin TTS emotional models with syllable-level emotion strength control as in [12] to evaluate the proposed method. Both the objective and subjective experiments show the proposed method is more effective than the ranking-function-based methods [11,12]. We also find introducing the proposed method into cross-speaker emotion transfer scenarios enhances emotion intensity in synthesized speech. Moreover, by assigning manual emotion strength values to synthesize speech, we find the proposed method has stronger ability of emotion strength control over both non-transferred emotional TTS and cross-speaker transferred ones. Overall, the main contributions of this work are summarized below:

- The proposed EG constructs gated emotion representations to bridge the gap in the ranking-function-based method [11,12] between utterance-level training and fine-grained inference, achieving better performance of emotion expressiveness.
- To our best knowledge, this is the first work to model fine-grained emotion strengths in speech synthesis in an end-to-end paradigm.
- The emotion strength control of the proposed method is verified to work for both non-transferred emotional TTS and cross-speaker emotion transferred TTS.

## 2   Methodology

### 2.1   Fine-Grained Emotion Strengths from Ranking Function

Most of recent works [11,12] employ the relative attributes method [9,10] to train an utterance-level ranking function $f_{rank}$ defined in Eq. 1 to annotate fine-grained emotion strengths for speech segments. By treating emotion strength as an attribute of speech, the ranking function $f_{rank}$ aims to learn the relative difference of emotion strengths between a category of emotional speech (such as joy) and neutral speech. In details, assuming the training set for learning the ranking function is $T$ represented in $R^n$ by utterance-level emotional features $\{x_u\}$ where $u$ is the index of utterances and $T = N \cup E$, where $N$ and $E$ are the neutral and joy emotion set respectively. The goal of relative attributes is to learn the ranking function Eq. 1:

$$f_{rank}(\boldsymbol{x_u}) = \boldsymbol{w}\boldsymbol{x_u} \tag{1}$$

which satisfies the maximum number of the following constraints:

$$\begin{aligned} \forall(i,j) \in O : f_{rank}(\boldsymbol{x_i}) > f_{rank}(\boldsymbol{x_j}) \\ \forall(i,j) \in S : f_{rank}(\boldsymbol{x_i}) = f_{rank}(\boldsymbol{x_j}) \end{aligned} \tag{2}$$

$O$ is the ordered set that is composed of sample pairs $(i,j)$ with different categories, e.g., $i \in E$ and $j \in N$. $S$ is the similar set which contains sample pairs from the same category. Any sample pair from $O$ has different emotion strengths and any sample pair from $S$ has the similar emotion strengths. This setup is based on the presumption that the emotion strengths of joy are greater than neutral ones. $\boldsymbol{w}$ in $f_{rank}$ is learned through Newton's method [10,11].

As described in [12], the $f_{rank}$ is first trained at utterance level and gets fine-grained emotional features to output ranking values by equation: $s_t = f_{rank}(\boldsymbol{x_t})$. Then those ranking values are normalized into $[0,1]$ for each emotion category and regard as fine-grained emotion strengths.

Once the utterance-level ranking function is trained, fine-grained emotion strengths can be obtained by feeding fine-grained speech segments, e.g., speech segments of phonemes or syllables, into the ranking function and all the ranking results are normalized into $[0,1]$ for each emotion category. The extracted emotion strengths are linearly projected and added into text encoder outputs in [11,12] to guide the emotion strengths in the synthesized speech.

## 2.2   The Proposed Method

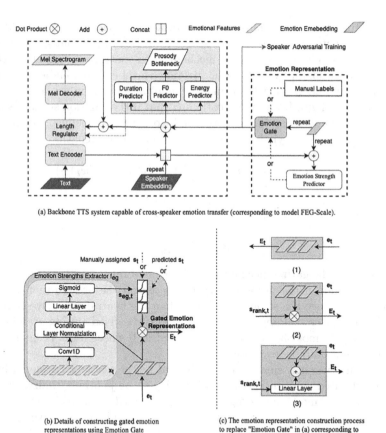

(a) Backbone TTS system capable of cross-speaker emotion transfer (corresponding to model FEG-Scale).

(b) Details of constructing gated emotion representations using Emotion Gate

(c) The emotion representation construction process to replace "Emotion Gate" in (a) corresponding to models Base (1), FRA-Scale (2), FRA-Transform (3)

**Fig. 1.** The overall architecture of emotional TTS system and its variants to evaluate the proposed method

Unlike the above two-stage processing which firstly train a ranking function and secondly use extracted emotion strengths to condition a TTS model, the proposed EG, demonstrated in Fig. 1 (b), merges the two stages into an unified component. The EG consists of an emotion strength extractor $f_{eg}$ described in Eq. 3 and a gating mechanism as shown in Fig. 1 (b). Those two parts work together to construct gated emotion representations which condition the TTS model to train and synthesize emotional speech.

The emotion strength extractor $f_{eg}$ takes the same emotional features $x_t$ of fine-grained speech segments as inputs similar to [11]. Within the extractor, a conditional layer normalization (CLN) [14,15] conditioned by emotion embedding is located between Conv1D and Linear layer, since we presume different emotion category has different scale and bias for hidden features to predict emotion strengths. Finally, a Sigmoid function is employed to construct a gating mechanism and output a gating value or emotion strength $s_{eg,t}$ for each segment.

The gating mechanism, illustrated in Eq. 5, scales the emotion embedding $e_t$ using the extracted emotion strengths $s_{eg,t}$ from Eq. 4 to output gated emotion representations. The similar gating mechanism to control emotion strength can also be found in [3,13] by scaling the utterance-level emotion embedding, while we apply it at fine-grained level and the emotion strengths are learned in an end-to-end paradigm. It's worth mentioning that if the emotion strengths are at syllable level where each syllable corresponds to several phonemes, the phonemes belonging to the same syllable share the same emotion strength.

$$f_{eg}(x_t) = Sigmoid(Linear(CLN(Conv1D(x_t)))) \tag{3}$$

$$s_{eg,t} = f_{eg}(x_t) \tag{4}$$

$$E_t = s_{eg,t} e_t \tag{5}$$

Apart from the gating mechanism and the emotion strength extractor, there is another design in the proposed method based on the presumption that any emotional utterance has stronger strength than the neutral ones. By setting the neutral emotion embedding to a $0$ vector which is fixed during training, the norm of neutral emotion representation is kept as 0 whose norm is smaller than any other emotion representations.

As for the training, since EG is connected to the whole TTS model by the constructed gated emotion representations, it can be jointly optimized with all losses of the TTS model including mel-spectrogram L1 reconstruction loss $\mathcal{L}_{mel}$, f0 loss $\mathcal{L}_{f0}$, energy loss $\mathcal{L}_{energy}$, duration loss $\mathcal{L}_{dur}$ and speaker adversarial loss $\mathcal{L}_{adv}$ [16] used in cross-speaker emotion transfer training. The $\mathcal{L}_{f0}$, $\mathcal{L}_{energy}$ and $\mathcal{L}_{dur}$ are bottleneck features mean-squared-error (MSE) loss in [8]. As shown in Fig. 1 (a), the fine-grained emotion strengths during inference can be manually assigned with a sequence of values within [0, 1] or predicted by an emotion strength predictor that shares the same structure as that of the variance predictor in [6] and is trained together with the whole TTS model to fit the extracted emotion strengths $s_{eg,t}$ by MSE loss $\mathcal{L}_s$. To sum up, the overall loss of the TTS model with the proposed EG is illustrated in Eq. 6 where $\alpha$, $\beta$, $\gamma$ are tunable weights for prosody bottleneck features loss, emotion strength predictor loss and speaker adversarial loss respectively.

$$\mathcal{L} = \mathcal{L}_{mel} + \alpha(\mathcal{L}_{f0} + \&\mathcal{L}_{energy} + \mathcal{L}_{dur}) + \beta\mathcal{L}_{adv} + \gamma\mathcal{L}_s \qquad (6)$$

## 3  Experiments

### 3.1  The Model Architecture of Baseline Emotional TTS

To evaluate the effectiveness of the proposed method in both intra-speaker and cross-speaker scenarios. A baseline cross-speaker emotional text-to-speech system is built according to [8]. Compared with the other SOTA emotional text-to-speech systems, [8] is capable of transferring emotion expression across speakers and achieves better performance than the other strong benchmarks [7]. Here, some of the components of the original model [8] are replaced by ones from other works as shown in Fig. 1 (a). First, the text encoder in DurIAN [3] is utilized to efficiently encode text information. Second, the non-autoregressive Transformer decoder in Fastspeech [4] is used to convert encoder outputs into 80-dim mel-spectrogram. For bottleneck features [8]: jointly trained variance predictors in [6] are adopted to predict duration, f0 and energy. In addition to the bottleneck design in [8], a speaker adversarial training task [16] is applied onto emotion representation to further disentangle the emotion and speaker attributes. To summarize, the whole model predicts mel-spectrogram by getting input from text information in phoneme sequence, speaker embedding and emotion representation conditioned by emotion strengths.

### 3.2  Tasks for Experiments and Models Setup

Two experimental tasks are conducted to evaluate the proposed method. *Task-I*: Building emotional TTS models based on an emotional database of single speaker A. The goal of Task-I is to compare the emotion expressiveness and emotion controllability of the proposed method with ranking function based methods and the baseline model. *Task-II*: Building multi-speaker emotional TTS models capable of transferring emotion from speaker A to speaker B. The goal of Task-II is to evaluate the performance of cross-speaker emotion transfer of models built with and without the proposed method as well as the emotion strength controlling effectiveness. The models built for the experiments are described as below:

- **Base**: Replace EG component in Fig. 1 (a) with directly adding the emotion embedding onto the encoder outputs as in Fig. 1 (c)(1) without any emotion strength conditions.

- **FRA-Scale**: Replace EG component in Fig. 1 (a) with ranking-function-based emotion strengths to scale emotion embedding as in Fig. 1 (c)(2), where FRA denotes fine-grained ranking-function-based emotion strengths.
- **FRA-Transform**: Replace EG component in Fig. 1 (a) with the representation in Fig. 1 (c)(3) which linearly transforms the ranking-function-based emotion strengths and adds them onto encoder outputs exactly as in [11,12].
- **FEG-Scale**: As demonstrated in Fig. 1 (a), TTS model with the proposed EG as shown in Fig. 1 (b).

### 3.3  Basic Setups

For both tasks, an internal female emotional database of speaker A is used. There are 3200 neutral utterances in the database and each of the rest 4 emotions (vigilance, disgust, joy, sad) has 600 utterances. Speaker B in Task-II is from another internal female neutral database which contains 6000 neutral utterances. All audios in the databases, with the length of each utterance ranging from 2 to 10 s, are in Mandarin and recorded in 44.1 kHZ while downsampled to 24 kHZ. For features pre-processing, 80-dimensional mel-spectrograms are used as acoustic features which are extracted using Hanning window with frame shift of 10 ms and frame length of 42.7 ms. Then, Kaldi toolkit [17] performs forced-alignment and retrieves duration of each phoneme. The extraction of f0 and energy follows the same setup in [6]. The emotion strength in our experiments is at syllable-level which is proved to be more effective for controlling emotion in Mandarin TTS [12]. As in [11,12], the emotional features are extracted using openSMILE tool [18] from syllable-level speech segments where the duration of each syllable is obtained by aggregating all the phoneme durations of the syllable. During the training, batch size is set to 24 and Adam Optimizer is used with learning rate equal to 0.0001. After cross-validations, both the $\alpha$ and $\beta$ in $\mathcal{L}$ are set to 0.05 and the adversarial weight $\gamma$ is set to 0.025. Finally, a multi-speaker HiFi-GAN [19] is trained to synthesize audio samples.

### 3.4  Task-I: Evaluating the Proposed Method for Non-transferred Emotional Speech Synthesis

**Emotion Strengths Extracted from Reference Speech.** In this part, similar to [11], we evaluate the synthesized speech with parallel transfer where emotion strengths are extracted from reference audio with the same text as the target one. The intuition behind it is that stronger emotion strength condition leads to better imitation of reference audios [11]. In this task, Mel-cepstral distortion (MCD) and A/B tests are used to evaluate the synthesized speech objectively and subjectively.

First, 30 test sentences for each emotion category in the database of speaker A are held out and synthesized with parallel transfer to compute MCD values. The synthesized speech is aligned with reference audio by dynamic time warping (DTW). The results in Table 1 show that all the models with syllable-level emotion strength modelling achieve lower MCD values compared to *Base* in which the emotion is represented only by global emotion embedding, with the proposed *FEG-Scale* achieving the best result. Also, it can be observed that the MCD values of *FRA-Scale* are higher than those of *FRA-Transform*, which indicates it is not the scaling but the end-to-end way to learn emotion strengths together with gated emotion representations that bring gains to performance.

**Table 1.** Results of MCD test (the lower the better)

|               | Parallel transfer | Predicted emotion strengths |
|---------------|-------------------|-----------------------------|
| Base          | 5.292             | 5.292                       |
| FRA-Scale     | 5.217             | 5.228                       |
| FRA-Transform | 5.121             | 5.165                       |
| **FEG-Scale** | **5.031**         | **5.084**                   |

**Table 2.** Results of preferences in A/B tests for parallel transfer

|                                    | Left (%) | Same (%) | Right (%) |
|------------------------------------|----------|----------|-----------|
| **FEG-Scale** vs. FRA-Transform    | **53**   | 23       | 24        |
| **FEG-Scale** vs. Base             | **53.3** | 20       | 26.7      |

Second, we conduct A/B tests in which participants are required to choose the audio that they believe is more similar to the reference audio in emotion expressiveness. 15 native Mandarin speakers participated in preference selections and 10 test sentences for each emotion category are randomly selected from the held out sentences to synthesize. *FRA-Scale* is dropped out since it does not perform better than the implementation in [11] which is *FRA-Transform*. The subjective evaluation results in Table 2 are consistent with the ones in objective MCD tests in Table 1, which show the superiority of the proposed method (*FEG-Scale*) over both *Base* and *FRA-Transform* for parallel transfer.

**Emotion Strengths from Emotion Strength Predictor.** In real applications, we also care about the performance of the proposed method when the emotion strengths are predicted. Therefore, the MCD and A/B tests of the same setup are conducted to evaluate generated speech with predicted emotion strengths as well. As shown in Table 1, for all models except *Base* without emotion strength modelling, the MCD values of generated speech with predicted emotion strengths only witness minor degradation compared to the ones with parallel transfer. Furthermore, even with predicted emotion strengths, all models built with emotion strength get lower MCD values than *Base*, with *FEG-Scale* still achieving the best result. Subjectively, an A/B test of the same setup is also conducted to compare the model with the proposed EG (*FEG-Scale*) and the model without it (*Base*). From the results in Table 3, *FEG-Scale* is more preferred than *Base* though the gap is small, which is reasonable since even minor precision loss in the predicted emotion strengths can lead to audible changes in emotion expressiveness for pair tests.

**Table 3.** Results of A/B test with predicted emotion strengths

|  | Left (%) | Same (%) | Right (%) |
|---|---|---|---|
| FEG-Scale (pred) vs. Base | **31.25** | 40 | 28.75 |

To sum up, Both objective and subjective results show the superiority of *FEG-Scale* which performs the best with parallel transfer and retains its advantage over other models with predicted emotion strengths.

### 3.5   Task-II: Evaluating the Proposed Method for Cross-Speaker Emotion Transfer

The experimental results in Task-I show the effectiveness of the proposed method in single-speaker emotional TTS. However, most of the speakers are non-emotional in databases and need cross-speaker emotion transfer to synthesize emotional speech. To do experiments with the proposed EG in such cases, in this task, models are built with and without EG to synthesize speech of speaker B whose emotion is transferred from speaker A.

To evaluate the effects of introducing EG into cross-speaker emotion transfer, we synthesize 10 held-out utterances for each emotion category with predicted emotion strengths and conduct two subjective tests with the same 15 participants: an A/B test over emotion intensity of generated speech and a mean opinion score (MOS) test to evaluate the overall naturalness. Next, to evaluate the effects of the proposed method over timbre similarity of generated speech to the target speaker, similar to [15], a speaker verification model [20] is trained with 12000-speaker data to extract 256-dim speaker embedding from synthesized speech and the actual recordings of the speaker B. 30 held-out utterances for each emotion category are generated to extract speaker embedding and compute cosine similarity with ones extracted from 10 recordings respectively. The cosine similarities for each emotion category are averaged and larger cosine similarity indicates more similar timbre to the target speaker B [15].

By looking at the results in Table 4, compared to *Base*, *FEG-Scale* has dominantly stronger emotion intensity for cross-speaker emotion transferred speech beside the sad ones. Although the sad speech from *Base* is advantageous to *FEG-Scale* in emotion intensity, it has the most dissimilar timbre as a cost according to the results of timbre similarity test in Table 5. Furthermore, for the rest of emotion apart from sad emotion, *FEG-Scale* and *Base* share the same level of timbre similarity to the target speaker. For naturalness evaluation, the differences of MOS scores between *FEG-Scale* and *Base* in Table 5 are all less than 0.1

**Table 4.** Results of A/B tests of emotion intensity on speaker B with emotion transfer

| Emotion | (FEG-Scale) Left (%) | Same (%) | (Base) Right (%) |
|---|---|---|---|
| vigilance | **73.33** | 13.33 | 13.33 |
| disgust | **50** | 36.67 | 13.33 |
| joy | **50** | 30 | 20 |
| sad | 13.33 | 33.33 | **53.33** |

**Table 5.** MOS of naturalness with 95% confidence interval and cosine similarity (timbre similarity)

| | MOS Scores | | Timbre Similarity | |
|---|---|---|---|---|
| | Base | FEG-Scale | Base | FEG-Scale |
| vigilance | 3.98 ± 0.1 | **4.02 ± 0.1** | **0.7817** | 0.7806 |
| disgust | **4.06 ± 0.09** | **4.06 ± 0.08** | 0.7908 | **0.8044** |
| joy | **3.9 ± 0.1** | 3.89 ± 0.09 | 0.8022 | **0.8095** |
| sad | **3.53 ± 0.16** | 3.47 ± 0.16 | 0.7289 | **0.7474** |
| neutral | 3.94 ± 0.09 | **3.99 ± 0.09** | **0.8450** | 0.8426 |
| AVG | 3.88 | **3.89** | 0.7897 | **0.7969** |

and the averaged scores for both models are very close. To summarize, we can conclude that introducing the proposed EG into cross-speaker emotion transfer does not bring degradation to naturalness or timbre of generated speech while enhancing the emotion intensity of generated speech and providing a strong emotion strength controllability which will be discussed in the following part.

### 3.6 Analysis of Manually Assigning Emotion Strengths for Both Task-I and Task-II

To evaluate the emotion strength controllability of the proposed method, we create manual emotion strength labels and expect the synthesized speech to follow the designed emotion expressiveness. Joy utterances with the same texts are synthesized using models in Task-I given syllable-level emotion strengths increasing from 0 to 1 and decreasing from 1 to 0, as shown in Fig. 2 (a) and (b). For synthesized speech from *FEG-Scale* in (a), both the F0 trend and pattern changes of mel-spectrogram clearly reflect the assigned emotion strengths, whereas the samples from *FRA-Transform* in (b) are relatively flat, which indicates superior emotion controllability of the proposed method. The analysis is also conducted in cross-speaker emotion transfer scenario. By comparing Fig. 2 (a) and (c), we find the proposed method owns the same level of emotion controllability in both non-transferred speech and speech with cross-speaker emotion transfer. We recommend to listen to samples of emotion control for all emotion categories in our demo page.

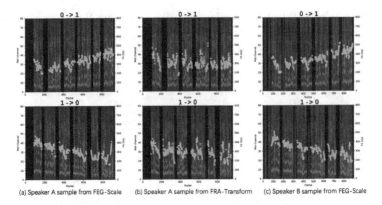

**Fig. 2.** Plots of speech with manually assigned emotion strengths increasing from 0 to 1 and decreasing from 1 to 0.

## 4    Conclusions

The proposed EG provides an end-to-end way to model fine-grained emotion strengths and construct gated emotion representations to guide emotional speech synthesis. During inference, the emotion strengths can be either predicted by an emotion strength predictor or manually assigned. First, the end-to-end paradigm facilitates modelling the controllable emotion in speech synthesis, which does not need training a separate emotion strength annotator. Second, according to our experimental results, the proposed method is superior to the ranking-function-based method in terms of emotion expressiveness. Third, beside the stronger emotion intensity the proposed method brings to generated speech with cross-speaker emotion transfer, it shares the same level of fine-grained emotion controllability for both non-transferred and emotion transferred cases.

## References

1. Yan, Y., et al.: Adaptive text to speech for spontaneous style. In: Proceedings of Interspeech 2021, pp. 4668–4672 (2021). https://doi.org/10.21437/Interspeech.2021-584
2. Wang, Y., et al.: Style tokens: unsupervised style modeling, control and transfer in end-to-end speech synthesis, March 2018
3. Yu, C., et al.: DurIAN: duration informed attention network for speech synthesis. In: Interspeech, pp. 2027–2031 (2020)
4. Ren, Y., et al.: FastSpeech: fast, robust and controllable text to speech. In: Proceedings of the 33rd International Conference on Neural Information Processing Systems, pp. 3171–3180 (2019)
5. Wang, Y., et al.: Tacotron: towards end-to-end speech synthesis, pp. 4006–4010, August 2017. https://doi.org/10.21437/Interspeech.2017-1452
6. Ren, Y., et al.: FastSpeech 2: fast and high-quality end-to-end text to speech. arXiv preprint arXiv:2006.04558 (2020)

7. Whitehill, M., Ma, S., McDuff, D., Song, Y.: Multi-reference neural TTS stylization with adversarial cycle consistency. In: Interspeech 2020, pp. 4442–4446. ISCA (2020). https://doi.org/10.21437/Interspeech.2020-2985

8. Pan, S., He, L.: Cross-speaker style transfer with prosody bottleneck in neural speech synthesis. In: Proceedings of Interspeech 2021, pp. 4678–4682 (2021). https://doi.org/10.21437/Interspeech.2021-979

9. Zhu, X., Yang, S., Yang, G., Xie, L.: Controlling emotion strength with relative attribute for end-to-end speech synthesis, pp. 192–199, December 2019. https://doi.org/10.1109/ASRU46091.2019.9003829

10. Parikh, D., Grauman, K.: Relative attributes. In: Proceedings of ICCV, pp. 503–510. IEEE (2011)

11. Lei, Y., Yang, S., Xie, L.: Fine-grained emotion strength transfer, control and prediction for emotional speech synthesis. CoRR abs/2011.08477 (2020). https://arxiv.org/abs/2011.08477

12. Lei, Y., Yang, S., Wang, X., Xie, L.: MsEmoTTS: multi-scale emotion transfer, prediction, and control for emotional speech synthesis. IEEE/ACM Trans. Audio Speech Lang. Process. **30**, 853–864 (2022). https://doi.org/10.1109/TASLP.2022.3145293

13. Li, T., Yang, S., Xue, L., Xie, L.: Controllable emotion transfer for end-to-end speech synthesis. In: 2021 12th International Symposium on Chinese Spoken Language Processing (ISCSLP), pp. 1–5 (2021). https://doi.org/10.1109/ISCSLP49672.2021.9362069

14. Chen, M., et al.: AdaSpeech: adaptive text to speech for custom voice. In: International Conference on Learning Representations (2021). https://openreview.net/forum?id=Drynvt7gg4L

15. Min, D., Lee, D.B., Yang, E., Hwang, S.J.: Meta-stylespeech: multi-speaker adaptive text-to-speech generation. In: ICML (2021)

16. Shang, Z., Huang, Z., Zhang, H., Zhang, P., Yan, Y.: Incorporating cross-speaker style transfer for multi-language text-to-speech. In: Proceedings of Interspeech 2021, pp. 1619–1623 (2021). https://doi.org/10.21437/Interspeech.2021-1265

17. Povey, D., et al.: The kaldi speech recognition toolkit, p. 4 (2011)

18. Eyben, F., Wöllmer, M., Schuller, B.: openSMILE the Munich versatile and fast open-source audio feature extractor, pp. 1459–1462, January 2010. https://doi.org/10.1145/1873951.1874246

19. Kong, J., Kim, J., Bae, J.: HiFi-GAN: generative adversarial networks for efficient and high fidelity speech synthesis. In: Larochelle, H., Ranzato, M., Hadsell, R., Balcan, M.F., Lin, H. (eds.) Advances in Neural Information Processing Systems, vol. 33, pp. 17022–17033. Curran Associates, Inc. (2020)

20. Desplanques, B., Thienpondt, J., Demuynck, K.: ECAPA-TDNN: emphasized channel attention, propagation and aggregation in TDNN based speaker verification. In: Proceedings of Interspeech 2020, pp. 3830–3834 (2020). https://doi.org/10.21437/Interspeech.2020-2650

# Violence Detection Through Fusing Visual Information to Auditory Scene

Hongwei Li, Lin Ma, Xinyu Min, and Haifeng Li[✉]

Faculty of Computing, Harbin Institute of Technology, Harbin, China
lihongwei@stu.hit.edu.cn

**Abstract.** In the field of audio and video detection, violence detection is a crucial task with significant theoretical and practical implications. In order to solve the present issue of the lack of violent audio datasets, we first created our own audio violent dataset named VioAudio. Then, we proposed a CNN-ConvLSTM network model for audio violence detection, which obtained an accuracy of 91.5% on VioAudio and a MAP value of 16.47% on the MediaEval 2015 dataset. Meanwhile, this paper integrated self-attention mechanisms and visual information into CNN-ConvLSTM network in order to address the issue of modality singularity in violence detection, and then confirmed them on MediaEval2015 dataset. The experimental results demonstrate that after fusing visual and auditory information, the CNN-LSTM network model greatly enhanced recognition accuracy, attaining a 31.25% MAP value, which is 1.94% higher than the best result. The method proposed in this paper considerably increased the accuracy of violence detection and offered fresh perspectives on how to integrate multimodal information to identify violence.

**Keywords:** Violence Detection · Auditory and Visual Information Fusion · Convolution Neural Network · Long-Short Term Memory Network

## 1   Introduction

The variety of audio and video available on the Internet is growing as a result of the growth of multimedia information technology. However, some videos or audio might have violent senses, which are unsuitable for widespread distribution. The exploding amount of multimedia cannot be reviewed by the traditional manual method [1]. Therefore, it becomes more crucial to automatically recognize violent senses. The use of multimedia-oriented violence detection with artificial intelligence has a wide range of potential applications [2].

Currently, most researchers use visual or auditory mode alone to detect violence. In the detection of auditory violence, Cheng et al. used Hidden Markov Model (HMM) and Gaussian Mixture Model (GMM) to model all kinds of sound

Supported by the National Natural Science Foundation of China [U20A20383]; the National Key Research and Development Program of China [2020YFC0833204]; the Key Research and Development Program of Heilongjiang [GY2021ZB0206].

separately, so as to effectively detect abnormal events such as explosion, gunshot, scream [3]. Gil-Pita et al. compared the effects of different audio features on the task of violence detection, and demonstrated that Mel Frequency Cepstral Coefficient (MFCC) was the most useful feature for the task, while the energy and pitch also performed well [4]. Penet et al. used factor analysis to model the variability of audio events in movies. The system built by Penet et al. has better generalization ability for the detection of violent audio such as screaming [5]. Sarman et al. extracted the acoustic features such as MFCC and zero-crossing rate from the audio and classified the violent senses by random forest method [6].

In the detection of video violence, Moreira [7] et al. proposed a TRoF feature descriptor to represent the local action features in the video, coded it with Fisher Vector, and sent the features to the Support Vector Machine (SVM) classifier to effectively detect dynamic violence in the video. Xu [8] et al. used MoSIFT algorithm to extract low dimensional features of video, and used kernel density estimation method to select features. In order to obtain distinguishing features, the author also uses sparse coding to further process features, achieving 94.3% accuracy on Hockey dataset. In recent years, the rise of deep learning represented by Convolution neural network has accelerated the research of visual violence detection. Serrano et al. proposed a two-dimensional CNN model using Hough forest features, and achieved very good results on the hockey and movies datasets, with accuracy rates of 94.6% and 99% respectively. Sudhakaran [9] et al. used Convolution neural networks (CNN) to extract features at the frame level from videos and integrate them through different Long Short Term Memory (LSTM) units to capture violent acts such as fighting in video.

From the existing research, we can see that the current research on violence detection has achieved very good results. But there are also some problems that cannot be ignored. First of all, video violence detection has many public datasets, but audio violence detection has no public datasets, which limits the further development of violence audio detection. Secondly, most of the researches focus on the single mode of audio or video, lacking the research on multimodal violence detection. The existing research shows that it is an effective method to combine multiple features of multiple modes in the field of multimedia event detection. Some scholars began to attempt detect multimodal violence. Jian et al. proposed a method of combining audio violence classifier and video classifier based on weak supervised learning. The accuracy of violence detection is higher than that of single mode [10]. Demarty et al. extracted linear prediction coefficients, line spectrum pair parameters, 196-dimensional acoustic features such as MFCC, 11-dimensional color histogram, and 81-dimensional gradient-oriented histogram, and fused them into a neural network for violent classification [11]. Zajdel et al. proposed a CASSANDRA system that captured complementary information in video and audio as detection features and used dynamic Bayesian networks to detect aggressive behavior in public [12]. The current method of multimodal fusion is relatively simple, mostly through simple union of different modal features or voting by multiple classifiers, lacking a deeper level of fusion [13].

In order to solve the present issue of the lack of violent audio datasets, this paper first created audio violent dataset named VioAudio. Then, a CNN-ConvLSTM network model was proposed for audio violence detection. Audio signal was converted into spectrogram through short-time Fourier transform, as well as the spectrogram was used to train the network and detect audio violence. Meanwhile, this paper integrated self-attention mechanisms and visual information into CNN-ConvLSTM network in order to address the issue of modality singularity in violence detection. The deep extraction and fusion of visual and auditory information is accomplished in this way.

The structure of this paper is as follows: Sect. 2 mainly introduces the methods used in this paper, including the construction of network models and training methods. The Sect. 3 verifies the proposed method through experiments. The experimental results validate that network model is effective, and the module combination boosts the performance compared to those of existing methods. Finally, in Sect. 4, the conclusion and future research directions are given.

## 2    Methods

### 2.1    CNN-ConvLSTM Model

For auditory violence detection, traditional framing feature extraction method can result in too large feature dimension, which will cause feature redundancy. Feature redundancy reduces model accuracy and increases detection time. In recent years, with the development of deep learning and hardware acceleration devices, deep neural network has made unprecedented progress in image recognition, video behavior recognition, speech recognition and many other fields [14]. Convolution neural network (CNN) has achieved considerable success in the fields of image classification and speech recognition because of its ability to extract features adaptively. The Convolution layer is used for local connectivity and parameter sharing and can be viewed as the process of observing local features by kernels and extracting the useful information. Long short-term memory (LSTM) effectively processes sequential signals and retains most information of the signal. In this paper, the CNN model was used to extract features from audio signals and the LSTM was used to model the extracted features in time series for violence detection.

Audio signal is a kind of one-dimensional signal, which is difficult to extract features directly through CNN. Therefore, the audio signal was converted to two-dimensional spectrogram as the input of the network by short-time Fourier transform. Using spectrograms as network input has tremendous advantages. In general, the spectrogram contains most information of the audio signal. Traditional audio features use a variety of artificially designed filter banks to extract features after Fourier transform, which results in loss of information in frequency domain, especially in high frequency area. In order to account for computation, traditional audio features must use a very large frame shift, which undoubtedly results in loss of information in time domain. These problems can be effectively avoided by directly using spectrograms as network input. Then, using spectrograms as network input is more in line with the processing mode of human brain for audio information. External sound waves are converted into frequency vibrations in the cochlea, which are then transmitted to the auditory nerve. At last, the time granularity of the spectrogram is larger than that of the traditional frame feature granularity, which can better preserve the long-term audio correlation.

In this study, the original audio signal was segmented, and then converted into spectrogram, which was sent to CNN network for audio feature extraction. Then, the extracted audio features were sent into the ConvLSTM network to model the timing signal. The diagram of violence audio detection system based on spectrogram is shown in Fig. 1.

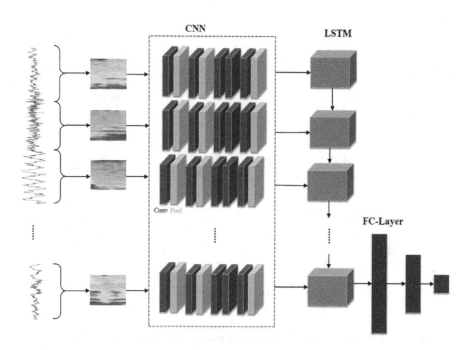

**Fig. 1.** CNN-ConvLSTM Network Model.

The CNN part of the network contains five convolution layers, three pooling layers, and the specific network parameters are shown in Table 1.

**Table 1.** CNN Network Parameters.

| Data dimension | Convolution kernel |
|---|---|
| $227 \times 227 \times 3$ | $Conv1 : (96, 11 \times 11 \times 3, 4)$ |
| $55 \times 55 \times 96$ | $Pool1 : (3 \times 3, 2)$ |
| $27 \times 27 \times 96$ | $Conv2 : (256, 5 \times 5 \times 96, 2)$ |
| $27 \times 27 \times 256$ | $Pool2 : (3 \times 3, 2)$ |
| $13 \times 13 \times 256$ | $Conv3 : (384, 3 \times 3 \times 256, 1)$ |
| $13 \times 13 \times 384$ | $Conv4 : (384, 3 \times 3 \times 384, 1)$ |
| $13 \times 13 \times 384$ | $Conv5 : (256, 3 \times 3 \times 384, 1)$ |
| $13 \times 13 \times 256$ | $Pool3 :: (3 \times 3, 2)$ |

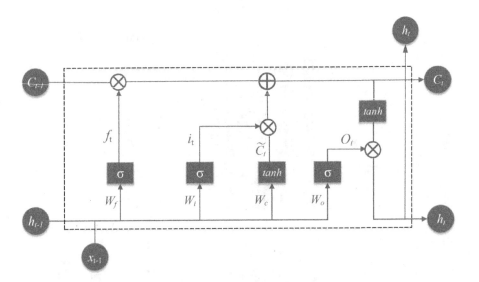

**Fig. 2.** The basic structure of LSTM.

The ConvLSTM was used to model features extracted from CNN. ConvL-STM is a deformation of LSTM. Its advantages are: 1) The occurrence of violence is a continuous process, and LSTM network has the ability to naturally remember information at multiple times; 2) Second, it can avoid the problem that the time length of each sample is not equal. The basic structure of LSTM is shown in Fig. 2.

LSTM combines the the current memory $\tilde{C}_T$ and long-term memory $C_{t-1}$ and forms a new unit state $C_t$. The forget gate can save the information long ago, while the input gate can prevent the current unimportant information from entering the memory. The output gate controls the effect of long-term memory on current output.

The learning of weight and threshold parameters were realized by traditional error back-propagation (EPB) algorithm. Cross entropy was applied as the loss function. $o_i^j$ denotes the predicted value of the networked and $y_i^j$ denotes the actual value, where $i$ represents the $i-th$ output neuron, and $j$ represents the $j-th$ training sample. The batch size is $N$ and the number of output neurons is $M$. Then the loss function can be expressed as eq1.

$$Loss = -\frac{1}{N} \sum_{i=1}^{M} \sum_{j=1}^{M} y_i^j \times log(o_i^j) \tag{1}$$

## 2.2 Attention Module

The basic idea of the attention mechanism is to allow the model to ignore irrelevant information and pay more attention to the key information we want it to focus on. The combination of deep learning and attention mechanism has focused on the mask to form the attention mechanism [15]. The principle of masking is to identify the key features in the data through another new layer of weights. By training, the deep neural network learns the areas of interest in the data and forms attention. The essence is to learn a set of weight distributions that can be applied to the original data.

In order to make ConvLSTM pay more attention on meaningful features and improve the accuracy of the model, this study introduces the attention mechanism into ConvLSTM. Then the corresponding input gate $i_t$, output gate $o_t$, forget gate $f_t$ and memory unit $c_t$ is calculated as follows.

$$i_t = \sigma(AttConv(X_t) + W_{hi} \times H_{t-1} + b_i) \tag{2}$$

$$o_t = \sigma(AttConv(X_t) + W_{ho} \times H_{t-1} + b_o) \tag{3}$$

$$f_t = \sigma(AttConv(X_t) + W_{hf} \times H_{t-1} + b_f) \tag{4}$$

$$H_t = o_t \circ tanh(C_t) \tag{5}$$

$$C_t = f_t \circ C_{t-1} + i_t \circ tanh(AttConv(X_t) + W_{hc} \times H_{t-1} + b_c) \tag{6}$$

Among them, $W_{hi}$, $W_{hf}$, $W_{ho}$ denote as Weight Matrix, $\sigma(*)$ denotes as sigmoid function, $\circ$ denotes as Hadamard product, $AttConv(*)$ denotes as Convolution attention function.

## 2.3    Audio-Visual Information Fusion

In the previous paper, we constructed a CNN-ConvLSTM model on the auditory channel. The spectrum of audio signals was used as the input of the CNN network, and the extracted features were sent to the ConvLSTM network for recognition. Research has shown that combining multiple modal features in the field of multimedia event detection is an effective method. Therefore, the CNN-ConvLSTM network is further improved in this study: two deep convolution networks are used to extract video and audio features, and then two channel features are fed into the ConvLSTM network. The diagram of multimodal violence detection system based on deep learning and multimodal feature fusion is shown in Fig. 3.

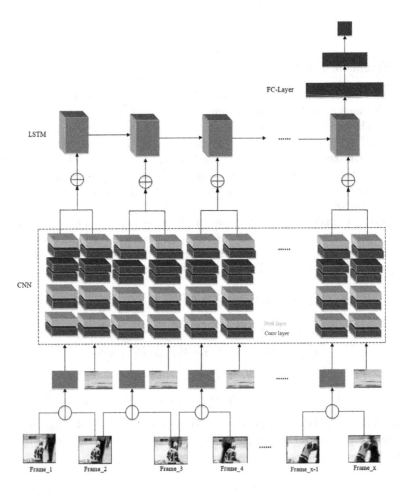

**Fig. 3.** Audio Video Fusion Model Based on CNN-ConvLSTM.

In the visual channel, the difference map of adjacent video frames is input into the visual CNN as network input. This is because visual violence is mostly continuous action, and the difference between adjacent frames can better represent the violence scene, which can shield the unchanged background. At the same time, the audio signal corresponding to the video frame is converted into a spectrogram and fed into the auditory CNN to extract the features of the audio signal. Finally, the two types of features at same time are fused and fed into ConvLSTM network to model the time series information to determine whether there are violent senses in the segment. The network still uses error back-propagation algorithm to learn parameters and cross-entropy as the loss function.

## 3    Experiments and Results

### 3.1    Datasets

**Violence Audio Dataset.** At present, there is no publicly trusted dataset available for the field of audio violence detection, so we build Violence Audio Dataset (VioAudio). The audio data sources in this dataset can be divided into two parts: one part is the audio from some video clips in the MediaEval 2015 Violence Detection Database. MediaEval is a competition dedicated to multimedia access and information retrieval algorithms, including speech recognition, multimedia content analysis, emotional recognition, violence detection and other multimedia tasks. In this paper, based on the labels of the original database, we have manually filtered and clipped 300 violent audio samples and 500 non-violent audio samples. Another part of VioAudio comes from 36 domestic movies. A total of 200 violent audio clips were selected through manual score screening.

Ultimately, VioAudio included 1,000 violent audio samples, each lasting 4–6 s. It contains 500 non-violent audio samples, including voices of conversation, music, laughter, singing and applause; 500 violent audio samples, including gunshots, explosions and screams.

**MediaEval 2015 Dataset.** In the Violence Detection experiment, we used a movie dataset from MediaEval 2015. The dataset includes 10,900 video samples, each 8–12 s long, from 199 real Hollywood movies. There were 502 violent and 10398 non-violent samples.

Violent videos are officially defined as videos that an eight-year-old cannot watch because of physical conflicts in the images. In the officially available datasets, the types of violence involved are explosions, screams, fights, gunshots, knife-holding assaults, and so on.

### 3.2    Audio Violence Detection

This section used CNN-ConvLSTM networks embedded in self-attention mechanisms for violent audio detection. The validity of the proposed network model

**Table 2.** The hyper-parameter settings of the network model.

| Hyper-parameter | Audio | Audio and Video |
|---|---|---|
| Learning Rate | $10^{-5}$ | $10^{-5}$ |
| LR decay rate | 0.5 | 0.5 |
| Batch | 16 | 16 |
| Hidden Size | 100 | 128 |
| Loss Function | Cross Entropy | Cross Entropy |
| Activation Function | ReLU | ReLU |
| Optimized Acceleration | Adam | Adam |

is verified first. The hyper-parameter settings of the network model are shown in Table 2.

During the training of the network, the maximum number of iterations is set to 100. The network's Loss and Accuracy changes with the number of iterations on the VioAudio dataset as shown in Fig. 4. From the change of Loss value on training set and test set, with the increase of iteration number, Loss value shows a downward trend as a whole, indicating that the network converges gradually.

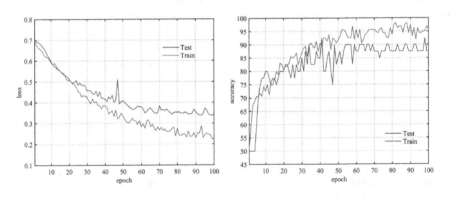

**Fig. 4.** Curve of Loss and Accuracy Changing with Iteration Times.

To verify the validity of the proposed method, this paper compares the recognition results of the CNN-ConvLSTM network with other methods, as shown in Table 3.

The results show that the detection method based on traditional acoustic features and machine learning method was still very effective, with 89.3% accuracy on the VioAudio audio dataset. Single CNN and single ConvLSTM networks were less effective than the detection algorithms based on traditional acoustic features. It is not enough to extract the acoustic features by CNN alone. The combination of CNN and ConvLSTM can better achieve the detection of violent audio. The attention mechanism further improves the detection accuracy of the network.

**Table 3.** Results of different algorithms in VioAudio dataset.

| Method | Accuracy |
|---|---|
| Acoustic features + SVM | 89.3% |
| Spectrogram +CNN | 86.7% |
| Spectrogram + ConvLSTM | 88.5% |
| Spectrogram + CNN + ConvLSTM | 91.5% |
| Spectrogram + Attention + CNN + ConvLSTM | 92.8% |

### 3.3  Audio-Visual Violence Detection

In the previous section, we built a CNN-ConvLSTM network on the auditory channel, taking the spectrogram of audio signal as the input of the network, and achieved good experimental results. On this basis, this section used two CNN networks to extract video and audio features respectively, and then fuses the features of the two channels into the ConvLSTM network to achieve violence event detection based on visual and auditory information fusion. The hyperparameter settings of the network are shown in Table 1. Except for the initial network structure and the number of hidden nodes, the network is basically unchanged. This section uses MediaEval 2015 Dataset to verify the network results. The experimental results are shown in Table 4.

**Table 4.** The results in the MediaEval 2015 Dataset.

| | Precision | Recall | F1 |
|---|---|---|---|
| Audio | 0.46 | 0.73 | 0.56 |
| Audio-Video | 0.51 | 0.84 | 0.63 |

The results show that the method of audio-video fusion is better than P single channel. The increase of recall rate indicates that when the information of both video and audio channels is considered in the detection of violence, more samples with violence can be detected and the miss rate is reduced. The improvement of Precision indicates that the audio-video feature fusion analysis can reduce the rate of misjudgment.

Then, we compared the MAP value with other open methods. MAP is the official evaluation index used in the MediaEval 2015 competition. The formula for calculating this index is as follows.

$$MAP = \frac{1}{N} \sum_{c=1}^{N} AP_c \tag{7}$$

$$AP_c = \frac{1}{R} \sum_{i=1}^{M} (I_i \times \frac{R_i}{i}) \tag{8}$$

where $R$ is the number of all positive samples in the test set, $M$ is the total number of samples in the test set. If the $i-th$ sample is positive, $I_i = 1$. $R_i$ is the number of positive samples in the top $i$ samples, and $N$ is the number of queries. The comparison results are shown in Table 5.

**Table 5.** The results in the MediaEval 2015 Dataset.

|             | methods               | MAP    |
|-------------|-----------------------|--------|
| Audio       | ICL-TUM-PASSAU [16]   | 14.9%  |
|             | TCS-ILAB [17]         | 6.38%  |
|             | CNN-ConvLSTM*         | 16.47% |
| Audio-Video | RUCMM [18]            | 21.6%  |
|             | NII-UIT [19]          | 26.8%  |
|             | CNN-ConvLSTM*         | 31.54% |

The results show that the CNN-ConvLSTM model could extract the depth features of audio and model the time series on auditory channel. The results of our method were what better than those of other teams. In the aspect of audio-video fusion, CNN was used to extract the features from video difference map and audio spectrogram, ConvLSTM was used to model and fuse the features. The experimental results were significantly better than the current best results.

## 4 Conclusion

Violence detection is an important research direction in the field of multimedia information processing, with a very wide range of application scenarios. In this paper, an audio violence dataset named VioAudio was constructed to solve the problem of the scarcity of dataset. Then, a CNN-ConvLSTM network model was built to detect audio violence. This paper explored the fusion methods that combine visual and audio features in the violence detection task. We combined the feature of the inter-frame difference map with the feature of the audio spectrogram, and used ConvLSTM network to model the time series information, which effectively improves the accuracy of violent event detection.

However, there are still many improvements. In the audio-video feature fusion, there may be asynchronization between audio and video information. To solve this problem, we will study the alignment method of audio and video features, and then model the aligned features to build a violence detection system. Secondly, the existing datasets are mainly for movie clips. We will build a dataset of violence detection based on monitoring data to enhance the practical significance of violence detection.

**Acknowledgements.** This work was supported by the National Natural Science Foundation of China [U20A20383]; the National Key Research and Development Program of China [2020YFC0833204]; the Key Research and Development Program of Heilongjiang [GY2021ZB0206].

# References

1. Gao, Y., Liu, H., Sun, X., et al.: Violence detection using oriented violent flows. Image Vision Comput. **48–49**, 37–41 (2016)
2. Zhang, T., Jia, W., He, X., et al.: Discriminative dictionary learning with motion weber local descriptor for violence detection. IEEE Trans. Circ. Syst. Video Technol. **27**(3), 696–709 (2017)
3. Cheng, W.-H., Chu, W.-T., Wu, J.-L.: Semantic context detection based on hierarchical audio models. In: Proceedings of the 5th ACM SIGMM International Workshop on Multimedia Information Retrieval - MIR 2003, vol. 109. ACM Press, Berkeley (2003). Accessed 14 Oct 2022
4. García-Gómez, J., Bautista-Durán, M., Gil-Pita, R., Mohino-Herranz, I., Rosa-Zurera, M.: Violence detection in real environments for smart cities. In: García, C.R., Caballero-Gil, P., Burmester, M., Quesada-Arencibia, A. (eds.) UCAmI/IWAAL/AmIHEALTH -2016. LNCS, vol. 10070, pp. 482–494. Springer, Cham (2016). https://doi.org/10.1007/978-3-319-48799-1_52
5. Penet, C., Demarty, C.-H., Gravier, G., et al.: Variability modelling for audio events detection in movies. Multimedia Tools Appl. **74**(4), 1143–1173 (2015)
6. Sarman, S., Sert, M.: Audio based violent scene classification using ensemble learning. In: 2018 6th International Symposium on Digital Forensic and Security (ISDFS), pp. 1–5. IEEE, Antalya (2018). Accessed 14 Oct 2022
7. Moreira, D., Avila, S., Perez, M., et al.: Temporal robust features for violence detection. In: Applications of Computer Vision, pp. 391–399. IEEE (2017)
8. Xu, L., Gong, C., Yang, J., et al.: Violent video detection based on MoSIFT feature and sparse coding. In: IEEE International Conference on Acoustics, Speech and Signal Processing, pp. 3538–3542. IEEE (2014)
9. Sudhakaran, S., Lanz, O.: Learning to detect violent videos using convolutional long short-term memory. In: 2017 14th IEEE International Conference on Advanced Video and Signal Based Surveillance (AVSS), pp. 1–6 (2017). https://doi.org/10.1109/AVSS.2017.8078468
10. Lin, J., Wang, W.: Weakly-supervised violence detection in movies with audio and video based co-training. In: Muneesawang, P., Wu, F., Kumazawa, I., Roeksabutr, A., Liao, M., Tang, X. (eds.) PCM 2009. LNCS, vol. 5879, pp. 930–935. Springer, Heidelberg (2009). https://doi.org/10.1007/978-3-642-10467-1_84
11. Demarty, C.-H., Penet, C., Ionescu, B., Gravier, G., Soleymani, M.: Multimodal violence detection in hollywood movies: state-of-the-art and benchmarking. In: Ionescu, B., Benois-Pineau, J., Piatrik, T., Quénot, G. (eds.) Fusion in Computer Vision. ACVPR, pp. 185–208. Springer, Cham (2014). https://doi.org/10.1007/978-3-319-05696-8_8
12. Zajdel, W., Krijnders, J.D., Andringa, T., et al.: CASSANDRA: audio-video sensor fusion for aggression detection. In: 2007 IEEE Conference on Advanced Video and Signal Based Surveillance, pp. 200–205. IEEE, London (2007). Accessed 14 Oct 2022
13. Pang, W.-F., He, Q.-H., Hu, Y., et al.: Violence detection in videos based on fusing visual and audio information. In: ICASSP 2021–2021 IEEE International Conference on Acoustics, Speech and Signal Processing (ICASSP), pp. 2260–2264. IEEE, Toronto (2021). Accessed 14 Oct 2022
14. Samuel, R.D.J., Fenil, E., Manogaran, G., et al.: Real time violence detection framework for football stadium comprising of big data analysis and deep learning through bidirectional LSTM. Comput. Netw. **151**, 191–200 (2019)

15. Rendón-Segador, F.J., Álvarez-García, J.A., Enríquez, F., et al.: ViolenceNet: dense multi-head self-attention with bidirectional Convolution LSTM for detecting violence. Electronics **10**(13), 1601 (2021)
16. Trigeorgis, G., Coutinho, E., Ringeval, F., et al.: The icl-tum-passau approach for the mediaeval 2015 "affective impact of movies" task. In: MediaEval (2015)
17. Chakraborty, R., Maurya, A.K., Pandharipande, M., et al.: TCS-ilab - mediaeval 2015: affective impact of movies and violent scene detection. In: MediaEval (2015)
18. Jin, Q., Li, X., Cao, H., et al.: RUCMM at mediaeval 2015 affective impact of movies task: fusion of audio and visual cues. In: MediaEval (2015)
19. Lam, V., Le, S.P., Le, D.-D., et al.: NII-uit at mediaeval 2015 affective impact of movies task. In: MediaEval (2015)

# Mongolian Text-to-Speech Challenge Under Low-Resource Scenario for NCMMSC2022

Rui Liu[1]([✉]), Zhen-Hua Ling[2], Yi-Fan Hu[1], Hui Zhang[1], and Guang-Lai Gao[1]

[1] Inner Mongolia University, Hohhot, China
`liurui_imu@163.com`, `{cszh,csggl}@imu.edu.cn`
[2] National Engineering Laboratory for Speech and Language Information Processing, University of Science and Technology of China, Hefei, People's Republic of China
`zhling@ustc.edu.cn`

**Abstract.** Mongolian Text-to-Speech (TTS) Challenge under Low-Resource Scenario is a special session for National Conference on Man-Machine Speech Communication 2022 (NCMMSC2022), termed as NCMMSC2022-MTTSC. A Mongolian TTS dataset was provided to participants this year, and a low-resource Mongolian TTS task was designed. Specifically, the task is to synthesize high-quality Mongolian speech with given Mongolian scripts. Thirteen teams submitted their results for final evaluation. Mean opinion score (MOS) listening tests were conducted online to measure the naturalness, intelligibility of the synthetic speech. In addition, the word error rate (WER) of automatic speech recognition was further treated as the objective metric for intelligibility evaluation. The evaluation results show that the top system achieved comparable naturalness and intelligibility with the ground truth speech.

**Keywords:** Mongolian · Text-to-Speech (TTS) · Low-Resource · NCMMSC2022

## 1 Introduction

Text-to-Speech (TTS), that is a standard technology in human-computer interaction, aims to convert the input text to human-like speech [1]. Mongolian TTS Challenge under Low-Resource Scenario is a special session for National Conference on Man-Machine Speech Communication 2022 (NCMMSC2022), termed as NCMMSC2022-MTTSC. The NCMMSC2022-MTTSC was organized by the Inner Mongolia University, the University of Science and Technology of China (USTC) and the other members of the committee[1]. The majority of previous TTS challenges have used speech datasets for mainstream languages, such as Mandarin Chinese and English. For example, Blizzard Challenges[2] 2008–2010 and 2019–2020 adopt Mandarin Chinese data and Blizzard Challenges 2005–2013 and 2016–2018 use English data. Note that the TTS research for the minority

---

[1] http://mglip.com/challenge/NCMMSC2022-MTTSC/index.html.
[2] http://festvox.org/blizzard/index.html.

L. Zhenhua et al. (Eds.): NCMMSC 2022, CCIS 1765, pp. 221–226, 2023.
https://doi.org/10.1007/978-981-99-2401-1_20

language gradually attracted wide attention. To this end, Blizzard Challenges 2013–2015 use several India languages and 2021 adopt European Spanish data as the official training corpus. Despite the progress, there are also some languages that have not attracted enough attention, such as Mongolian language [2].

Mongolian is the most famous and widely spoken language of the Mongolian language family. In addition, Mongolian is the main national language in the Inner Mongolia Autonomous Region of China and is mainly used in Mongolian-inhabited areas of China, Mongolia, and the Siberian Federal District of the Russian Federation. Recently, the researchers began a series study of Mongolian TTS [3]. Specifically, deep learning techniques were first introduced to Mongolian TTS [4], and DNN-based acoustic models trained on 5-hours training data were used instead of HMM acoustic models to achieve high-quality synthesized speech. Li et al. [5] further implemented a Tacotron-based Mongolian TTS system. To accelerate the inference speed and improve the speech fidelity, Liu et al. [6] proposed a pure non-autoregressive neural Mongolian TTS model, called MonTTS, which consists of the FastSpeech2-based acoustic model and the HiFi-GAN vocoder. The above mentioned works provide a solid foundation for the research of Mongolian TTS technology. However, there is still a lack of in-depth research on Mongolian TTS, especially in a low-resource scenario that aims to synthesize high-quality Mongolian speech with limited training data.

The NCMMSC2022-MTTSC is the first time that a minority language, i.e., Mongolian, has been used for TTS challenge in China. This challenge also promotes the development of intelligent information processing in minority languages within China. This paper will present the details of the speech dataset, tasks, participating systems, evaluations and results of challenge.

## 2  Voices to Build

### 2.1  Speech Dataset

A Mongolian speech dataset kindly provided by Inner Mongolia University was released for voice building. The dataset contains recorded speech from a professional female native Mongolian speaker together with text transcriptions. The texts were from various domains, including daily life, sport, education, travelling, etc. The speech was recorded in a studio of school of computer science with quiet environment. The total duration of the waveform files, which were sampled at 22.05 kHz with a sampling accuracy of 16 bit, amounts to around 2 h.

### 2.2  Task

We design a Mongolian TTS task under a low-resource scenario using the released dataset. Each participating team should build a voice from the provided 2-h Mongolian data to synthesize the given Mongolian text, following the challenge rules (See Foonote 1). The submitted synthetic speech should be 16 bit depth, and at any standard sampling rate (e.g., 16 kHz, 22.05 kHz, 44.10 kHz, or 48 kHz). For evaluation, teams were required to synthesize 200 test sentences (disjoint from the training data) that contained only Mongolian text.

Regarding the use of external data, the NCMMSC2022-MTTSC allowed that each participant can use any data of any language, whether freely-available or not, to conduct the pre-training etc. Participants were asked to report their data usage instruction when submitting synthetic speech and in their paper.

# 3  Participants

There are 13 teams submitted their results. Note that there is no benchmark system for NCMMSC2022-MTTSC. Following Blizzard challenges, all systems are identified using letters in these published results. Specifically, letter A denotes natural or ground truth speech. Letters B to T were assigned to the systems submitted by participants. Each participating team is free to choose whether to reveal their system identifier in their workshop paper.

We summarized the detailed structure of all systems in Table 1. We see that all systems adopted a neural approach, and the great majority employed VITS,

**Table 1.** The participating teams and their institutions. The system identifier of natural speech (the first row) is letter A. The method descriptions are summarised based on the questionnaires and the workshop papers from participants.

| Team Name | Institution | Input Type | Acoustic Model | Vocoder | Transfer Learning |
|---|---|---|---|---|---|
| Natural Speech | N/A | N/A | N/A | N/A | N/A |
| all u need | University of Science and Technology of China, Hefei | Latin | FullConv | Griffin-Lim | No |
| Mnemosyne | Microsoft Azure Speech, Beijing | Latin | Conformer based FastSpeech2 | HiFinet2 | Yes |
| sigma | VXI Global Solutions, Shanghai | Latin | VITS | N/A | Yes |
| TJUCCA_TTS | Tianjin University, Tianjin (**no paper submission**) | Phoneme | Tacotron2 | HiFi-GAN | Yes |
| RoyalFlush | Zhejiang Hithink RoyalFlush AI Research Institute, HangZhou | Latin | Tacotron2 | HiFi-GAN | Yes |
| 火之源 | Inner Mongolia University, Hohhot | Latin | FastSpeech2 | HiFi-GAN | Yes |
| IOA-THINKIT | Institute of Acoustics, and University of Chinese Academy of Sciences, Beijing (**no paper submission**) | Latin | VITS | N/A | Yes |
| DBLAB | OPPO | Latin | VITS | N/A | Yes |
| 在线_AI特工队 | China Mobile Online Services Co., Ltd., Luoyang | Latin | VITS | N/A | No |
| qdreamer | Suzhou Qimeng People Network Technology Company, Suzhou | Latin | VITS | N/A | Yes |
| FlySpeech | Audio, Speech and Langauge Processing Group, Northwestern Polytechnical University, XiAn | Phoneme | Delightful TTS | HiFi-GAN | Yes |
| Cyber | Chengdu Rongwei Software Service Co., Ltd. Chengdu (**no paper submission**) | Phoneme | Flow-based model | HiFi-GAN | Yes |
| Y | Mobvoi (**no paper submission**) | Phoneme | VITS | N/A | No |

which is a state-of-the-art fully end-to-end model. The classic Tacotron2 and FastSpeech2 models were also favored by many teams. Just one team build the acoustic model with convolutional neural network (CNN). Neural vocoder was also adopted by many teams, of which the majority (6 out of 13) was HiFi-GAN.

## 4    Evaluations and Results

### 4.1    Evaluation Materials

We released 200 sentences as the testing data for the listening test. All participants used their own system to synthesize 200 sentences for the final subjective and objective rating.

### 4.2    Evaluation Metrics

The evaluation results for NCMMSC2022-MTTSC consisted of three metrics in terms of naturalness and intelligibility, respectively, as follows:

- **Naturalness mean opinion score (N-MOS)** was used to test the speech quality for all teams in terms of naturalness subjectively.
- **Intelligibility mean opinion score (I-MOS)** was used to test the speech quality for all teams in terms of intelligibility subjectively.
- **Word error rate (WER)** was used to test the speech quality for all teams in terms of intelligibility objectively.

For N-MOS and I-MOS, the organizers recruited 20 listeners that were all native speakers of Mongolian and all instructions and other text on the listening test webpages were includes Chinese and Mongolian. For WER, the organizers calculated the WER by leveraging a Mongolian speech recognition interface[3] from Inner Mongolia University.

### 4.3    Results

We report the evaluation results for all metrics in Fig. 1. In all subfigures of Fig. 1, a consistent system ordering is adopted, which is the descending order of mean values for the corresponding metric. The mean values are calculated from the listeners' scores for each metric. Please note that this ordering only aims to make the plots more readable by using the same system ordering across all plots for each task and can not be interpreted as a ranking, because the ordering does not indicate which systems are significantly better than others.

As shown in Fig. 1(a), system C achieved significantly better naturalness than all other submitted systems. The N-MOS ratings of natural speech (system A) and system C were 4.886 and 4.468 respectively, and the difference between them is minimal compared to other systems. All other systems scored below 4.3.

---

[3] http://asr.mglip.com/.

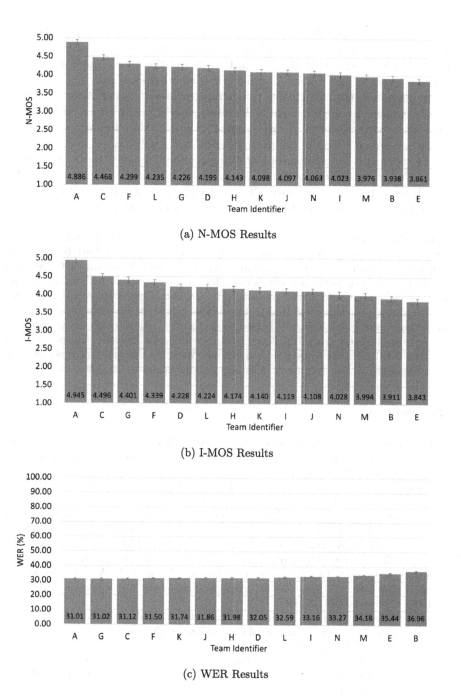

(a) N-MOS Results

(b) I-MOS Results

(c) WER Results

**Fig. 1.** N-MOS, I-MOS and WER results for NCMMSC2022-MTTSC. A is natural speech, the remaining letters denote the systems submitted by participants.

As shown in Fig. 1(b), system C still beat other systems and achieve highest I-MOS score. The I-MOS scores of system A and system C were 4.945 and 4.496 respectively. Regarding objective intelligibility metric, Fig. 1(c) reports the WER. We found that the system G achieved the lowest value with 31.02, which is closer to the system A with 31.01 than other systems.

In a nutshell, the participants using transfer learning generally achieve good results, which proves that transfer learning plays a positive role in improving the speech synthesis effect in low-resource scenarios. More important, from the submitted papers[4], we can see that system C has conducted an in-depth study on the pronunciation phenomenon of Mongolian and designed special processing rules. Therefore, for agglutinative language, the addition of language-related knowledge is helpful to improve TTS model performance. It is hoped that the above findings will contribute to the future study of Mongolian TTS.

**Acknowledgements.** This work was partially supported by the High-level Talents Introduction Project of Inner Mongolia University (No. 10000-22311201/002) and the Young Scientists Fund of the National Natural Science Foundation of China (No. 62206136). We wish to thank a number of additional contributors without whom running the challenge would not be possible. Prof. Zhenhua Ling at University of Science and Technology of China provide all instruction related to the challenge selflessly. Yifan Hu at Inner Mongolia University helped to prepare the official challenge website and the training and test data. Hui Zhang at Inner Mongolia University helped to conduct the WER calculation. Pengkai Yin at Inner Mongolia University is responsible for gathering listening test volunteers. Thanks to all participants and listeners.

# References

1. Ling, Z.H., et al.: Deep learning for acoustic modeling in parametric speech generation: a systematic review of existing techniques and future trends. IEEE Signal Process. Maga. **32**(3), 35–52 (2015)
2. Janhunen, J.: Mongolic languages. In: The Encyclopedia of Language & Linguistics, pp. 231–234. Elsevier Scientific Publ. Co. (2006)
3. Liu, R., Sisman, B., Bao, F., Yang, J., Gao, G., Li, H.: Exploiting morphological and phonological features to improve prosodic phrasing for Mongolian speech synthesis. IEEE/ACM Trans. Audio Speech Lang. Process. **29**, 274–285 (2020)
4. Liu, R., Bao, F., Gao, G., Wang, Y.: Mongolian text-to-speech system based on deep neural network. In: Tao, J., Zheng, T.F., Bao, C., Wang, D., Li, Y. (eds.) NCMMSC 2017. CCIS, vol. 807, pp. 99–108. Springer, Singapore (2018). https://doi.org/10. 1007/978-981-10-8111-8_10
5. Li, J., Zhang, H., Liu, R., Zhang, X., Bao, F.: End-to-end mongolian text-to-speech system. In: 2018 11th International Symposium on Chinese Spoken Language Processing (ISCSLP), pp. 483–487. IEEE (2018)
6. Liu, R., Kang, S., Gao, G., Li, J., Bao, F.: MonTTS: a real-time and high-fidelity Mongolian TTS model with pure non-autoregressive mechanism. J. Chin. Inf. Process. **36**(7), 86 (2022)

---

[4] We encourage readers to visit the official website of NCMMSC2022-MTTSC to check the submitted papers.

# VC-AUG: Voice Conversion Based Data Augmentation for Text-Dependent Speaker Verification

Xiaoyi Qin[1], Yaogen Yang[1], Yao Shi[1], Lin Yang[2], Xuyang Wang[2], Junjie Wang[2], and Ming Li[1(✉)]

[1] Data Science Research Center, Duke Kunshan University, Kunshan, China
{xiaoyi.qin,yaogen.yang,yao.shi,ming.li369}@dukekunshan.edu.cn
[2] AI Lab of Lenovo Research, Beijing, China
{yanglin13,wangxy60,wangjj9}@lenovo.com

**Abstract.** In this paper, we focus on improving the performance of the text-dependent speaker verification system in the scenario of limited training data. The deep learning based text-dependent speaker verification system generally needs a large-scale text-dependent training data set which could be both labor and cost expensive, especially for customized new wake-up words. In recent studies, voice conversion systems that can generate high quality synthesized speech of seen and unseen speakers have been proposed. Inspired by those works, we adopt two different voice conversion methods as well as the very simple re-sampling approach to generate new text-dependent speech samples for data augmentation purposes. Experimental results show that the proposed method significantly improves the Equal Error Rate performance from 6.51% to 4.48% in the scenario of limited training data. In addition, we also explore the out-of-set and unseen speaker voice conversion based data augmentation.

**Keywords:** speaker verification · voices conversion · text-dependent · data augmentation

## 1 Introduction

Speaker verification technology aims to determine whether the test utterance is indeed spoken by the enrollment speaker. In recent years, x-vectors [23] demonstrate state-of-the-art results in the speaker verification field. Multiple different backbone architectures, e.g. TDNN [23], ResNet [2], and their variants [18], etc. are proposed for the front-end feature extraction.

Futhermore, the research works of deep learning based speaker verification also enjoy those publicly open and free speech databases, e.g., AISHELL2 [7], Librispeech [17], Voxceleb1&2 [4,16] in the text-independent field, and RSR2015 [15], HIMIA [19], MobvoiHotwords in the text-dependent field, etc. Methods in [10,24] achieve a good performance in the text-dependent speaker verification task if a large amount of text-dependent training data are available. However, it is both labor expensive and time consuming to collect the database. With the

© The Author(s), under exclusive license to Springer Nature Singapore Pte Ltd. 2023
L. Zhenhua et al. (Eds.): NCMMSC 2022, CCIS 1765, pp. 227–237, 2023.
https://doi.org/10.1007/978-981-99-2401-1_21

rise of smart home and Internet of Things applications, there are great demands for text-dependent speaker verification, with customized wake-up words. It is almost impossible to collect the corresponding text-dependent speech data for each customized wake-up word.

In recent studies, the speech signals generated by the multi-speaker Text-to-Speech (TTS) and one-to-many or many-to-many voice conversion (VC) systems are getting harder to be distinguished between real-person voice and synthesized voice [5,27,29]. So, it is natural to adopt TTS or VC as a data augmentation strategy for speaker verification under the limited training data scenario [12]. The multi-speaker TTS system could create a large amount of speech data from multiple target speakers with different lexical contents. However, in the context of text-dependent cases, since the input text is the same, the synthesized speech data are very similar even for different target speakers. Moreover, different from multi-speaker TTS, the VC system can generate data with various kinds of styles all with the same text-dependent content. Therefore, the VC approaches are more appropriate than TTS as the data augmentation method for text-dependent speaker verification.

This paper aims to improve the text-dependent speaker verification system's performance with a limited number of speakers and training data.

- Limited training data for each speaker. The number of text-dependent utterances of each speaker is less than 10.
- Limited speakers for training. The number of speakers is less than 500.

Targeting the aforementioned scenarios, we propose to train a voice conversion model with limited existing text-dependent data to generate more new text-dependent data. We use two different voice conversion methods as our data augmentation systems. The first one is a Mel-to-Mel voice conversion system [26] using the conditional Seq-to-Seq neural network framework with dual speaker embeddings as the inputs while the other one is a PPP-to-Mel system that converts the phoneme posterior probability(PPP) features [11] with target speaker embedding into Mel-spectrograms [28]. Furthermore, in the limited speaker number case, we adopt the pitch shift(speed perturbation with re-sampling) strategy to augment more speakers. Besides, we also attempt to use the out-of-set unseen speakers' embeddings to generate the text-dependent data from out-of-set speakers. In order to compare TTS and VC based data augmentation methods in the text-dependent speaker verification task, we also train a popular one-hot multi-speaker TTS framework. The ResNet34-GSP [2] model is adopted as the speaker verification system to evaluate different systems.

The paper is organized as follows. Section 2 describes the related works about voice conversion and speaker verification we adopted in this paper. The proposed methods and strategies are described in Sect. 3. Section 4 shows the experimental results. Finally, the conclusion is provided in Sect. 5.

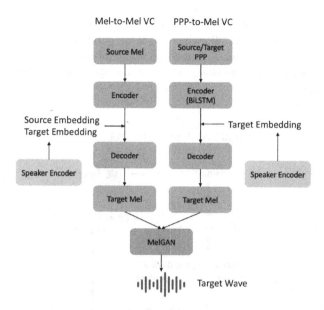

**Fig. 1.** The architectures of two voice conversion systems used in this work

## 2    Related Works

### 2.1    Speaker Verification System

In this paper, we adopt the same structure as [2]. The network structure contains three main components: a front-end pattern extractor, an encoder layer, and a back-end classifier. The ResNet34 [9] structure is employed as the front-end pattern extractor, which learns a frame-level representation from the input acoustic feature. The global statistic pooling (GSP) layer, which computes the mean and standard deviation of the output feature maps, can project the variable length input to the fixed-length vector. The output of a fully connected layer following after the pooling layer is adopted as the speaker embedding layer. The ArcFace [6] ($s = 32$, $m = 0.2$) which could increase intra-speaker distances while ensuring inter-speaker compactness is used as a classifier. The detailed configuration of the neural network is the same with [21]. The cosine similarity serves as the back-end scoring method.

### 2.2    Voice Conversion System

**Mel-to-Mel VC System.** Firstly, we introduce a many-to-many voice conversion model using the conditional sequence-to-sequence neural network framework with dual speaker embedding [26]. The model is trained on many different source-target speaker pairs, which requires the speaker embeddings from both the source speaker and the target speaker as the auxiliary inputs. To improve speaker similarity between reference speech and converted speech, we use a feed-back constraint mechanism [3], which adds an auxiliary speaker identity loss in the network. This model is named as the Mel-to-Mel VC system because the model directly maps the source speaker Mel-spectrogram to target speaker Mel-spectrogram.

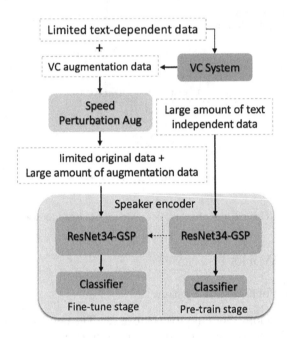

**Fig. 2.** The pipeline of Data Augmentation based on Voice Conversion in Text-Dependent Speaker Verification.

**PPP-to-Mel VC System.** Besides, we also introduced another VC system. The model is proposed in [28]. First, we use a DNN based auto-speech recognition (ASR) acoustic model, trained on the AISHELL-2 database, to obtain the target speaker phoneme posterior probabilities(PPP) features as the voice conversion

**Table 1.** The dataset usage of training VC and TTS systems.

| Model | Dataset | Training Spk/Utt Num |
| --- | --- | --- |
| ASV | HIMIA | 340/3060 |
| VC (PPP-to-Mel) | HIMIA | 340/3060 |
| VC (Mel-to-Mel) | HIMIA | 340/3060 |
| TTS | DIDI-speech | 500/53425 |

model's input. The model's output is the target Mel-spectrogram feature. In the testing, the source PPP will be assumed to be exactly the same as the target PPP to generate the results. This system is named as the PPP-to-Mel VC system in this paper. The PPP-to-Mel VC system architecture is similar to the Mel-to-Mel VC system expect that there is no feedback constraint. Besides, since the system's input is the target speaker feature rather than the source speaker feature, the input PPP feature is selected randomly from the limited training data.

Figure 1 shows the architectures of two voice conversion systems. The speaker encoder component is the same as the aforementioned ResNet34-GSP model. The vocoder MelGAN [14] is used to reconstruct the time-domain waveform from the predicted Mel-spectrogram.

## 3    Methods

In this section, the VC data augmentation strategies and the speed perturbation method are introduced in detail. The pipeline of our proposed data augmentation strategy is shown in Fig. 2. Those methods are all focused on the limited text-dependent data scenario. In this experiment, we adopt the HIMIA database with 340 speakers [19]. 9 utterances of each speaker in the HIMIA database are randomly chosen as the limited text-dependent data to train the baseline system. Therefore, only 3060 utterances (total have 340 * 9 = 3060 utterances) are used to train the VC conversion and fine-tune speaker verification models. The close-talk text-dependent data of the FFSVC20 challenge [21] are chosen as the test data. The trial file can be download from trial_file[1]. Since 3060 sentences with only 'ni hao,mi ya' text are not enough to train a TTS system, we use DiDi-speech [8] with 500 speaker to train a multi-speaker TTS system. The dataset usage of training VC and TTS system show in the Table 1.

### 3.1    Pre-training and Fine-tuning

According to our previous works [19,20], fine-tuning is an effective transfer learning approach to improve the speaker verification system performance in the limited training data scenario. In this work, we pre-trained the deep speaker verification network with a large-scale text-independent mix-dataset. There are in total 3742 speakers in the pre-training dataset, including AISHELL-2 [7], SLR68[2] and SLR62[3] from openslr.org. These three databases are also considered as out-of-set unseen speaker data for the VC augmentation system. The model was trained for 200 epochs in the pre-training stage, with an initial learning rate of 0.1. The network was optimized by stochastic gradient descent (SGD). All weights in the network remain trainable with an initial learning rate of 0.01 during the fine-tuning stage.

---

[1] https://github.com/qinxiaoyi/VCaug_ASV.

[2] https://openslr.org/68/.

[3] https://openslr.org/62/.

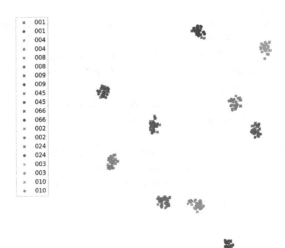

**Fig. 3.** Speaker embedding visualization by t-SNE for the in-set data. The voice conversion data is generate by PPP-to-Mel VC system. The * stands for the original data and the • stands for voice conversion data

### 3.2   Data Augmentation Based on the VC System

The training data of both VC systems is only 3060 utterances with 'ni hao, mi ya' text (Fig. 3).

**Data Augmentation Using the Mel-to-Mel VC System.** For training the Mel-to-Mel VC system, the loss function of the many-to-many voice conversion model is

$$\mathcal{L}_{total} = \mathcal{L}_{mel\_before} + \mathcal{L}_{mel\_after} + \mathcal{L}_{stop\_token\_loss}$$
$$+ 5 * \mathcal{L}_{embedding\_loss} + \mathcal{L}_{regular\_loss} \tag{1}$$

The loss function is also described in detail in [3]. To make the speaker embedding of the voice generated by the voice conversion model close to the target speaker embedding, we increased the weight of embedding loss and set it to 5.

After that, we generated 200 utterances for each target speaker based on a trained Mel-to-Mel VC system. For every target speaker, the source speech of VC's input was random chosen from the other 339 speaker utterances. The embeddings generated by the VC system were computed the cosine similarity with target speaker embedding to handle the outlier. The data with similarity greater than 0.6 are retained.

The limited text-dependent training data (3060 utts) are adopted as source speech for the out-of-set unseen speaker augmentation. Each out-of-set unseen speaker has 20 VC generated text-dependent utterances. After that, the generated data with cosine similarity less than 0.3 are filtered out. Since the out-of-set

**Table 2.** The performance of the text dependent speaker verification systems under different data augmentation methods. The 9*utt* denotes the limited training data scenario, each speaker only has 9 utterances; the VC $AUG_{in}$ and the VC $AUG_{out}$ denotes the voice conversion data from in-set and out-of-set speakers respectively; the Pitch shift AUG denotes the SoX *speed* function based pitch shift augmentation method.

| Model | Training data | Spk/Utt Num. | EER[%] | $mDCF_{0.1}$ |
|---|---|---|---|---|
| Pre-train model | AISHELL2 +SLR62 +SLR68 | 3472/518864 | 6.51 | 0.265 |
| Fine-tune model | 9 Utts per spk (baseline system) | 340/3060 | 7.63 | 0.331 |
| | + Pitch shift AUG | 1020/9180 | 5.76 | 0.248 |
| | + VC $AUG_{in}$ (Mel-to-Mel) | 340/26160 | 6.36 | 0.304 |
| | + VC $AUG_{in}$ (PPP-to-Mel) | 340/29089 | 5.16 | 0.249 |
| | + VC $AUG_{out}$ (Mel-to-Mel) | 3210/48890 | 6.08 | 0.295 |
| | + VC $AUG_{in}$ (Mel-to-Mel) + Pitch shift AUG | 1020/76978 | 5.19 | 0.241 |
| | + VC $AUG_{in}$(PPP-to-Mel) + Pitch shift AUG | 1020/87267 | **4.48** | **0.212** |
| | + TTS (DiDi) | 792/7323 | 6.01 | 0.292 |

voice conversion is a challenging task, the threshold is not very strict (the most out of set embedding similarity is less than 0.5).

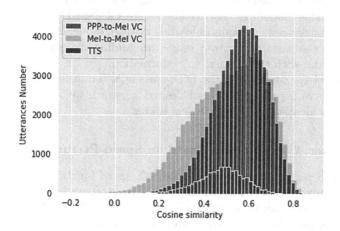

**Fig. 4.** Histogram of cosine similarity score on the in-set experiment.

**Data Augmentation Using the PPP-to-Mel VC System.** The procedure the PPP-to-Mel VC augmentation method is the same as the Mel-to-Mel VC system, and the loss function is the same as [22].

For the in-set speaker augmentation scenario, the word error rate (WER) and Cosine similarity are adopted as objective metrics to measure the VC and TTS systems. Figure 4 and Table 3 shows the quality of synthesized speech from different VC systems on in-set speakers' data. Each VC system generates 68000

**Table 3.** The WER[%] and cosine similarity for different system on the in-set experiment.

| Model | Cosine/Utt Num (average/all) | Utt Num (>0.6) | WER [%] |
|---|---|---|---|
| PPP-to-Mel(9utt) | 0.555/68000 | 26029 | 9.11 |
| Mel-to-Mel(9utt) | 0.510/68000 | 23100 | 10.28 |
| TTS (DiDi) | 0.475/10000 | 4263(>0.5) | – |

text-dependent utterances $(340 * 200 = 68000$, each speaker generate 200 utterances). Comparing with the Mel-to-Mel system, the PPP-to-Mel system's average speaker similarity is higher. Moreover, as shown in Table 3, the WER of the PPP-to-Mel VC system is less than Mel-to-Mel in retained utterances data. Therefore, the speech quality of the PPP-to-Mel system is higher in terms of these objective metrics.

### 3.3 Data Augmentation Based on the TTS System

We also train a one-hot multi-speaker TTS system to generate the augmented data. The system is based on Tacotron-2 [22] with GMMv2 [1] attention. For the multi-speaker modeling, a naive embedding-table based strategy is employed, where 128 dimensional embeddings learned through model optimization are concatenated to the encoder's output sequence, guiding the attention mechanism and the decoder with target speaker's information.

The model is trained from the DiDi-speech [8] database with 500 speakers. For each pair of target speaker and desired keyword, we synthesize 20 speech samples with identical voice and lexical content.

### 3.4 Speaker Augmentation Based on Speed Perturbation

We use speed perturbation based on the SoX *speed* function that modifies the pitch and tempo of speech by resampling. This strategy has been successfully used in the speech and speaker recognition tasks [13,25]. The limited text-dependent dataset is expanded by creating data created two versions of the original signal with speed factors of 0.9 and 1.1. The new classifier labels are generated at the same time since speech samples after pitch shift are considered to be from new speakers.

## 4    Experimental Results

Table 2 shows the results of different data augmentation strategies. The evaluation metrics are Equal Error Rate (EER) and minimum Detection Cost Function (mDCF) with $P_{target} = 0.1$. The baseline system employs the original limited text-dependent dataset (9 utts per speaker) to fine tune the pre-trained model. Since the size of in-set speaker dataset is too small, the system performance is degraded significantly. On the other hand, since the pitch shift AUG expand the number of speakers, the EER of the system has been improved by 10% relatively. The VC AUG with the PPP-to-Mel system also reduces the EER by relatively 20%. Moreover, it is observed that the system with both pitch shift AUG and VC AUG achieves the best performance. Experimental results show that, in the scenario of limited training data, the proposed method significantly reduces the EER from 6.51% to 4.48%, and the performance of the $mDCF_{0.1}$ also improves from 0.265 to 0.212.

Without the Pitch shift Aug, the VC $AUG_{in}$ (PPP-to-Mel) have the lower EER and $mDCF_{0.1}$ than TTS Aug under the less speakers. Since all the synthesized speech sentences have the similar tone regarding the TTS Aug system, the VC Aug is more suitable than TTS Aug in the text-dependent speaker verification task.

Furthermore, since the speech quality and speaker similarity of synthesized speech from the PPP-to-Mel VC system are better than the Mel-to-Mel VC system, a better result is achieved by using the PPP-to-Mel VC system for data augmentation. Nevertheless, the Mel-to-Mel VC system explores the direction of out-of-set unseen speaker augmentation and achieves some improvement. The results obtained show that the VC $Aug_{in}$ method is feasible, while the VC $Aug_{out}$ method still needs to be explored in the future.

## 5    Conclusion

This paper proposes two voice conversion based data augmentation methods to improve the performance of text-dependent speaker verification systems under the limited training data scenario. The results show that VC-AUG and pitch-shift strategy are feasible and effective. In the future works, we will further explore the methods and strategies for voice conversion based data augmentation with unseen or even artificiality created speakers.

**Acknowledgement.** This research is funded in part by the Synear and Wang-Cai donation lab at Duke Kunshan University. Many thanks for the computational resource provided by the Advanced Computing East China Sub-Center.

## References

1. Battenberg, E., et al.: Location-relative attention mechanisms for robust long-form speech synthesis. In: Proceedings of ICASSP, pp. 6194–6198 (2020)
2. Cai, W., Chen, J., Li, M.: Exploring the encoding layer and loss function in end-to-end speaker and language recognition system. In: Proceedings of Odyssey, pp. 74–81 (2018)
3. Cai, Z., Zhang, C., Li, M.: From speaker verification to multispeaker speech synthesis, deep transfer with feedback constraint. In: Proceedings of Interspeech (2020)
4. Chung, J.S., Nagrani, A., Zisserman, A.: Voxceleb2: deep speaker recognition. In: Proceedings of Interspeech (2018)

5. Das, R.K., et al.: Predictions of subjective ratings and spoofing assessments of voice conversion challenge 2020 submissions. In: Proceedings of Joint Workshop for the Blizzard Challenge and Voice Conversion Challenge 2020, pp. 99–120. https://doi.org/10.21437/VCC_BC.2020-15
6. Deng, J., Guo, J., Xue, N., Zafeiriou, S.: ArcFace: additive angular margin loss for deep face recognition. In: Proceedings of CVPR, pp. 4685–4694 (2019). https://doi.org/10.1109/CVPR.2019.00482
7. Du, J., Na, X., Liu, X., Bu, H.: AISHELL-2: transforming mandarin ASR research into industrial scale. arXiv:1808.10583 (2018)
8. Guo, T., et a.: Didispeech: a large scale mandarin speech corpus. arXiv:2010.09275 (2020)
9. He, K., Zhang, X., Ren, S., Sun, J.: Deep residual learning for image recognition. In: Proceedings of CVPR, pp. 770–778 (2016)
10. Heigold, G., Moreno, I., Bengio, S., Shazeer, N.: End-to-end text-dependent speaker verification. In: Proceedings of ICASSP, pp. 5115–5119 (2016). https://doi.org/10.1109/ICASSP.2016.7472652
11. Zheng, H., Cai, W., Zhou, T., Zhang, S., Li, M.: Text-independent voice conversion using deep neural network based phonetic level features. In: Proceedings of ICPR, pp. 2872–2877 (2016). https://doi.org/10.1109/ICPR.2016.7900072
12. Huang, Y., Chen, Y., Pelecanos, J., Wang, Q.: Synth2aug: cross-domain speaker recognition with tts synthesized speech. In: 2021 IEEE Spoken Language Technology Workshop (SLT), pp. 316–322 (2021). https://doi.org/10.1109/SLT48900.2021.9383525
13. Ko, T., Peddinti, V., Povey, D., Khudanpur, S.: Audio augmentation for speech recognition. In: Proceedings of Interspeech, pp. 3586–3589. (2015)
14. Kumar, K., et al.: MelGAN: generative adversarial networks for conditional waveform synthesis. Adv. Neural Inf. Process. Syst. **32**, 1–12 (2019)
15. Larcher, A., Lee, K.A., Ma, B., Li, H.: Text-dependent speaker verification: classifiers, databases and RSR2015. Speech Commun. **60** (2014). https://doi.org/10.1016/j.specom.2014.03.001
16. Nagrani, A., Chung, J.S., Zisserman, A.: Voxceleb: a large-scale speaker identification dataset. In: Proceedings of Interspeech, pp. 2616–2620 (2017)
17. Panayotov, V., Chen, G., Povey, D., Khudanpur, S.: Librispeech: an ASR corpus based on public domain audio books. In: Proceedings of ICASSP, pp. 5206–5210 (2015). https://doi.org/10.1109/ICASSP.2015.7178964
18. Povey, D., et al.: Semi-orthogonal low-rank matrix factorization for deep neural networks. In: Proceedings of Interspeech, pp. 3743–3747 (2018). https://doi.org/10.21437/Interspeech.2018-1417
19. Qin, X., Bu, H., Li, M.: HI-MIA: a far-field text-dependent speaker verification database and the baselines. In: Proceedings of ICASSP, pp. 7609–7613 (2020)
20. Qin, X., Cai, D., Li, M.: Far-field end-to-end text-dependent speaker verification based on mixed training data with transfer learning and enrollment data augmentation. In: Proceedings of Interspeech (2019)
21. Qin, X., et al.: The INTERSPEECH 2020 far-field speaker verification challenge. In: Proceedings of Interspeech, pp. 3456–3460 (2020). https://doi.org/10.21437/Interspeech.2020-1249
22. Shen, J., et al.: Natural TTS synthesis by conditioning wavenet on MEL spectrogram predictions. In: Proceedings of ICASSP, pp. 4779–4783 (2018)
23. Snyder, D., Garcia-Romero, D., Sell, G., Povey, D., Khudanpur, S.: x-vectors: robust DNN embeddings for speaker recognition. In: Proceedings of ICASSP, pp. 5329–5333 (2018)

24. Wan, L., Wang, Q., Papir, A., Moreno, I.L.: Generalized end-to-end loss for speaker verification. In: Proceedings of ICASSP, pp. 4879–4883 (2018). https://doi.org/10.1109/ICASSP.2018.8462665

25. Yamamoto, H., Lee, K.A., Okabe, K., Koshinaka, T.: Speaker augmentation and bandwidth extension for deep speaker embedding. In: Proceedings of Interspeech, pp. 406–410 (2019)

26. Yang, Y., Li, M.: The sequence-to-sequence system for the voice conversion challenge 2020. In: Proceedings of Interspeech 2020 satellite workshop on Spoken Language Interaction for Mobile Transportation System (2020)

27. Yi, Z., et al.: Voice conversion challenge 2020 - intra-lingual semi-parallel and cross-lingual voice conversion. In: Proceedings of Joint Workshop for the Blizzard Challenge and Voice Conversion Challenge 2020, pp. 80–98. https://doi.org/10.21437/VCC_BC.2020-14

28. Zhao, G., Ding, S., Gutierrez-Osuna, R.: Foreign accent conversion by synthesizing speech from phonetic posteriorgrams. In: Proceedings of Interspeech, pp. 2843–2847 (2019). https://doi.org/10.21437/Interspeech.2019-1778

29. Zhou, X., Ling, Z.H., King, S.: The blizzard challenge 2020. In: Proceedings of Joint Workshop for the Blizzard Challenge and Voice Conversion Challenge 2020, pp. 1–18 (2020). https://doi.org/10.21437/VCC_BC.2020-1

# Transformer-Based Potential Emotional Relation Mining Network for Emotion Recognition in Conversation

Yunwei Shi[1] and Xiao Sun[2,3(✉)]

[1] Anhui University, Hefei, Anhui, China
e20201097@stu.ahu.edu.cn
[2] Hefei University of Technology, Hefei, Anhui, China
sunx@hfut.edu.cn, sunx@iai.ustc.edu.cn
[3] Hefei Comprehensive National Science Center, Hefei, Anhui, China

**Abstract.** Emotion recognition in conversation (ERC) has attracted much attention due to its widespread applications in the field of human communication analysis. Compared with the vanilla sentiment analysis of the single utterance, the ERC task which aims to judge the emotion labels of utterances in the conversation requires modeling both the contextual information and the speaker dependency. However, previous models are limited in exploring the potential emotional relation of the utterances. To address the problem, we propose a novel transformer-based potential emotion relation mining network (TPERMN) to better explore the potential emotional relation and integrate the emotional clues. First, we utilize the global gated recurrence unit to extract the situation-level emotion vector. Then, different speaker GRU are assigned to different speakers to capture the intra-speaker dependency of the utterances and obtain the speaker-level emotion vector. Second, a potential relation mining transformer called PERformer is devised to extract the potential emotional relation and integrate emotional clues for the situation-level and speaker-level emotion vector. In PERformer, we combine graph attention and multi-head attention mechanism to explore the deep semantic information and potential emotional relation. And an emotion augment block is designed to enhance and complement the inherent characteristics. After multi-layer accumulation, the updated representation is obtained for emotion classification. Detailed experiments on two public ERC datasets demonstrate our model outperforms the state-of-the-art models.

**Keywords:** Emotion recognition in conversation · Transformer encoder · Natural language processing

## 1 Introduction

Emotion intelligence is an important part of human intelligence and plays an important role in our daily life [13]. Recently, with the continuous application of emotion recognition in conversation (ERC) in the fields of opinion mining, social

L. Zhenhua et al. (Eds.): NCMMSC 2022, CCIS 1765, pp. 238–251, 2023.
https://doi.org/10.1007/978-981-99-2401-1_22

media analysis, and dialogue systems, ERC task has attracted much attention from researchers and companies.

ERC task aims to identify the emotion of utterances in the conversation. Different from the vanilla sentiment analysis of single utterance, ERC task not only needs to model the context information but also needs to consider the speaker dependency [2]. Contextual information contributes to emotional judgment, which plays an important role in ERC tasks. In addition, due to the influence of the emotion dynamics [4], speaker dependency also needs to be considered.

Recent works can be divided into recurrence-based models and graph-based models. The recurrence-based models e.g., DialogueCRN [5], DialogueRNN [11], COSMIC [1] use recurrence neural networks to process the utterance features in sequence. However, the recurrence-based models are limited by their structure and can not learn the semantic information of utterances well. Some early recurrence-based models e.g., HiGRU [7], BIERU [9] do not consider the speaker dependencies, which also leads to the mediocre results. Graph-based models e.g., KET [18], DialogueGCN [2], DAG-ERC [14] can better reflect the struct of conversation and achieve the better results. However, these models are limited in exploring the potential emotional relation of utterances in the conversation.

In order to identify the emotion of utterances effectively, it is necessary to model the context and speaker information. Also, we can't ignore the role of the utterance's potential emotional connections. The emotions of other utterances have potential effects to the current utterance. These effects can be called "emotional clues", which can help the current utterance to make correct emotional judgments as clues.

As shown in Fig. 1, speaker A couldn't find his sandwich, so he asked B whether he saw his sandwich. It can be seen from $U_2$ that B ate the sandwich. $U_5$ shows that A did not want others to eat his sandwich. $U_7$ has the potential emotion relation with other utterances. It is difficult to judge the angry label only by $U_7$. By integrating the emotional clues and mining the potential relation of utterance can we make a correct judgment.

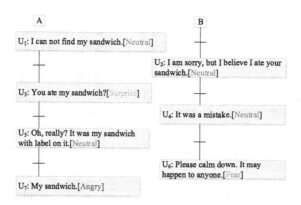

**Fig. 1.** An example of the conversation segment from the MELD dataset.

To address the problem, we propose a novel transformer-based potential emotion relation mining network (TPERMN) to better explore contextual information, speaker dependency, and potential emotion relation. The model is mainly composed of two modules. In the emotion extraction module, we first utilize the global gated recurrent unit (GRU) to capture the situation-level emotion vector. Afterward, aiming to capture the speaker dependency, the speaker GRUs are assigned to different speakers to interact the utterances spoken by the same speakers. Then, we restore the output of the speaker GRUs as the order of the conversation to get the speaker-level emotion vector. After this stage, we obtain the situation-level and the speaker-level emotion vector. In the potential relation mining module, we introduce the **P**otential **E**motional **R**elation mining transformer called PERformer to explore the potential relation of emotions and integrate the emotional clues. Here, we first utilize the graph attention mechanism and the multi-head attention to obtain semantic information of utterances and explore the potential relationship. Then the emotion augment block is designed to make the outputs of the two attention mechanisms complement each other and enhance their inherent characteristics.

The contributions of our work are as follows:

- We propose a novel emotion extraction module which contains the global GRU and speaker GRUs to extract the situation-level and speaker-level emotion vector respectively.
- We present the parallel potential emotional relation mining transformer (PERformer), namely, situation-level PERformer and speaker-level PERformer, which can better explore the potential emotional relation and integrate the emotional clues.
- Extensive experiments on the two public benchmarks show that our model achieves the state-of-the-art result.

## 2    Related Work

### 2.1    Emotion Recognition in Conversation

**Recurrence-Based Models.** bc-LSTM [12] uses content-dependent LSTM with an attention mechanism to capture contextual information. HIGRU [7] obtains contextual information through the attention mechanism and hierarchical GRU. ICON [3] adds a global GRU based on CMN [4] to improve the previous model. DialogueRNN [11] uses three GRU to model dialog dynamics and uses the attention mechanism to capture contextual information.

**Graph-Based Models.** KET [18] combines external commonsense and transformer to model the context through a multi-layer attention mechanism. CTNET [10] uses the transformer to model the intra- and cross-model interactions. DialogueGCN [2] connects the utterance by a certain size window and uses the relation graph attention network to aggregation information. DAG-ERC [14] models conversation by a directed acyclic graph.

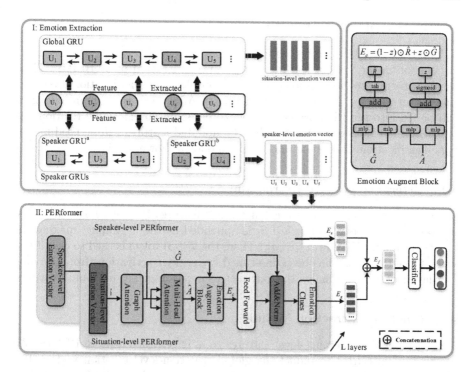

**Fig. 2.** The framework of our potential emotion relation mining network (TPERMN).

## 3    Task Definition

In ERC task, conversation is defined as $\{u_1, u_2, ..., u_N\}$, where $N$ denotes the utterance numbers. Each utterance $u_i = \{w_1, w_2, ..., w_n\}$ formed by $n$ words. And we define $U_\lambda$ to represent the utterances spoken by the speaker $\lambda$. $u_i^\lambda$ means that utterance $i$ is spoken by speaker $\lambda$. $\lambda \in S$, where $S$ is the speaker set. And each utterance has an emotion label $y_i \in E$, where $E$ is the emotion label set. Our task is to predict the emotion label $y_i$ of a given utterance $u_i$ in a conversation by the given context information and the speaker information.

## 4    Proposed Method

An ERC model requires advanced reasoning capability to understand the conversation context and integrate emotion clues which lead to accurate classification results.

We assume that the emotion of the utterances in conversation depends on three factors:

– The contextual information.
– The speaker dependency.

– The potential emotional relation of utterances.

Our TPERMN is modeled as follows:

We obtain the textual features of utterances extracted by RoBERTa. Then, we input them into the emotion extraction module to get the situation-level emotion vector and speaker-level emotion vector. Afterward, these representations will be fed into the PERformer to explore the potential emotional relation and integrate the emotional clues. Finally, the updated features will be used for the final classification. The framework of the model is shown in Fig. 2.

## 4.1  Utterance Feature Extraction

Following Ghosal et al. [1], we get utterance features by RoBERTa-Large. For utterance $u_i$, a special token $[CLS]$ is appended at the beginning, making the input a form of $[CLS], x_1, x_2, ..., x_n$. Then, the $[CLS]$ token is used to classify its lable. Once the model has been fine-tuned, we extract activations from the final four layers of $[CLS]$. Finally, the four vectors are averaged as utterance features.

## 4.2  Emotion Extraction Module

After getting the utterance-level features $u_i \in \mathbb{R}^d$ extracted by RoBERTa. We use the global $GRU^g$ to model the global situation-level emotion vector. The global situation-level emotion vector $G = \{g_1, ..., g_N\}$ can be formulated as follows:

$$g_i, h_i^g = \overrightarrow{GRU}^g(u_i, h_{i-1}^g), \tag{1}$$

where $g_i \in \mathbb{R}^{2d}$ is the situation-level emotion vector representation, $h_i^g \in \mathbb{R}^d$ is the $i$-th hidden state of the global GRU, $d$ is the hidden dimension of utterances.

We utilize different speaker GRUs to capture the intra-speaker dependency. For each speaker $\lambda \in \{a, b\}$, $GRU_\lambda^s$ is utilized to model the emotional inertia of speaker $\lambda$. The speaker-level emotion vector $s_i \in \mathbb{R}^{2d}$ is computed as follows:

$$s_i^\lambda, h_i^\lambda = \overrightarrow{GRU}_\lambda^s(u_i^\lambda, h_{i-1}^\lambda), \lambda \in \{a, b\}, \tag{2}$$

where $u_i^\lambda \in \mathbb{R}^d$ are the utterances spoken by speaker $\lambda$, $a$, $b$ are different speakers in conversation, $h_i^\lambda \in \mathbb{R}^d$ is the $i$-th hidden state of the speaker GRU.

After that, we restore the output of the speaker GRUs as the order of the conversation to get the speaker-level emotion representation $S = \{s_1, ..., s_N\}$, $s_i \in \mathbb{R}^{2d}$.

## 4.3  PERformer Module

After the emotion extraction module, we get the situation-level emotion vector $G = \{g_1, ..., g_N\}$, $g_i \in \mathbb{R}^{2d}$ and the speaker-level emotion vector $S = \{s_1, ..., s_N\}$, $s_i \in \mathbb{R}^{2d}$. Then, these representations are fed into the PERformer respectively. In later section, we will mainly use the situation-level emotion vector $G = \{g_1, ..., g_N\}$, $g_i \in \mathbb{R}^{2d}$ to illustrate the structure of PERformer.

**Graph Attention.** Graph attention Network (GAT) [15], leveraging self-attention to assign different importance to neighbor nodes. In our work, we utilize the graph attention mechanism to mining the deep semantic information of the emotion vector.

$$\alpha_{ij}^l = \frac{\exp(\mathcal{F}(g_i^l, g_j^l))}{\sum_{j \in \mathcal{N}_i} \exp(\mathcal{F}(g_i^l, g_j^l))}, \tag{3}$$

$$\mathcal{F}(g_i^l, g_j^l) = LeakyReLU\left(\mathbf{a}^\top [W_h g_i^l \| W_h g_j^l]\right), \tag{4}$$

$$\hat{g}_i^l = \sigma\left(\sum_{j \in \mathcal{N}_i} \alpha_{ij}^l W_h g_j^l\right), \tag{5}$$

where $l$ is the current layer, $\alpha_{ij}^l$ is the weight calcuate via an attention process, $\mathcal{N}_i$ is the neighbor nodes of $i$, we construct a fully-connect graph for the conversation and $j$ are the rest utterances. $LeakyReLU$ is an activation function, $W_h \in \mathbb{R}^{2d \times 2d'}$, $\mathbf{a} \in \mathbb{R}^{4d}$ are the trainable weight matrixs. Then, we get the updated representation in $l$ layer $\hat{G}^l = \{\hat{g}_1^l, \ldots, \hat{g}_N^l\}$, where $\hat{g}_i^l \in \mathbb{R}^{2d}$.

**Multi-head Attention.** We further adapt the multi-head attention mechanism to combine different attentions in parallel, enabling the model to jointly attend to important potential relations and emotional clues from various embeddings at different positions. The similarity score of different utterances can be seen as potential emotion relation. The process of aggregation can be seen as the process of integrating emotional clues. Specially, in the $l$-th layer, PERformer first transforms $\hat{G}^l$ into multiple subspaces with different linear transformations,

$$Q_h, K_h, V_h = \hat{G}^l W_h^Q, \hat{G}^l W_h^K, \hat{G}^l W_h^V \tag{6}$$

where $\hat{G}$ is projected to three matrixs: query matrix $Q_h$, key matrix $K_h$ of dimension $d_k$, and value matrix $V_h$, $W_h^Q, W_h^K, W_h^V$ are trainable parameters. Then, the self-attention is applied to calculate the potential relation,

$$\hat{A}_h^l = softmax\left(\frac{Q_h K_h^\top}{\sqrt{d_k}}, \dim{=}1\right) V_h, \tag{7}$$

where $\hat{A}_h^l$ is the output of $h$-th head in self-attention mechanism. Then, we concatenate the output of these attention heads to exlplore the potential relation in parallel,

$$\hat{A}^l = [\hat{A}_1^l; \hat{A}_2^l; \ldots; \hat{A}_m^l]W^O, \tag{8}$$

where $\hat{A}^l \in \mathbb{R}^{2d}$ is the final output of multi-head attention in $l$ layer, $m$ is the number of parallel heads, $W^O$ is a fully connected layer to connect $h$ heads.

**Emotion Augment Block.** We design an emotion augment block based on the gating mechanism to make comprehensive use of the attention mechanisms and enhance their inherent characteristics. The method has been shown effective in various past works [8]. The formulas are as follows:

$$\hat{R}^l = tanh(W_{mr}^l \hat{G}^l + W_{nr}^l \hat{A}^l + b_{mr}^l), \tag{9}$$

$$z^l = \sigma(W_{mz}^l \hat{G}^l + W_{nz}^l \hat{A}^l + b_{mz}^l), \tag{10}$$

$$E_c^l = (1 - z^l) \odot \hat{R}^l + z^l \odot \hat{G}^l, \tag{11}$$

where $l$ is the current layer, $z^l$ is the update gate, $\hat{R}^l$ is a candidate of updated representation, $\sigma$ is sigmoid activation, $tanh$ is a hyperbolic tangent activation, $\odot$ is the Hadamard product, $w_{mr}^l$, $w_{nr}^l$, $w_{mz}^l$, and $w_{nz}^l$ are trainable weight matrix, $b_{mz}^l$, $b_{mz}^l$ are trainable bias, $E_c^l \in \mathbb{R}^{2d}$ is the augmented emotion clues. Then, a feed-forward network is deployed to produce $E_g^l$ in $l$ layer,

$$F_g^l = max(0, W_{mf}^l E_c^l + b_{mf}^l)W_{nf}^l + b_{nf}^l, \tag{12}$$

$$E_g^l = LayerNorm(E_c^l + F_g^l), \tag{13}$$

where $W_{mf}^l, W_{nf}^l$ are two trainable matrixs, $b_{mf}^l$, $b_{nf}^l$ are trainable biases, and $E_g^l \in \mathbb{R}^{2d}$ is the output of situation-level PERformer in $l$ layer. Then, the output of the $l$-th layer will continue to be calculated as the input of $l+1$ layer.

### 4.4  Emotion Classifier

After stacking $L$ layers, we get the final integrated emotion clues. For the situation- and speaker-level representation $G$ and $S$, the above processes can be summarized as $E_g = PRRformer(G)$, $E_s = PRRformer(S)$ respectively, where $E_g$, $E_s \in \mathbb{R}^{N \times 2d}$. Then we connect them to get the finally representation $E_f = \{[E_f]_1, \ldots, [E_f]_N\}$, where $[E_f]_i = [[E_g]_i \oplus [E_s]_i] \in \mathbb{R}^{4d}$. Finally, we change the dimension to emotion classes and use the cross-entropy function to calculate loss. The formulas are as follow:

$$P_i = Softmax(W_e[E_f]_i + b_e), \tag{14}$$

$$y_i = Argmax(P_i[k]), \tag{15}$$

$$\mathcal{L}(\theta) = -\sum_{i=1}^{M} \sum_{j=1}^{c(i)} log P_{i,j}[y_{i,j}]. \tag{16}$$

where $M$ is the number of conversations, $c(i)$ is the number of utterances in $i$-th conversation, $P_{i,j}$ is the probability of the label of utterance $j$ in conversation $i$, $y_{i,j}$ is the label of utterance $j$ in conversation $i$, $\theta$ is the trainable parameters.

### 4.5  Datasets

## 5  Experiments

### 5.1  Datasets

We conduct detailed experiments on two public datasets. The specific information of the datasets is shown in Table 1.

**Table 1.** Data distribution of the two datasets.

| Dataset | # Conversations | | | # Uterrances | | |
|---------|-------|-----|------|-------|-----|------|
|         | Train | Val | Test | Train | Val | Test |
| IEMOCAP | 120   |     | 31   | 5810  |     | 1623 |
| MELD    | 1038  | 114 | 280  | 9989  | 1109 | 2610 |

**IEMOCAP:** The conversations in IEMOCAP comes from the performance based on script by two actors. There are in total six emotion classes *neutral, happiness, sadness, anger, frustrated,* and *excited.*

**MELD:** A multi-speaker dataset collected from *Friends* TV series. MELD has more than 1400 dialogues and 13000 utterances. There are seven emotion classes including *neutral, happiness, surprise, sadness, anger, disgust,* and *fear.*

### 5.2 Implementation Details

In our experiment, the dimension of utterance features is 1024, we set the hidden dimension $d$ to 100. The learning rate is set to $\{0.0001, 0.001\}$ and dropout is set to $\{0.1, 0.2\}$ for IEMOCAP and MELD dataset respectively. The batch size is set to $\{16, 64\}$ and $l_2$ is set to $\{0.001, 0.002\}$ respectively. As for the potential emotional relation mining transformer (PERformer), we set the layer of PERformer to 4, 5 respectively and we set the number of the multi-head attention to 4, 8 respectively. The epoch is set to 50 and each training and testing process runs on Tesla P100.

### 5.3 Evaluation Metrics

Following Ghosal et al. [1], we use the weighted-average F1 score as the evaluation metrics. Besides, following Majumder et al. [11], we use Acc. to calculate the classification accuracy.

### 5.4 Comparing Methods and Metrics

**bc-LSTM** [12] introduces the bidirectional LSTM and attention mechanism to model the contextual information.

**CMN** [4] models the utterances spoken by different speakers respectively, and uses the attention mechanism to fuse context information. ICON [3] adds a global GRU based on CMN to improve the previous model.

**DialogueRNN** [11] uses three RNN to model dialog dynamics which include the speaker, the context, and the emotion of the consecutive utterances and uses the attention mechanism to capture contextual information.

**AGHMN** [6] uses the BiGRU fusion layer to model the correlation of historical utterances. In addition, it uses the attention mechanism to update the internal state of GRU.

**CoGCN** [17] models utterance features and initialized speaker features as vertex features to construct graph structure.

**A-DMN** [16] models self and inter-speaker dependency respectively and combines them to update the memory.

**BIERU** [9] introduces a compact and fast emotional recurrent unit(ERU) to obtain contextual information.

**DialogueGCN** [2] connects utterances in a certain window size and uses the relation graph attention network to aggregation information.

**COSMIC** [1] proposes a new framework that introduces different commonsense.

**CTNET** [10] uses different transformers to model the intra- and inter-modal information and the speaker embedding mechanism is used to add speaker information.

**DAG-ERC** [14] models conversation by a directed acyclic graph(DAG) and uses the contextual information unit to enhance the information of historical context. DAG-ERC achieves superior performance as a strong baseline.

**Table 2.** Comparison of results of specific emotion categories on the IEMOCAP dataset.

| Methods | IEMOCAP | | | | | | | | | | | | | |
|---|---|---|---|---|---|---|---|---|---|---|---|---|---|---|
| | Happy | | Sad | | Neutral | | Angry | | Excited | | Frustrated | | W-Avg | |
| | Acc. | F1 | Acc. | F1 | Acc. | F1 | Acc. | F1 | Acc. | F1 | Acc. | F1 | Acc. | F1 |
| bc-LSTM | 37.5 | 43.4 | 67.7 | 69.8 | 64.2 | 55.8 | 61.9 | 61.8 | 51.8 | 59.3 | 61.3 | 60.2 | 59.2 | 59.1 |
| CMN | 20.1 | 28.7 | 62.9 | 68.3 | 56.0 | 57.4 | 58.8 | 60.4 | 68.2 | 66.7 | **74.3** | 63.2 | 60.7 | 59.8 |
| ICON | 22.2 | 29.9 | 58.8 | 64.6 | 62.8 | 57.4 | 64.7 | 63.0 | 58.9 | 63.4 | 67.2 | 60.8 | 59.1 | 58.5 |
| DialogueRNN | 25.7 | 33.2 | 75.1 | 78.8 | 58.6 | 59.2 | 64.7 | 65.3 | 80.3 | 71.9 | 61.2 | 58.9 | 63.4 | 62.8 |
| AGHMN | 48.3 | 52.1 | 68.3 | 73.3 | 61.6 | 58.4 | 57.5 | 61.9 | 68.1 | 69.7 | 67.1 | 62.3 | 63.5 | 63.5 |
| DialogueGCN | 40.6 | 42.8 | **89.1** | **84.5** | 61.9 | 63.5 | 67.5 | 64.2 | 65.5 | 63.1 | 64.2 | **67.0** | 65.3 | 64.2 |
| A-DMN | 43.1 | 50.6 | 69.4 | 76.8 | 63.0 | 62.9 | 63.5 | 56.5 | **88.3** | 77.9 | 53.3 | 55.7 | 64.6 | 64.3 |
| BiERU | 54.2 | 31.5 | 80.6 | 84.2 | 64.7 | 60.2 | 67.9 | 65.7 | 62.8 | 74.1 | 61.9 | 61.3 | 66.1 | 64.7 |
| COSMIC | – | – | – | – | – | – | – | – | – | – | – | – | – | 65.3 |
| CTNET | 47.9 | 51.3 | 78.0 | 79.9 | 69.0 | 65.8 | **72.9** | **67.2** | 85.3 | **78.7** | 52.2 | 58.8 | 68.0 | 67.5 |
| DAG-ERC | – | – | – | – | – | – | – | – | – | – | – | – | – | 68.0 |
| Ours | **56.9** | **57.1** | 79.2 | 81.3 | **70.3** | **69.4** | 71.2 | 64.7 | 76.6 | 75.6 | 61.2 | 64.1 | **69.56** | **69.59** |

**Table 3.** Comparison of results of specific emotion categories on the MELD dataset.

| Methods | MELD | | | | | | | |
|---|---|---|---|---|---|---|---|---|
| | Anger | Disgust | Fear | Joy | Neutral | Sadness | Surprise | W-Avg F1 |
| CMN | 44.7 | 0.0 | 0.0 | 44.7 | 74.3 | 23.4 | 47.2 | 55.5 |
| bc-LSTM | 43.4 | **23.7** | 9.4 | 54.5 | 76.7 | 24.3 | 51.0 | 59.3 |
| ICON | 44.8 | 0.0 | 0.0 | 50.2 | 73.6 | 23.2 | 50.0 | 56.3 |
| DialogueRNN | 43.7 | 7.9 | 11.7 | 54.4 | 77.4 | 34.6 | 52.5 | 60.3 |
| CoGCN | 46.8 | 10.6 | 8.7 | 53.1 | 76.7 | 28.5 | 50.3 | 59.4 |
| A-DMN | 41.0 | 3.5 | 8.6 | 57.4 | 78.9 | 24.9 | 55.4 | 60.5 |
| DAG-ERC | – | – | – | – | – | – | – | 63.6 |
| COSMIC | – | – | – | – | – | – | – | 65.2 |
| Ours | **53.4** | 16.1 | **20.3** | **65.3** | **79.2** | **42.4** | **59.5** | **65.8** |

## 5.5    Compared with the State-of-the-art Method

In this section, we will compare our model with other benchmark models. The results are shown in Table 2 and 3.

**IEMOCAP:** As shown in Table 2, our model achieves the best results on the W-Avg Acc. and W-Avg F1 score. As for the performance on specific emotions, our model is relatively balanced. On *happy* and *neutral*, our model achieves the best performance. DAG-ERC is the start-of-the-art model which combines the advantage of recurrence-based and graph-based models by using the directed acyclic graph. Compared with DAG-ERC, our model improves by 1.59% on the W-Avg F1 score. CTNET uses the transformer encoder to model the information of different modals. Compared with CTNET, our model makes improvements on the vanilla transformer encoder to mine the potential emotional relation and integrate emotional clues. Besides, our model uses different GRUs to model the speaker dependency is more reasonable. Thus, our model improves by 1.56%. 2.09% on Acc. and F1 respectively.

**MELD:** As shown in Table 3, our model achieves the best result of 65.8% on the W-Avg F1. COSMIC is a strong baseline that introduces extra common-sense knowledge to improve the results, and the structure of COSMIC is similar to DialogueRNN. Compared with COSMIC, our model improves by 0.6% on the F1 score. As for the specific emotional classification, our model performs better compared to the baseline model. On the emotion of *disgust* and *fear*, most models perform poorly or even can not distinguish. But our model achieves 16.1% and 20.3% on the F1 score respectively, which shows the excellent ability to understand the conversation context. Compared with A-DMN, on the emotion of *anger, joy, sadness,* and *surprise*, our model improves by 12.4%, 7.9%, 17.5%, and 4.1% respectively.

## 5.6    Ablation Study

In this section, we will analyze the results of the ablation study on both IEMO-CAP and MELD datasets as shown in Table 4 and 5. There are two main mod-

**Table 4.** The ablation study of our model.

| Emotion Extraction | | PERformer | | IEMOCAP | MELD |
|---|---|---|---|---|---|
| situation-level — | speaker-level | situation-level — | speaker-level | W-Avg F1 | |
| ✓ | ✓ | ✓ | ✓ | 69.59 | 65.82 |
| ✓ | ✗ | ✓ | ✗ | 65.76 | 65.09 |
| ✗ | ✓ | ✗ | ✓ | 68.21 | 64.55 |
| ✓ | ✓ | ✗ | ✗ | 65.70 | 64.97 |
| ✓ | ✓ | ✓ | ✗ | 66.57 | 65.14 |
| ✓ | ✓ | ✗ | ✓ | 68.87 | 65.22 |
| ✗ | ✗ | ✗ | ✗ | 54.55 | 62.02 |

Table 5. The ablation study of PERformer.

| Method | IEMOCAP | MELD |
|---|---|---|
| w/o emotion augment blcok | 68.97 | 65.57 |
| w/o graph attention layer | 68.05 | 65.49 |
| w/o graph attention layer and emotion augment blcok | 67.87 | 65.24 |
| **OUR** | **69.59** | **65.82** |

ules in our paper: the emotion extraction module (EE) and the potential relation mining transformer (PRMformer). Both modules are composed of speaker-level and situation-level units.

**The Effort of Different Level Modules:** As shown in the first three rows in Table 4, when the speaker-level emotion extraction modele (speaker GRUs) and speaker-level PERformer are removed simultaneously, the results decrease by 3.83% and 0.73% respectively. And when the situation-level emotion extraction modele (global GRU) and situation-level PERformer are removed simultaneously, the results decrease by 1.38% and 1.27% on the IEMOCAP and MELD datasets respectively. This result shows that both the situation-level and speaker-level modules have unique functions, and they can complement each other's missing information to make the final result better. The best results are obtained when they are used simultaneously. In addition, the speaker-level modules is more important for the IEMOCAP dataset, and the situation-level modules is more important for the MELD dataset. The results are caused by the IEMOCAP is a two-speaker dataset and the MELD is a multi-speaker dataset. In the emotion extraction module, our model only divided the utterances spoken by two different speakers. In addition, the utterances in MELD are short and highly dependent on the context information.

**The Effort of PERformer:** As shown in the fourth row of Table 4, when the PERformer is removed, the results are sharply dropped by 3.89%, 0.85%. These are caused by our PERformer can better mine the potential emotion relation and integrate the emotional clues. When the speaker-level and situation-level PERformer are removed respectively, the results also decline to a certain extent.

As shown in Table 5, we will analyze the effort of the internal components of the PERformer. When the graph attention mechanism is removed, the F1 decreases by 1.54% and 0.33%, respectively on the IEMOCAP and MELD datasets. And when the emotion augment block is removed, the F1 decreases by 0.62% and 0.25%. When two components are used at the same time, our model achieves the best results.

## 5.7   Analysis on Parameters

In this section, we will analyze the multi-head number $M$ and the PERformer layer $L$. The results are shown in Fig. 3. We choose $L$ from $\{1,...,8\}$ and $M$ from $\{1,...,12\}$, respectively.

As shown in Fig. 3(a) and 3(b). On IEMOCAP, when $L$ increases from 1 to 4, F1 always increases. When $L = 4$, the best result is obtained for 69.59%. However, F1 score declines after $L = 4$. On MELD, the best result is achieved when $L=5$. The phenomenons show that increasing the number of layers will improve the experimental results, but too deep layers will extract more useless potential relations.

As shown in Fig. 3(c) and 3(d). The increase in the number of heads $M$ will improve the ability of our model to mine the potential emotional relation and integrate emotional clues. On the IEMOCAP dataset, when $M = 4$, our model achieves the best result. After this point, the F1 begins to decrease. On MELD, when $M = 8$, our model achieves the best result. The results indicate that too many attention heads will not improve the experimental results, but will learn redundant information.

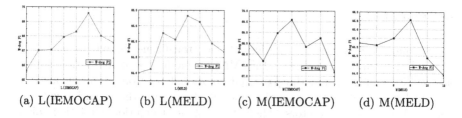

(a) L(IEMOCAP)    (b) L(MELD)    (c) M(IEMOCAP)    (d) M(MELD)

**Fig. 3.** Parameter analysis of the module layers ($L$) and the multi-head number ($M$).

## 5.8   Error Study

In this section, we will analyze the misclassification through the confusion matrix. The columns and rows on confusion matrix indicate the true and predicted labels, respectively.

As shown in Fig. 4, our model has a good performance in all kinds of emotion discrimination and the accuracy is high in most emotion categories. For example, our model correctly judges 194 *sad* emotion with a total of 245 on the IEMOCAP dataset, and the accuracy rate of our model is 79.18%. On the emotion of *anger*, our model correctly judges 121 *anger* emotion with a total of 170 on the IEMOCAP dataset, and the accuracy rate of our model is 71.17%. However, on the IEMOCAP dataset, our model is insufficient in distinguishing similar emotions. *happy* and *excited* are similar emotions and our model is easy to misjudge them. In addition, *frustrated* is easy to be misjudged as *natural* and *angry*.

On the MELD dataset, there are 1256 shapes of *neutral* labels in the test set. Because of this uneven distribution, our model will misjudge other emotions as *natural*. As can be seen from Fig. 4, the number of other emotions misjudged as *neutral* is the largest. For the *fear* and *disgust* labels, there are only 50 and 68 in the test set, respectively. The accuracy of a judgment is not high due to too few samples.

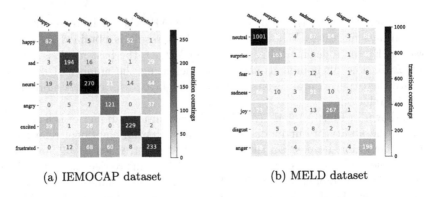

(a) IEMOCAP dataset                    (b) MELD dataset

**Fig. 4.** Confusion matrix on the IEMOCAP and MELD dataset.

## 6  Conclusion

In this paper, we propose a model entitled transformer-based potential emotion relation mining network (TPERMN) to model the conversation context, speaker information, and the potential emotional relation simultaneously. We first utilize the emotion extraction module to obtain the situation-level and speaker-level emotion vectors. Then, the potential emotional relation mining transformer (PERformer) is introduced to explore the potential relationship and integrate emotional clues. In addition, we propose two parallel PERformers called situation-level PERformer and speaker-level PERformer to model these representations respectively. The method complements each other to improve the performance of the final emotional clues. Empirical results on two ERC datasets show that our model is effective in solving the ERC task. However, our model has shortcomings in similar emotional judgment. Our feature work will focus on the application of multi-modals features.

**Acknowledgment.** This research is supported by the Major Project of Anhui Province under Grant (No.202203a05020011).

## References

1. Ghosal, D., Majumder, N., Gelbukh, A., Mihalcea, R., Poria, S.: Cosmic: commonsense knowledge for emotion identification in conversations. In: EMNLP, pp. 2470–2481 (2020)
2. Ghosal, N., Poria, N., Gelbukh, A.: Dialoguegcn: a graph convolutional neural network for emotion recognition in conversation. In: EMNLP, pp. 154–164 (2019)
3. Hazarika, D., Poria, R., Cambria, E., Zimmermann, R.: Icon: interactive conversational memory network for multimodal emotion detection. In: EMNLP, pp. 2594–2604 (2018)
4. Hazarika, D., Poria, S., Zadeh, A., Cambria, E., Morency, L.P., Zimmermann, R.: Conversational memory network for emotion recognition in dyadic dialogue videos. In: NAACL, pp. 2122–2132 (2018)

5. Hu, D., Wei, L., Huai, X.: Dialoguecrn: contextual reasoning networks for emotion recognition in conversations. arXiv preprint arXiv:2106.01978 (2021)
6. Jiao, W., Lyu, M.R., King, I.: Real-time emotion recognition via attention gated hierarchical memory network. In: AAAI, pp. 8002–8009 (2020)
7. Jiao, W., Yang, H., King, I., Lyu, M.R.: Higru: hierarchical gated recurrent units for utterance-level emotion recognition. In: NAACL, pp. 397–406 (2019)
8. Lei, J., Wang, L., Shen, Y., Yu, D., Berg, T.L., Bansal, M.: MART: memory-augmented recurrent transformer for coherent video paragraph captioning. In: ACL, pp. 2603–2614 (2020)
9. Li, W., Shao, W., Ji, S., Cambria, E.: Bieru: bidirectional emotional recurrent unit for conversational sentiment analysis. Neurocomputing **467**, 73–82 (2022)
10. Lian, Z., Liu, B., Tao, J.: Ctnet: conversational transformer network for emotion recognition. IEEE ACM Trans. Audio Speech Lang. Process. **28**, 985–1000 (2021)
11. Majumder, N., Poria, S., Hazarika, D., Mihalcea, R., Gelbukh, A., Cambria, E.: Dialoguernn: an attentive rnn for emotion detection in conversations. In: AAAI, pp. 6818–6825 (2019)
12. Poria, S., Cambria, E., Hazarika, D., Majumder, N., Zadeh, A., Morency, L.P.: Context-dependent sentiment analysis in user-generated videos. In: ACL, pp. 873–883 (2017)
13. Poria, S., Majumder, N., Mihalcea, R., Hovy, E.: Emotion recognition in conversation: research challenges, datasets, and recent advances. IEEE Access **7**, 100943–100953 (2019)
14. Shen, W., Wu, S., Yang, Y., Quan, X.: Directed acyclic graph network for conversational emotion recognition. In: ACL/IJCNLP, pp. 1551–1560 (2021)
15. Velickovic, P., Cucurull, G., Casanova, A., Romero, A., Liò, P., Bengio, Y.: Graph attention networks. In: ICLR (2017)
16. Xing, S., Mai, S., Hu, H.: Adapted dynamic memory network for emotion recognition in conversation. IEEE Trans. Affect. Comput. **13**, 1426–1439 (2020)
17. Zhang, D., Wu, L., Sun, C., Zhou, G.: Modeling both context- and speaker-sensitive dependence for emotion detection in multi-speaker conversations. In: IJCAI, pp. 5415–5421 (2019)
18. Zhong, P., Wang, D., Miao, C.: Knowledge-enriched transformer for emotion detection in textual conversations. In: EMNLP-IJCNLP, pp. 165–176 (2019)

# FastFoley: Non-autoregressive Foley Sound Generation Based on Visual Semantics

Sipan Li[1], Luwen Zhang[1], Chenyu Dong[1], Haiwei Xue[1],
Zhiyong Wu[1,2(✉)], Lifa Sun[3], Kun Li[3], and Helen Meng[1,2]

[1] Tsinghua-CUHK Joint Research Center for Media Sciences, Technologies and
Systems, Shenzhen International Graduate School, Tsinghua University,
Shenzhen, China
{lsp20,zlw20,dcy20,xhw22}@mails.tsinghua.edu.cn
[2] Department of Systems Engineering and Engineering Management, The Chinese
University of Hong Kong, Hong Kong, Hong Kong SAR, China
{zywu,hmmeng}@se.cuhk.edu.hk
[3] SpeechX Ltd., Shenzhen, China
{lfsun,kli}@speechx.cn

**Abstract.** Foley sound in movies and TV episodes is of great importance to bring a more realistic feeling to the audience. Traditionally, foley artists need to create the foley sound synchronous with the content occurring in the video using their expertise. However, it is quite laborious and time consuming. In this paper, we present FastFoley, a Transformer based non-autoregressive deep-learning method that can be used to synthesize a foley audio track from the silent video clip. Existing cross-model generation methods are still based on autoregressive models such as long short-term memory (LSTM) recurrent neural network. Our FastFoley offers a new non-autoregressive framework on the audio-visual task. Upon videos provided, FastFoley can synthesize associated audio files, which outperforms the LSTM based methods in time synchronization, sound quality, and sense of reality. Particularly, we have also created a dataset called Audio-Visual Foley Dataset(AVFD) for related foley work and make it open-source, which can be downloaded at https://github.com/thuhcsi/icassp2022-FastFoley.

**Keywords:** Foley · Cross-Modal · Audio-Visual · Transformer · Video Sound Generation

## 1 Introduction

Sound is the earliest way for us to contact this world. Heartbeat, breathing, knocking, running water and footsteps..., it can be said that sound is the first step for us to communicate with this world. It extends from information to emotion and supports the commonality of people's psychology and senses.

S. Li and L. Zhang—Equal contributions.

© The Author(s), under exclusive license to Springer Nature Singapore Pte Ltd. 2023
L. Zhenhua et al. (Eds.): NCMMSC 2022, CCIS 1765, pp. 252–263, 2023.
https://doi.org/10.1007/978-981-99-2401-1_23

Sound effects mean the sound added to the vocal cords to enhance the realism or atmosphere of a scene. As sound is passively accepted by users, it is easier to use sound effects to convey information than visual communication. Foley is a kind of sound effect, which is usually created by foley designers using props in a noiseless environment. For digital media applications such as movies, games, and children's picture books, adding foley sound corresponding to the video content on top of the original vocals and background music greatly enhance the audience's immersive experience [2]. But the design of foley relies heavily on the personal understanding and creativity of foley designers themselves. The foley sound creation is extremely restricted by the environment and inspiration, and also very time-consuming and laborious. Now, some teams are trying to automate this process, e.g. AutoFoley [8]. However, the existing methods are mostly based on autoregressive models, which need a long time to train and are not robust enough. Non-autoregressive Transformer can offer a better ability to fit the visual features of the front-end and the audio features of the back-end, which can avoid over-fitting that often occurs in LSTM.

Generating foley sound from videos is a task of cross-modal generation. There are already a lot of works on audio-visual tasks, such as audio-visual representation [5,18]. Synthesizing audio for video has recently attracted considerable attention. Recent works have explored the relationship between auditory and visual modalities, such as Visual2Sound collected an unconstraint dataset (VEGAS) that includes 10 types of sounds recorded in the wild and proposed a SampleRNN-based method to directly generate a waveform from a video [30]; Audeo [22] and FoleyMusic [6] focus on generating music from the image information including musicians playing instruments; Listen to the Image [14] aims to assist the blind in sensing the visual environment better by translating the visual information into a sound pattern, and proposed a new method to evaluate the sensing ability by machine, not human; REGNET [3] focuses on learning how to only keep the sound of the object in the video contents; AutoFoley [8] proposed two different streams of networks for synthesizing foley sound, one is traditional CNN and LSTM, the other is based on Temporal-Relation-Network (TRN); FoleyGAN [9] is the follow-up work of AutoFoley, which introduced GAN [10] to generate more realistic sound; CMCGAN [12] offered a novel model based on CycleGAN, as a uniform framework for the cross-modal visual-audio mutual generation. Similar to AutoFoley, we consider generating foley sound from videos.

In this paper, a Transformer-based non-autoregressive sound generation network is proposed as an automatic tool for foley sound synthesis, named Fast-Foley. The FastFoley task is similar to text-to-speech (TTS) that aims to synthesize acoustic speech from text. In our task, we consider visual information as the input to generate various kinds of foley sound that is synchronized with the motions in the input silent video. Different from the existing cross-model generation methods where autoregressive models such as LSTM are used, our proposed method offers a new non-autoregressive framework on the auto-visual task. Experimental results demonstrate that our proposed method outperforms the LSTM-based methods in time synchronization, sound quality and sense of reality. In addition, we also created a dataset called Audio-Visual Foley Dataset (AVFD) for the related foley work and make it open-source for public access.

## 2    Method

In this section, we first introduce the overview of our method and then describe the sound generation pipeline in the following subsections. As shown in Fig. 1, the proposed framework consists of three major parts: 1) Audio and visual features extraction. 2) Acoustic model which transforms the visual features to the corresponding spectrograms. 3) Sound generation based on converting the spectrograms to the waveforms.

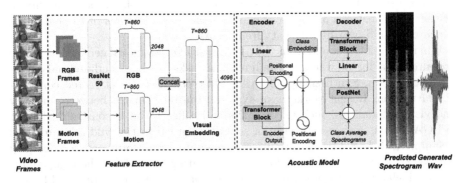

**Fig. 1.** The structure of proposed FastFoley network.

### 2.1    Audio and Visual Features Extraction

The original video clips and audios are all 5 s. The Short-Time Fourier Transform (STFT) [11] is used to get the spectrograms from audio files as the audio features. The sampling rate of the audio files is 44.1 kHz. When computing STFT, Hanning window is used, with window size 1024 and window shift 256. Therefore the dimensions of the spectrogram are 513 * 860.

Before moving towards the visual feature extraction step, aligning the video frames with the audio feature frames by interpolating the original video is of great importance to make the sound and video frames synchronous. The original videos are with the frame rate of 25 or 30 FPS (frames per second). To match with the 860 frames with 5 s of the audio features, the videos have been adjusted from their inherent FPS to 172 FPS after interpolation by using m-interpolation filter in the FFmpeg [24] package that can generate intermediate frames instead of simple duplication.

The information of a video not only includes the texture and color of the objects but also the motion information that can be extracted from consecutive video frames. The motion information is quite important to model the synchronicity between video and audio frames. Previous work [8] has shown that ResNet-50 is better than VGG-19, hence we extract the motion information as well as the texture and color information separately using a pre-trained ResNet-50 [13] and concatenate them together as the visual features. For extraction the motion information, the current, previous and next video frames are firstly gray-scaled and input to ResNet-50 as three channels. By doing so, the motion information can be represented better because at each step it contains the consecutive motion, contributing to more continuous visual features. Meanwhile, we input the current frame with RGB channels to ResNet-50 to extract the texture and color information.

## 2.2   Acoustic Model

Basically, the proposed model is similar to the acoustic model in TTS. Previous neural TTS models such as Tacotron [26] first generate Mel-spectrograms autoregressively from text and then synthesize speech from the generated Mel-spectrograms using a separately trained vocoder. As for foley sound generation, [3,8] both take an autoregressive model as the acoustic model, [3] which makes it a model looks like Tacotron.

As shown in Fig. 2, the proposed acoustic model is based on a non-autoregressive model with Feed-Forward Transformer blocks [25], which can parallelize the process to generate spectrograms with extremely fast speed. As Fig. 2(a) shows, Transformer blocks are based on the attention mechanism, with which long-range dependencies of the input visual features can be learnt that contribute to the generation of foley sound. The sequential information between different visual features is also crucial for modeling the synchronicity between foley sound and visual input. To help the Transformer blocks distinguish different positions of the visual features, positional embeddings [25] are added to the encoded visual features.

In TTS, the length regulator [19] is essential so that the linguistic features can be aligned with the acoustic features. However, in the proposed model there is no longer a need for the length regulator as video frames and audio frames have already been aligned. The linear layer in encoder is used to transform the dimension of visual features to the dimension of encoder hidden. Additionally, class embedding [4] is added to the encoder output for the support of training with multiple sound classes, to ensure the matching between feature extraction and acoustic model. The average spectrograms of respective classes are viewed as a conditional input to the decoder output so that the acoustic model only predicts the residual between the average spectrograms and the ground-truth. The class embedding is a look-up table consisting of the class label of datasets. The dimention of class embedding is the number of classes. We expand the dimension of class embedding to the dimension of encoder output and then add them up. The average spectrograms are obtained by calculating the average value of each class's spectrogram of the corresponding audio files. The linear layer in decoder is used to transform the dimension of decoder hidden to the dimension of spectrogram. As is shown in Fig. 2(b), the PostNet contains 5 1D convlutional layers and BatchNorm layers. The input of PostNet is added to the output via residual connections. We reduce the difference between the ground-truth and the predicted sound feature for every time step by calculating the Smooth L1 Loss, which is found better than L1 Loss and L2 loss during experiments.

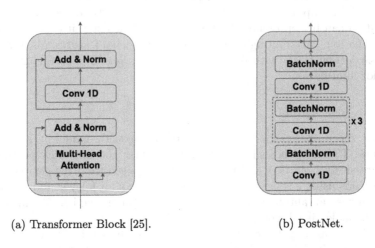

(a) Transformer Block [25].    (b) PostNet.

**Fig. 2.** Transformer Block and PostNet structure.

For sound generation from predicted spectrograms, the Inverse Short Time Fourier Transform (ISTFT) [11] method with Hanning window is performed because of its less computational complexity than neural-based vocoder. For phase reconstruction from the spectrogram, the iterative Griffin-Lim algorithm [11] is adopted when performing ISTFT.

## 3   Dataset

The goal of this work is to generate matched and synchronized sound effects according to video content, so the content of audio must be highly matched with the action in the video. Towards this, we found some datasets related to motion recognition, video understanding and audio-visual synchronization tasks, such as HMDB [16], UCF101 [21], VEGAS [30]. However, most of these datasets pay too much attention to video content but neglect the importance of sound. Elements such as human voice, background music, noise, and even the absence of sound have a great impact. We also found some datasets in the film field. For example, MovieNet [15] is a large-scale dataset for comprehensive movie understanding. But it is too artistic and talks more about the hierarchical and content structure of the film itself. We consider adding the artistic quality evaluation in the future, but it is not suitable for our current work.

In order to complete this foley task, we construct the Audio-Visual Foley Dataset (AVFD)[1] derived from AudioSet [7], UCF101 [21] and VEGAS [30].

AudioSet is a dataset that consists of 2,084,320 human-labeled 10-second sound clips drawn from YouTube videos. Although these video files own corresponding natural sound, AudioSet isn't absolutely ideal for our task because many videos and audios are loosely related [30]. So we screened out some suitable classes manually for further cleaning, such as JackHammer, Sawing, and so on. According to AudioSet ontology's JSON file, we know the id of each class. Consequently, we could obtain the YouTube URL of all videos belonging to this class. Then, we download all of them by YouTube-dl [1] and select the available video manually. As the downloaded video is the complete video in YouTube, the digital media editor uses FFmpeg [24] and Adobe Premiere Pro to extract the exact video clips according to the timestamp tag provided by AudioSet. In addition, we noticed that part of videos in UCF101 and VEGAS also meet the requirements (little noise, no vocals, no background music). So we have also screened out the available video from these datasets and integrated them together with the videos from AudioSet to construct the AVFD.

**Fig. 3.** The Audio-Visual Foley Dataset (AVFD) Video Classes.

Altogether, our Audio-Visual Foley Dataset(AVFD) contains a total of 5,565 videos from 12 different classes, as is shown in Fig. 3. 5 sound classes namely Hammer, BoxingSpeedBag, Gun, TableTennisShot, and Typing are selected from UCF101 and VEGAS. The other classes are obtained from AudioSet, namely Hammer (merged with the same Hammer class in UCF101), Chopping, Clapping, Footsteps, Powertool, JackHammer, Waterfall and Sawing. There are 464 videos

---

[1] https://github.com/thuhcsi/icassp2022-FastFoley.

in each class on average, and the duration of each video is within 10 s. We make this Audio-Visual Foley Dataset (AVFD) open-source for public access.

# 4 Experiments and Results

## 4.1 Implementation Details

Firstly we dynamically extract gray-scale feature maps and RGB feature maps and then feed them to ResNet-50 to extract visual features, which is used for training the acoustic model. In this way, the whole framework is in end-to-end mode. However, we found that dynamically extracting the feature maps leads to a fairly slow training process. Hence, we separate the training process into two stages: 1) Extract the visual features with ResNet-50 and compute the audio features, and save them as NumPy arrays. 2) Training the acoustic model with prepared audio and visual features.

At first stage, we compute audio features using librosa [17], an audio processing package from Python. The configurations are mentioned in Sect. 2.1. For every sound class, we compute an average spectrogram. For video interpolation, we use the m-interpolation filter in the FFmpeg video editing tool [24]. To get visual features, we separately take the current, previous frame and next frames as the three-channel input of ResNet-50 to get a 2048-d vector containing motion information, and use the RGB frame in the same way to get a 2048-d vector with texture and color information at each time step. The final visual embeddings are derived by concatenating the two vectors, leading to a 4096-d vector.

At second stage, the acoustic model is trained, in which the encoder and decoder are both made of one Feed-Forward Transformer block. For each Transformer block, the hidden dimension is 512, and the heads of multi-head attention are set as 2. The category embedding is a look-up table, which introduces the class information of data. We add it to the encoder output to support multi-class training.

The dataset is divided into 80% for training, 10% for validation to make sure the training process is not over-fitting, and the last 10% for testing. For training, we use minibatch gradient descent with the Adam optimizer. The minibatch size and learning rate are 16 and 0.001 respectively. Warm-up strategy is urse for the first thousand steps, and gradient clipping is used to avoid exploding. The model has been trained 3000 epochs, using an NVIDIA RTX 2080 Ti GPU.

## 4.2 Experiments and Results

**Baseline.** AutoFoley [8] has achieved good effects in foley sound generation, therefore we consider it as our baseline. However, the implementation and synthesized demos of AutoFoley have not been released, so we reproduce the AutoFoley framework with CNN and LSTM as the baseline and use the generated results for performance comparisons.

**Subjective Evaluation.** As the task is to generate the audio corresponding to the video contents, making the sound more realistic and more synchronous with the video is essential. We choose 16 pairs of videos with audio tracks synthesized by the proposed model and baseline and then shuffle the videos before conducting the experiments. The experiments are ABX tests, which contain four standards for selection: 1) Select the more realistic sample; 2) Select the sample with better quality; 3) Select the most synchronized sample; 4) Select the sample the participant prefers more in overall. The experiments are conducted by 33 participants. The results are shown in Table 1, where *NP* means no preference between the two methods. The average preference rates of the FastFoley model for all metrics are higher than the sum of baseline and *NP*, which explicitly demonstrates the proposed system receives a better preference rate in all standards.

**Objective Evaluation.** As the task of sound generation is similar to speech synthesis, we draw on some metrics that evaluate speech quality to objectively evaluate our generation results, such as MS-STFT-Loss (Multi-resolution-STFT-Loss) [28], SSIM (Structure Similarity Index Measure) [27], STOI (Short-Time Objective Intelligibility) [23], PESQ (Perceptual evaluation of speech quality) [20]. MS-STFT-Loss is proposed to apply the needs of waveform generation, a single STFT-Loss is defined as:

$$L_s(G) = E_{z \sim p(z), x \sim p_{data}}[L_{sc}(x, \hat{x}) + L_{mag}(x, \hat{x})] \qquad (1)$$

where $\hat{x}$ represents the generated sample, also known as $G(z)$, $L_{sc}$ represents spectral convergence loss, and $L_{mag}$ represents log STFT magnitude loss, respectively, as defined below:

$$L_{sc}(x, \hat{x}) = \frac{\| \ |STFT(x)| \ - \ |STFT(\hat{x})| \ \|_F}{\| \ \|STFT(x)\| \ \|_F} \qquad (2)$$

$$L_{mag}(x, \hat{x}) = \frac{1}{N} \| \ log|STFT(x)| - log|STFT(\hat{x})| \ \|_1 \qquad (3)$$

Multi-resolution STFT loss is multiple single-short-time Fourier transform losses using different analytical parameters (such as Fourier transform size FFT size, window size, frame shift). Combining multiple short-term Fourier transform losses under different analytical parameters shows that the proposed method is 0.1944, better than the baseline, which is 0.2961.

The SSIM(Structure Similarity Index Measure) metric measures the degree of distortion of a picture, as well as the similarity of two pictures, which is often used in the image generation task. Therefore, we extracted the spectrogram of the generated audios by the proposed method and baseline, and then extracted the ground truth spectrogram for experimentation. The result shows that the proposed method has gotten 0.6742 while the baseline has gotten 0.5373.

The STOI and PESQ are often used as metrics to measure the quality of synthesized speech, which actually need the synthesized speech contain the semantic information. Short-Time Objective Intelligibility (STOI) is an important indicator of speech comprehensibility. For a word in the speech signal, only can be understood and can not be understood two cases, from this point of view can be considered to be diverse, so the value range of STOI is quantified in 0 to 1, representing the percentage of words that are correctly understood, and the value of 1 means that the speech can be fully understood. Perceptual Evaluation of Speech Quality (PESQ) is one of the most commonly used indicators to evaluate speech quality, the calculation process includes preprocessing, time alignment, perceptual filtering, masking effect, etc., the value range is $-0.5$–$4.5$, the higher the PESQ value indicates that the tested voice has better auditory speech quality. However, in the foley sound generation task, which is specific, the sound generated and ground truth are without semantic information, only convey the acoustic information, so we defined the lower the STOI and PESQ metrics of the model, the better the sound generated. The results show that the proposed model has reach 0.1185 of STOI and 0.2978 of PESQ, while the baseline's are 0.1704 of STOI and 0.3383 (Table 2).

**Table 1.** ABX tests between the proposed method and baseline under the four standards (Reality, Quality, Synchronization, Prefer).

|  | Reality | Quality | Synchronization | Prefer |
|---|---|---|---|---|
| Baseline | 11.9% | 8.3% | 17.4% | 14.2% |
| NP | 18.2% | 22.3% | 21.2% | 9.5% |
| **FastFoley** | **69.9%** | **69.3%** | **61.4%** | **76.3%** |

**Table 2.** objective metrics measured by the proposed method and baseline.

|  | MS-STFT-Loss | SSIM | STOI | PESQ | Speed |
|---|---|---|---|---|---|
| Baseline | 0.2961 | 0.5373 | 0.1704 | 0.3383 | 0.45 iter/s |
| **FastFoley** | **0.1944** | **0.6742** | **0.1185** | **0.2978** | **3.32 iter/s** |

**Visualization.** Figure 4 explicitly shows the audio waveforms of two video clips, denoting the ground truth waveform extracted from the original video, the waveform synthesized by baseline, and the waveform synthesized by FastFoley respectively. The results show that FastFoley performs better. The left one (a) shows that compared with Ground-Truth, the waveform synthesized by the baseline may occur sound redundant and missing. Sound redundant may also happen to FastFoley. The right one (b) shows that both baseline and FastFoley presented sound temporal mismatch to some extent. However, the temporal mismatch occurring in the baseline is ahead of the real sound in time. While, in FastFoley, there exists hysteresis against the Ground-Truth. According to human perception, people always feel the sight first, then the hearing. Hence it is more reasonable that the synthesized waveforms by FastFoley are perceived better in subjective experiments than the baseline. Fig. 5 shows the generated spectrograms of baseline and the proposed method, which is a "Chopping" foley sound. The result demonstrates that the spectrogram generated by the proposed method is much more similar to ground-truth, while much noise is included in the spectrogram generated by the baseline.

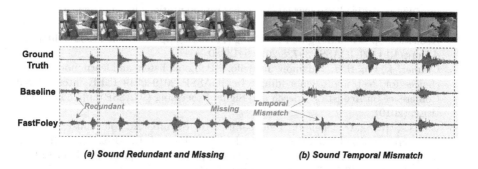

(a) Sound Redundant and Missing          (b) Sound Temporal Mismatch

**Fig. 4.** The visualization results of waveforms.

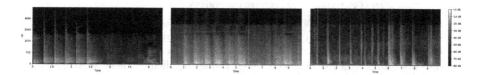

**Fig. 5.** Spectrograms of Ground Truth, Baseline, FastFoley.

## 5  Conclusion

In this paper, we proposed FastFoley to synthesize foley sound related to video contents. We also built a foley dataset, which contains various kinds of foley sound with audio-video pairs. Compared with the existing work, the results demonstrate that FastFoley outperforms subjectively in time synchronization, foley authenticity, and audio quality, and performs better objectively in MS-STFT-Loss, SSIM, STOI, PESQ. Due to the novelty of this task, we will try to introduce more effective front-end video feature engineering such as pre-trained or prompt Vision-Language models [29] in future work, also we will attempt to set up more reasonable objective experiments to evaluate the quality and the effects of the generated sound.

# References

1. Ajwani, P., Arolkar, H.: TubeExtractor: a crawler and converter for generating research dataset from YouTube videos. Int. J. Comput. Appl. **975**, 8887 (2019)
2. Ament, V.T.: The Foley Grail: The Art of Performing Sound for Film, Games, and Animation. Routledge (2014)
3. Chen, P., Zhang, Y., Tan, M., Xiao, H., Huang, D., Gan, C.: Generating visually aligned sound from videos. IEEE Trans. Image Process. **29**, 8292–8302 (2020)
4. Chien, C.M., Lin, J.H., Huang, C.Y., Hsu, P.C., Lee, H.Y.: Investigating on incorporating pretrained and learnable speaker representations for multi-speaker multi-style text-to-speech. In: ICASSP 2021–2021 IEEE International Conference on Acoustics, Speech and Signal Processing (ICASSP), pp. 8588–8592 (2021). https:// doi.org/10.1109/ICASSP39728.2021.9413880
5. Cramer, J., Wu, H.H., Salamon, J., Bello, J.P.: Look, listen, and learn more: design choices for deep audio embeddings. In: ICASSP 2019–2019 IEEE International Conference on Acoustics, Speech and Signal Processing (ICASSP), pp. 3852–3856. IEEE (2019)
6. Gan, C., Huang, D., Chen, P., Tenenbaum, J.B., Torralba, A.: Foley music: learning to generate music from videos. In: Vedaldi, A., Bischof, H., Brox, T., Frahm, J.-M. (eds.) ECCV 2020. LNCS, vol. 12356, pp. 758–775. Springer, Cham (2020). https://doi.org/10.1007/978-3-030-58621-8_44
7. Gemmeke, J.F., et al.: Audio set: an ontology and human-labeled dataset for audio events. In: 2017 IEEE International Conference on Acoustics, Speech and Signal Processing (ICASSP), pp. 776–780. IEEE (2017)
8. Ghose, S., Prevost, J.J.: AutoFoley: artificial synthesis of synchronized sound tracks for silent videos with deep learning. IEEE Trans. Multimedia (2020)
9. Ghose, S., Prevost, J.J.: FoleyGAN: visually guided generative adversarial network-based synchronous sound generation in silent videos. arXiv preprint arXiv:2107.09262 (2021)
10. Goodfellow, I., et al.: Generative adversarial nets. In: Advances in Neural Information Processing Systems, vol. 27 (2014)
11. Griffin, D., Lim, J.: Signal estimation from modified short-time Fourier transform. IEEE Trans. Acoust. Speech Signal Process. **32**(2), 236–243 (1984)
12. Hao, W., Zhang, Z., Guan, H.: CMCGAN: a uniform framework for cross-modal visual-audio mutual generation. In: Proceedings of the AAAI Conference on Artificial Intelligence, vol. 32 (2018)
13. He, K., Zhang, X., Ren, S., Sun, J.: Deep residual learning for image recognition. In: Proceedings of the IEEE Conference on Computer Vision and Pattern Recognition, pp. 770–778 (2016)
14. Hu, D., Wang, D., Li, X., Nie, F., Wang, Q.: Listen to the image. In: Proceedings of the IEEE/CVF Conference on Computer Vision and Pattern Recognition, pp. 7972–7981 (2019)
15. Huang, Q., Xiong, Yu., Rao, A., Wang, J., Lin, D.: MovieNet: a holistic dataset for movie understanding. In: Vedaldi, A., Bischof, H., Brox, T., Frahm, J.-M. (eds.) ECCV 2020. LNCS, vol. 12349, pp. 709–727. Springer, Cham (2020). https://doi.org/10.1007/978-3-030-58548-8_41
16. Kuehne, H., Jhuang, H., Garrote, E., Poggio, T., Serre, T.: HMDB: a large video database for human motion recognition. In: 2011 International Conference on Computer Vision, pp. 2556–2563. IEEE (2011)

17. McFee, B., et al.: librosa: audio and music signal analysis in Python. In: Proceedings of the 14th Python in Science Conference, vol. 8, pp. 18–25. Citeseer (2015)
18. Monfort, M., et al.: Spoken moments: learning joint audio-visual representations from video descriptions. In: Proceedings of the IEEE/CVF Conference on Computer Vision and Pattern Recognition, pp. 14871–14881 (2021)
19. Ren, Y., et al.: FastSpeech: fast, robust and controllable text to speech. arXiv preprint arXiv:1905.09263 (2019)
20. Rix, A.W., Beerends, J.G., Hollier, M.P., Hekstra, A.P.: Perceptual evaluation of speech quality (PESQ)-a new method for speech quality assessment of telephone networks and codecs. In: 2001 IEEE International Conference on Acoustics, Speech, and Signal Processing. Proceedings (Cat. No. 01CH37221), vol. 2, pp. 749–752. IEEE (2001)
21. Soomro, K., Zamir, A.R., Shah, M.: UCF101: a dataset of 101 human actions classes from videos in the wild. arXiv preprint arXiv:1212.0402 (2012)
22. Su, K., Liu, X., Shlizerman, E.: Audeo: audio generation for a silent performance video. arXiv preprint arXiv:2006.14348 (2020)
23. Taal, C.H., Hendriks, R.C., Heusdens, R., Jensen, J.: A short-time objective intelligibility measure for time-frequency weighted noisy speech. In: 2010 IEEE International Conference on Acoustics, Speech and Signal Processing, pp. 4214–4217. IEEE (2010)
24. Tomar, S.: Converting video formats with FFmpeg. Linux J. **2006**(146), 10 (2006)
25. Vaswani, A., et al.: Attention is all you need. In: Advances in Neural Information Processing Systems, pp. 5998–6008 (2017)
26. Wang, Y., et al.: Tacotron: towards end-to-end speech synthesis. arXiv preprint arXiv:1703.10135 (2017)
27. Wang, Z., Bovik, A.C., Sheikh, H.R., Simoncelli, E.P.: Image quality assessment: from error visibility to structural similarity. IEEE Trans. Image Process. **13**(4), 600–612 (2004)
28. Yamamoto, R., Song, E., Kim, J.M.: Parallel waveGAN: a fast waveform generation model based on generative adversarial networks with multi-resolution spectrogram. In: ICASSP 2020–2020 IEEE International Conference on Acoustics, Speech and Signal Processing (ICASSP), pp. 6199–6203. IEEE (2020)
29. Yao, Y., Zhang, A., Zhang, Z., Liu, Z., Chua, T.S., Sun, M.: CPT: colorful prompt tuning for pre-trained vision-language models. arXiv preprint arXiv:2109.11797 (2021)
30. Zhou, Y., Wang, Z., Fang, C., Bui, T., Berg, T.L.: Visual to sound: generating natural sound for videos in the wild. In: Proceedings of the IEEE Conference on Computer Vision and Pattern Recognition, pp. 3550–3558 (2018)

# Structured Hierarchical Dialogue Policy with Graph Neural Networks

Zhi Chen, Xiaoyuan Liu, Lu Chen$^{(\boxtimes)}$, and Kai Yu$^{(\boxtimes)}$

X-LANCE Lab, Department of Computer Science and Engineering, MoE Key Lab of
Artificial Intelligence, AI Institute, Shanghai Jiao Tong University, Shanghai, China
{zhenchi713,chenlusz,kai.yu}@sjtu.edu.cn

**Abstract.** Dialogue policy training for composite tasks, such as restaurant reservations in multiple places, is a practically essential and challenging problem. Recently, hierarchical deep reinforcement learning (HDRL) methods have achieved excellent performance in composite tasks. However, in vanilla HDRL, both top-level and low-level policies are all represented by multi-layer perceptrons (MLPs), which take the concatenation of all observations from the environment as the input for predicting actions. Thus, the traditional HDRL approach often suffers from low sampling efficiency and poor transferability. In this paper, we address these problems by utilizing the flexibility of graph neural networks (GNNs). A novel ComNet is proposed to model the structure of a hierarchical agent. The performance of ComNet is tested on composited tasks of the PyDial benchmark. Experiments show that ComNet outperforms vanilla HDRL systems with performance close to the upper bound. It not only achieves sample efficiency but also is more robust to noise while maintaining the transferability to other composite tasks.

**Keywords:** Hierarchical Deep Reinforcement Learning · Dialogue Policy · Graph Neural Network

## 1 Introduction

Different from normal one-dialogue-one-domain dialogue task [5], the composite dialogue task may involve multiple domains in a single dialogue. The agent must complete all subtasks (accomplish the goals in all domains) to get positive feedback. Consider the process of completing a composite task (e.g., multi-area restaurant reservation). An agent first chooses a subtask (e.g., reserve-Cambridge-restaurant), then makes a sequence of decisions to gather related information (e.g., price range, area) until all information required by users are provided, and this subtask is completed. Then chooses the next subtask (e.g., reserve-SF-restaurant) to complete. The state-action space will increase with the number of subtasks. Thus, dialogue policy learning for the composite task needs more exploration, and it needs to take more dialogue turn between agent and user to complete a composite task. The sparse reward problem is further magnified.

L. Zhenhua et al. (Eds.): NCMMSC 2022, CCIS 1765, pp. 264–277, 2023.
https://doi.org/10.1007/978-981-99-2401-1_24

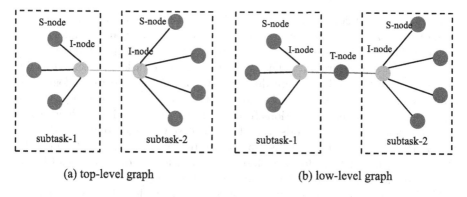

(a) top-level graph                    (b) low-level graph

**Fig. 1.** A composite dialogue task contains two subtasks, where (a) is the graph of the top-level policy and (b) is the graph of the low-level policy.

Solving composite tasks using the same method as the one solving single domain tasks may hit obstacles. The complexity of the composite task makes it hard for an agent to learn an acceptable strategy. While hierarchical deep reinforcement learning (HDRL) [1] shows its promising power, by introducing the framework of options over Markov Decision Process (MDP), the original task can be decomposed into two parts: deciding which subtask to solve and how to solve one subtask, thus simplifying the problem.

However, in previous works, multi-layer perceptrons (MLPs) are often used in DQN to estimate the Q-value. MLPs use the concatenation of the flatten dialogue state as its inputs. In this way, it cannot capture the structural information of the semantic slots in that state easily, which results in low sampling efficiency. In our work, we propose ComNet, which makes use of the Graph Neural Network (GNN) [3] to better leverage the graph structure in the observations (e.g., dialogue states) and being coherent with the HDRL method.

Our main contributions are three-fold: (1). We propose a new framework ComNet combining HDRL and GNN to solve the composite tasks while achieving sample efficiency. (2). We test ComNet based on PyDial [17] benchmark and show that our result over-performed the vanilla HDRL systems and is more robust to noise in the environment. (3). We test the transferability of our framework and find that an efficient and accurate transfer is possible under our framework.

## 2   Related Work

Reinforcement learning is a recently mainstream method to optimize statistical dialogue management policy under the partially observable Markov Decision Process (POMDP) [21]. One line of research is on single-domain task-oriented dialogues with flat deep reinforcement learning approaches, such as DQN [2,4,8], policy gradient [19,20] and actor critic [9,10,13]. Multi-domain task-oriented dialogue task is another line, where each domain learns a separate dialogue policy [5,6]. The performance of this one domain model is tested on different domains to highlight its transferability. The composite dialogue task is presented in [11]. Different from the multi-domain dialogue system, the composite dialogue task requires all the individual subtasks have to be accomplished. The composite dialogue task is formulated by options framework [14] and solved using hierarchical reinforcement learning methods [1,11,15]. All these works are built based on the vanilla HDRL, where the policy is represented by multi-layer perceptron (MLP). However, in this paper, we focus on designing a transferable dialogue policy for the composite dialogue task based on Graph Neural Network [12].

GNN is also used in other aspects of reinforcement learning to provide features like transferability or less over-fitting [18]. In dialogue system building, models like BUDS also utilize the power of graph for dialogue state tracking [16]. Previous works also demonstrated that using GNN to learn a structured dialogue policy can improve system performance significantly in a single-domain setting by creating graph nodes corresponding to the semantic slots and optimizing the graph structure [3]. However, we need to exploit the particularity of the tasks and change the complete framework for the composite dialogue.

## 3    Hierarchical Reinforcement Learning

Before introducing ComNet, we first present a short review of HRL for a composite task-oriented dialogue system. According to the *options* framework, assume that we have a dialogue state set $\mathcal{B}$, a subtask (or an option) set $\mathcal{G}$ and a primitive action set $\mathcal{A}$.

Compared to the traditional Markov decision process (MDP) setting where an agent can only choose a primitive action at each time step, the decision-making process of hierarchical MDP consists of (1) a top-level policy $\pi_b$ that selects subtasks to be completed, (2) a low-level policy $\pi_{b,g}$ that selects primitive actions to fulfill a given subtask. The top-level policy $\pi_b$ takes as input the belief state $b$ generated by the global state tracker and selects a subtask $g \in \mathcal{G}$. The low-level policy $\pi_{b,g}$ perceives the current state $b$ and the subtask $g$, and outputs a primitive action $a \in \mathcal{A}$. All subtasks share the low-level policy $\pi_{b,g}$.

In this paper, we take two Q-function to represent these two-level policies, learned by deep Q-learning approach (DQN) and parameterized by $\theta_e$ and $\theta_i$ respectively. Corresponding to two-level policies, there are two kinds of reward signal from the environment (the user): *extrinsic* reward $r^e$ and *intrinsic* reward $r^i$. The extrinsic rewards guide the dialogue agent to choose the right subtask order. The intrinsic rewards are used to learn an option policy to achieve a given subtask. The extrinsic reward and intrinsic reward are combined to help the dialogue agent accomplish a composite task as fast as possible. Thus, the extrinsic and intrinsic rewards are designed as follows:

**Intrinsic Reward.** At the end of a subtask, the agent receives a positive intrinsic reward of 1 for the success subtask or 0 for the failure subtask. To encourage

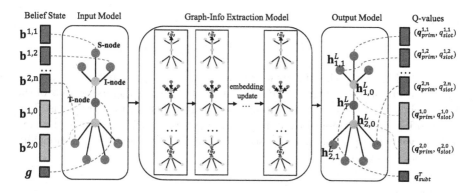

**Fig. 2.** The low-level policy model of ComNet contains three parts: input module, graph-info extraction module and output module. In the input module, for each node, ComNet fetches the corresponding elements from the observation. ComNet then computes the messages sent to neighbors in the graph and updates the embedding vector of each node. The main content of the output module is how to use the highest embedding vectors of all the nodes to calculate the corresponding Q-values. The subscript shapes like $(k, i)$ denotes the $i$-th node corresponding to subtask $k$. The top-level policy model of ComNet has a similar structure.

shorter dialogues, the agent receives a negative intrinsic reward of $-0.05$ at each turn.

**Extrinsic Reward.** Let $K$ be the number of subgoals. At the end of a dialogue, the agent receives a positive extrinsic reward of $K$ for the success dialogue or $0$ for the failure dialogue. To encourage shorter dialogues, the agent gets a negative extrinsic reward of $-0.05$ at each turn.

Assume we have a subtask trajectory of $T$ turns: $\mathcal{T}_k = (\mathbf{b}_0^k, a_0^k, r_0^k, \ldots, \mathbf{b}_T^k, a_T^k, r_T^k)$, where $k$ represents the k-th subtask $g_k$. The dialogue trajectory consists of a sequence of subtask trajectories $\mathcal{T}_0, \mathcal{T}_1 \ldots$ . According to Q-learning algorithm, the parameter $\theta_e$ of the top-level Q-function is updated as follows:

$$\theta_e \leftarrow \theta_e + \alpha \cdot (q_k - Q(g_k|\mathbf{b}_0^k; \theta_e)) \cdot \nabla_{\theta_e} Q(g_k|\mathbf{b}_0^k; \theta_e),$$

where

$$q_k = \sum_{t=0}^{T} \gamma^t r_t^e + \gamma^T \max_{g' \in \mathcal{G}} Q(g'|\mathbf{b}_T^k; \theta_e),$$

and $\alpha$ is the step-size parameter, $\gamma \in [0, 1]$ is a discount rate. The first term of the above expression of q equals to the total discounted reward during fulfilling subtask $g_k$, and the second estimates the maximum total discounted value after $g_k$ is completed.

The learning process of the low-level policy is in a similar way, except that intrinsic rewards are used. For each time step $t = 0, 1, \ldots, T$,

$$\theta_i \leftarrow \theta_i + \alpha \cdot (q_t - Q(a_t|\mathbf{b}_t^k, g_k; \theta_i)) \cdot \nabla_{\theta_i} Q(a_t|\mathbf{b}_t^k, g_k; \theta_i),$$

where

$$q_t = r_t^i + \gamma \max_{a' \in \mathcal{A}} Q(a'|\mathbf{b}_{t+1}^k, g_k; \theta_i).$$

In vanilla HDRL, the above two Q-functions are approximated using MLP. The structure of the dialogue state is ignored in this setting. Thus the task of the MLP policies is to discover the latent relationships between observations. This leads to longer convergence time, requiring more exploration trials. In the next section, we will explain how to construct a graph to represent the relationships in a dialogue observation.

## 4    ComNet

In this section, we first introduce the notation of the composite task. We then explain how to construct two graphs for two-level policies of a hierarchical dialogue agent, followed by the description of the ComNet.

### 4.1    Composite Dialogue

Task-oriented dialogue systems are typically defined by a structured *ontology*. The *ontology* consists of some properties (or slots) that a user might use to frame a query when fulfilling the task. As for composite dialogue state, which contains $K$ subtasks, each subtask corresponds to several slots. For simplification, we take the subtask $k$ as an example to introduce the belief state. There are two boolean attributes for each slot of the subtask $k$, whether it is *requestable* and *informable*. The user can request the value of the requestable slots and can provide specific value as a search constraint for the informable slots. At each dialogue turn, the dialogue state tracker updates a belief state for each informable slot.

Generally, the belief state consists of all the distributions of candidate slot values. The value with the highest belief for each informable slot is selected as a constraint to search the database. The information of the matched entities in the database is added to the final dialogue state. The dialogue state $\mathbf{b}^k$ of the subtask $k$ is decomposed into several *slot-dependent* states and a *slot-independent* state, represented as $\mathbf{b}^k = \mathbf{b}^{k,1} \oplus \mathbf{b}^{k,2} \oplus \cdots \oplus \mathbf{b}^{k,n} \oplus \mathbf{b}^{k,0}$. $\mathbf{b}^{k,j}(1 \leq j \leq n)$ is the $j$-th informable slot-related state of the subtask $k$, and $\mathbf{b}^{k,0}$ represents the slot-independent state of the subtask $k$. The whole belief state is the concatenation of all the subtask-related state $\mathbf{b}^k$, i.e. $\mathbf{b} = \mathbf{b}^1 \oplus \cdots \oplus \mathbf{b}^K$, which is the input of the top-level dialogue policy.

The output of the top-level policy is a subtask $g \in \mathcal{G}$. In this paper, we use a one-hot vector to represent one specific subtask. Furthermore, the whole belief state $\mathbf{b}$ and the subtask vector $g$ are fed into the low-level policy. The output of the low-level policy is a primitive dialogue action. Similarly, for each subtask $k$, the dialogue action set $\mathcal{A}^k$ can be divided into $n$ slot-related action sets $\mathcal{A}^{k,j}(1 \leq j \leq n)$, e.g. $request\_slot^{k,j}, inform\_slot^{k,j}, select\_slot^{k,j}$ and a one slot-independent action set $\mathcal{A}^{k,0}$, e.g. $repeat^{k,0}, reqmore^{k,0}, \ldots, bye^{k,0}$. The whole dialogue action space $\mathcal{A}$ is the union of all the subtask action spaces.

## 4.2   Graph Construction

As introduced in Sect. 4.1, the dialogue state $\mathbf{b}$ consists of $K$ subtask-related state, and each subtask-related state can further be decomposed into several slot-dependent states and a slot-independent state, which are logically inde-composable, named *atomic states*. The hierarchical format of the dialogue state can be naturally regarded as a graph. Each node in the graph represents the corresponding atomic state. To simplify the graph's structure, we choose the slot-independent nodes as the delegate of the nodes which correspond to the same subtask. All the slot-independent nodes are fully connected in the top-level graph, and the slot-dependent nodes are only connected to their delegate node. Unlike the input of top-level policy, the input of the low-level policy adds a new node named subtask node to represent the goal information produced by the top-level policy. In the low-level graph, the slot-independent nodes are all connected to the subtask node (or the global delegate nodes) instead of fully connecting.

## 4.3   ComNet as Policy Network

We now turn to ComNet, which parameterizes two-level policies with two Graph Neural Networks (GNNs). Before delving into details, we first introduce our notation. We denote the graph structure as $G = (V, E)$ with nodes $v_i (0 \leq i \leq n) \in V$ and directed edges $e_{ij} \in E$. The adjacency matrix $\mathbf{Z}$ denotes the structure of $G$. If there is a directed edge from $i$-th node $v_i$ to $j$-th node $v_j$, the element $z_{ij}$ of $\mathbf{Z}$ is 1, otherwise $z_{ij}$ is 0. We denote the out-going neighborhood set of node $v_i$ as $\mathcal{N}_{out}(v_i)$. Similarly, $\mathcal{N}_{in}(v_i)$ denotes the in-coming neighborhood set of node $v_i$. Each node $v_i$ has an associated node type $p_i$. Each edge $e_{ij}$ has an edge type $c_e$, which is determined by starting node type $p_i$ and ending node type $p_j$. In other words, two edges have the same type if and only if their starting node type and ending node type are both the same.

For top-level policy, it has two types of nodes: `slot-dependent` nodes (S-nodes) and `slot-independent` nodes (I-node). Since there is no edge between `slot-dependent` nodes, it has only four edge types. Similarly, for low-level policy, it has three types of nodes (`slot-dependent`, `slot-independent` and `subtask` (T-node)) and four edge types. The two level graphs show in Fig. 1.

Until now, the graphs of top-level policy and low-level policy are both well defined. ComNet, which has two GNNs, is used to parse these graph-format observations of the low-level policy and top-level policy. Each GNN has three parts to extract useful representation from initial graph-format observation: input module, graph-info extraction module and output module.

**Input Module.** Before each prediction, each node $v_i$ of top-level and low-level graphs will receive the corresponding atomic state $\mathbf{b}$ or subtask information $g$ (represented as $x_i$), which is fed into an input module to obtain a state embedding $\mathbf{h}_i^0$ as follows:

$$\mathbf{h}_i^0 = F_{p_i}(x_i),$$

where $F_{p_i}$ is a function for node type $p_i$, which may be a multi-layer perceptron (MLP). Normally, different slots have a different number of candidate values. Therefore, the input dimension of the slot-dependent nodes is different. However, the belief state of each slot is often approximated by the probability of *sorted* top $M$ values [7], where $M$ is usually less than the least value number of all the slots. Thus, the input dimension of nodes with the same type is the same.

**Graph-Info Extraction Module.** The graph-info extraction module takes $\mathbf{h}_i^0$ as the initial embedding for node $v_i$, then further propagates the higher embedding for each node in the graph. The propagation process of node embedding at each extraction layer shows as the following operations.

**Message Computation.** At $l$-th step, for every node $v_i$, there is a node embedding $\mathbf{h}_i^{l-1}$. For every out-going node $v_j \in \mathcal{N}_{out}(v_i)$, node $v_i$ computes a message vector as below,

$$\mathbf{m}_{ij}^l = M_{c_e}^l(\mathbf{h}_i^{l-1}),$$

where $c_e$ is edge type from node $v_i$ to node $v_j$ and $M_{c_e}^l$ is the message generation function which may be a linear embedding: $M_{c_e}^l(\mathbf{h}_i^{l-1}) = \mathbf{W}_{c_e}^l \mathbf{h}_i^{l-1}$. Note that the subscript $c_e$ indicates that edges of the same edge type share the weight matrix $\mathbf{W}_{c_e}^l$ to be learned.

**Message Aggregation.** After every node finishes computing message, The messages sent from the in-coming neighbors of each node $v_j$ will be aggregated. Specifically, the aggregation process shows as follows:

$$\overline{\mathbf{m}}_j^l = A(\{\mathbf{m}_{ij}^l | v_i \in \mathcal{N}_{in}(v_j)\}),$$

where $A$ is the aggregation function which may be a summation, average or max-pooling function. $\overline{\mathbf{m}}_j^l$ is the aggregated message vector which includes the information sent from all the neighbor nodes.

**Embedding Update.** Until now, every node $v_i$ has two kinds of information, the aggregated message vector $\overline{\mathbf{m}}_i^l$ and its current embedding vector $\mathbf{h}_i^{l-1}$. The embedding update process shows as below:

$$\mathbf{h}_i^l = U_{p_i}^l(\mathbf{h}_i^l, \overline{\mathbf{m}}_i^l),$$

where $U_{p_i}^l$ is the update function for node type $p_i$ at $l$-th extraction layer, which may be a non-linear operation, i.e.

$$\mathbf{h}_i^l = \delta(\lambda^l \mathbf{W}_{p_i}^l \mathbf{h}_i^l + (1 - \lambda^l)\overline{\mathbf{m}}_i^l),$$

where $\delta$ is an activation function, i.e. RELU, $\lambda^l$ is a weight parameter of the aggregated information which is clipped into $0 \frown 1$, and $\mathbf{W}_{p_i}^l$ is a trainable matrix. Note that the subscript $p_i$ indicates that the nodes of the same node type share the same instance of the update function, in our case the parameter $\mathbf{W}_{p_i}^l$ is shared.

**Output Module.** After updating node embedding $L$ steps, each node $v_i$ has a final representation $\mathbf{h}_i^L$, also represented as $\mathbf{h}_{k,i}^L$, where the subscript $k, i$ denotes the node $v_i$ corresponds to the subtask $k$.

**Top-Level Output:** The top-level policy aims to predict a subtask to be fulfilled. In the top-level graph, for a specific subtask, it corresponds to several S-nodes and one I-node. Thus, when calculating the Q-value of a specific subtask, all the final embedding of the subtask-related nodes will be used. In particular, for each subtask $k$, we perform the following calculating:

$$q_{top}^k = O_{top}(\sum_{v_i \in S-node} \mathbf{h}_{k,i}^L, \mathbf{h}_{k,0}^L),$$

where $O_{top}$ is the output function which may be a MLP and the subscripts $k, 0$ and $k, i$ denote the I-node and $i$-th S-node of the subtask $k$, respectively. In practice, we take the concatenation of $\sum_{v_i \in S-node} \mathbf{h}_{k,i}^L$ and $\mathbf{h}_{k,0}^L$ as the input of a MLP and outputs a scalar value. For all the subtask, this MLP is shared. When making a decision, all the $q_{top}^k$ will be concatenated, i.e. $\mathbf{q}_{top} = q_{top}^1 \oplus \cdots \oplus q_{top}^K$, then the subtask is selected according to $\mathbf{q}_{top}$ as done in vanilla DQN.

**Low-Level Output:** The top-level policy aims to predict a primitive dialogue action. As introduced in Sect. 4.1, a primitive dialogue action must correspond to a subtask. If we regard slot-independent nodes as a special kind of slot-dependent nodes, a primitive dialogue action can further correspond to a slot node. Thus, the Q-value of each dialogue action contains three parts of information: subtask-level value, slot-level value and primitive value. We use T-node embedding $\mathbf{h}_T^L$ to compute subtask-level value:

$$\mathbf{q}_{subt}^T = O_{subt}^T(\mathbf{h}_T^L),$$

where $O_{subt}^T$ is output function of subtask-level value, which may be a MLP. The output dimension of $O_{subt}^T$ is $K$ where each value distributes to a corresponding subtask. The nodes $v_i$ that belong to S-nodes and I-nodes will compute slot-level value and primitive value:

$$q_{slot}^{k,i} = O_{slot}^{p_i}(\mathbf{h}_{k,i}^L),$$
$$\mathbf{q}_{prim}^{k,i} = O_{prim}^{p_i}(\mathbf{h}_{k,i}^L),$$

where $O_{slot}^{p_i}$ and $O_{prim}^{p_i}$ are output functions of slot-level value and primitive value respectively, which may be MLPs in practice. Similarly, the subscript $p_i$ indicates that the nodes of the same node type share the same instance of the output functions. The Q-value of an action $\mathbf{a}_{k,i}$ corresponding to the slot node $v_i$ is $\mathbf{q}_{low}^{k,i} = (\mathbf{q}_{subt}^T)_k + q_{slot}^{k,i} + \mathbf{q}_{prim}^{k,i}$, where $+$ is element-wise operation and $(\mathbf{q}_{subt}^T)_k$ denotes the $k$-th value in $\mathbf{q}_{subt}^T$. When predicting a action, all the $\mathbf{q}_{low}^{k,i}$ will be concatenated, i.e. $\mathbf{q}_{low} = \mathbf{q}_{low}^{1,1} \oplus \cdots \oplus \mathbf{q}_{low}^{K,0}$, then the primitive action is chosen according to $\mathbf{q}_{low}$ as done in vanilla DQN.

**Discussion.** Note that although the parameters of the input module and graph-info extraction module are not shared between the top-level GNN and low-level GNN (shown as Fig. 2), there are many shared parameters in each single GNN. Assume that now the composite task is changed, and one subtask adds some new slot. We only need to create new nodes in each GNN. If the number of the edge type has not changed, the parameters of the GNN will stay the same after adding new nodes. This attribution of ComNet leads to transferability. Generally, if the node-type set and edge-type set of the composite task $Task_1$ are both subsets of another task $Task_2$'s, the ComNet policy learned in $Task_2$ can be directly used on $Task_1$.

Since the initial output of the same type of nodes has a similar semantic meaning, they share the parameters in ComNet. We hope to use the GNN to propagate the relationships between the nodes in the graph based on the connection of the initial input and the final outputs.

## 5    Experiments

In this section, we first verify the effectiveness of ComNet on the composite tasks of the PyDial benchmark. We then investigate the transferability of ComNet.

### 5.1    PyDial Benchmark

A composite dialogue simulation environment is required for the evaluation of our purposed framework. PyDial toolkit [17], which supports multi-domain dialogue simulation with error models, has laid a good foundation for our composite task environment building.

**Table 1.** The number of data constraints, the number of informative slots that user can request and the number of database result values vary in different composite tasks. Semantic error rate (SER) presents an ascending order in three environments.

| Composite Tasks | Constraints | Requests | Values |
|---|---|---|---|
| CR+SFR | 9 | 20 | 904 |
| CR+LAP | 14 | 30 | 525 |
| SFR+LAP | 17 | 32 | 893 |
| | **Env. 1** | **Env. 2** | **Env. 3** |
| **SER** | 0% | 15% | 30% |

We modified the policy management module and user simulation module to support 2-subtask composite dialogue simulation among three available subtasks, which are Cambridge Restaurant (CR), San Francisco Restaurant (SFR), and generic shopping task for laptops (LAP). We still preserve fully functional error simulation of different levels in Table 1. Note that in the policy management module, we discard the domain input provided by the dialogue state tracking (DST) module to make a fair comparison. We updated the user simulation module and evaluation management module to support the reward design in Sect. 3.

## 5.2   Implementation

We implement the following three composite task agents to evaluate the performance of our proposed framework.

- **Vanilla HDQN:** A hierarchical agent using MLPs as its models. This serves as the baseline for our comparison.
- **ComNet:** Our purposed framework utilizing the flexibility of GNNs. The complete framework is discussed in Sect. 4.
- **Hand-crafted:** A well-designed rule-based agent with a high success rate in composite dialogue without noise. This agent is also used to warm up the training process of the first two agents. Note that this agent uses the precise subtask information provided by DST, which is not fair comparing with the other two.

Here, we train models with 6000 dialogues or iterations. The total number of the training dialogues is broken down into milestones (30 milestones of 200 iterations each). At each milestone, there are 100 dialogues to test the performance of the dialogue policy. The results of 3 types of composite tasks in 3 environments are shown in Fig. 3.

## 5.3   Analysis

From Fig. 3, we can observe that ComNet outperforms the vanilla MLP policy in all nine settings (3 environments * 3 types of composite tasks) in both success rate and learning speed. In ComNet, the top-level policy and low-level policy are both represented by a GNN where the same of type nodes and the same type of edges share the parameters. It means that the same type of nodes shares the input space (belief state space). Thus the exploration space will greatly decrease. As shown in Fig. 3, ComNet learns to vary faster than vanilla MLP policy. Note that the hand-crafted agent performs well because it has cheated by peeking at the precise subtask information, which means the hand-crafted agent is solving the multi-domain tasks. This should be the upper bound for the performance of our model. Comparing with vanilla HDQN, our ComNet shows its robustness in all environment by a greater margin, which is helpful for dialogue system building when an ASR or a DST with high accuracy is not available.

We also compared the difference of the dialogues produced by both vanilla HDQN and ComNet after 6000-dialogue training, which is shown in Table 2. After that much training, it seems that the vanilla HDQN agent still cannot

**Table 2.** This table is an example of two different trained agents handling the same composite task with same user goal. On the left-hand side, vanilla HDQN succeed in finishing the first subtask but failed to produce meaningful action in the second. On the right-hand side, ComNet successfully full-filled two subtasks, thus succeed in the dialogue.

**User Goal**

| **CR**: restaurant | **SFR**: restaurant |
|---|---|
| [('pricerange', '=', 'dontcare'), ('area', '=', 'centre'), ('food', '=', 'mediterranean')] | [('pricerange', '=', 'cheap'), ('allowedforkids', '=', 'dontcare'), ('area', '=', 'civic center')] |
| **req**: ['name', 'description'] | **req**: ['name'] |

| Vanilla HDQN | ComNet |
|---|---|
| **CR domain** | |
| Sys : hello(help) | **CR domain** |
| User : inform(type= "restaurant", pricerange= "dontcare", area= "centre", food= "mediterranean") | Sys : hello(help) |
| | User : inform(type= "restaurant") |
| Sys : request(area) | Sys : request(area) |
| User : inform(area= "centre") | User : inform(area= "centre") |
| Sys : inform(food= "mediterranean", pricerange= "expensive", name= "la mimosa", area= "centre") | Sys : request(food) |
| | User : inform(food= "mediterranean") |
| User : thankyou() | Sys : request(pricerange) |
| Sys : inform(food= "mediterranean", pricerange= "expensive", name= "la mimosa", area= "centre") | User : inform(= "dontcare") |
| | Sys : inform(food= "mediterranean", pricerange= "expensive", name= "la mimosa", area= "centre") |
| User : request(description) | User : thankyou() |
| Sys : inform(name= "la mimosa", description= "not available") | Sys : reqmore() |
| **SFR domain** | User : request(description) |
| User : hello(type= "restaurant", pricerange= "cheap", allowedforkids= "dontcare", area= "civic center") | Sys : inform(name= "la mimosa", description= "not available") |
| | **SFR domain** |
| Sys : request(area) | User : inform(type= "restaurant", area= "civic center") |
| User : inform(area= "civic center") | Sys : request(allowedforkids) |
| Sys : request(area) | User : inform(goodformeal= "lunch") |
| User : inform(area= "civic center") | Sys : request(allowedforkids) |
| Sys : request(area) | User : inform(allowedforkids= "dontcare") |
| User : inform(area= "civic center") | Sys : request(food) |
| Sys : request(area) | User : inform(= "dontcare", pricerange= "dontcare") |
| User : inform(area= "civic center") | Sys : inform(goodformeal= "lunch", name= "sai jai thai restaurant", area= "civic center", food= "thai", allowedforkids= "1", pricerange= "cheap") |
| Sys : request(area) | User : bye() |
| User : bye() | Sys : bye() |
| Sys : bye() | |
| **\*FAILED, subtask-2 is not finished.** | **\*SUCCESS** |

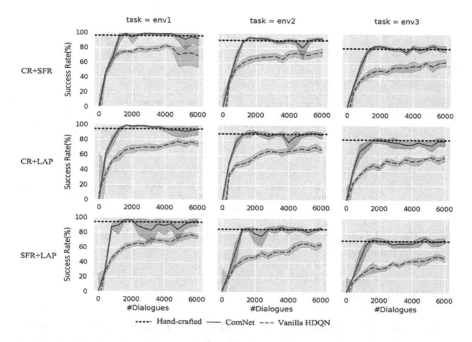

**Fig. 3.** The comparison between 3 kinds of agent. ComNet achieved performance close to the upper bound (hand-crafted) while there is still room for improvement for vanilla DQN.

choose a proper action in some specific dialogue state, which results in the loss of customer patience. On the other hand, ComNet also chose the same action, but it advanced the progress of the dialogue as soon as it got the information it needed, thus finished the task successfully. This also helps to find that ComNet is more sample efficient comparing to the vanilla framework.

## 5.4 Transferability

As we discussed in Sect. 4.3, another advantage of ComNet is that because of the flexibility of GNNs, ComNet is transferable naturally. To evaluate its transferability, we first trained 6,000 dialogues on the CR+SFR task. We then initiate the parameters of the policy models on the other two composite tasks using trained policy and continue to train and test the models. The result is shown in Fig. 4.

We can find that the transferred model learned on the CR+SFR task is compatible with the other two composite tasks. It demonstrates that ComNet can propagate the task-independent relationships among the graph nodes based on the connection of the initial nodes' inputs and final outputs. Under the framework of ComNet, it is possible to boost the training process for a new composite task by using pre-trained parameters of related tasks. After all, It is essential to solving the start-cold problems in the task-oriented dialogue systems.

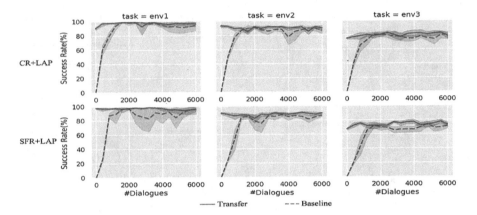

**Fig. 4.** The model pretrained on CR+SFR task is compared with the one started with randomized parameters.

## 6    Conclusion

In this paper, we propose ComNet, a structured hierarchical dialogue policy represented by two graph neural networks (GNNs). By replacing MLPs in the traditional HDRL methods, ComNet makes better use of the structural information of dialogue state by separately feeding observations (dialogue state) and the top-level decision into slot-dependent, slot-independent and subtask nodes and exchange message between these nodes. We evaluate our framework on the modified PyDial benchmark and show high efficiency, robustness, and transferability in all settings.

**Acknowledgments.** We would like to thank all anonymous reviewers for their insightful comments. This work has been supported by the China NSFC Projects (No.62120106006 and No.62106142), Shanghai Municipal Science and Technology Major Project (2021SHZDZX0102), CCF-Tencent Open Fund and Startup Fund for Youngman Research at SJTU (SFYR at SJTU).

## References

1. Budzianowski, P., et al.: Sub-domain modelling for dialogue management with hierarchical reinforcement learning. arXiv preprint arXiv:1706.06210 (2017)
2. Chang, C., Yang, R., Chen, L., Zhou, X., Yu, K.: Affordable on-line dialogue policy learning. In: Proceedings of the 2017 Conference on Empirical Methods in Natural Language Processing, pp. 2200–2209 (2017)
3. Chen, L., Tan, B., Long, S., Yu, K.: Structured dialogue policy with graph neural networks. In: Proceedings of the 27th International Conference on Computational Linguistics, pp. 1257–1268 (2018)

4. Chen, L., Zhou, X., Chang, C., Yang, R., Yu, K.: Agent-aware dropout DQN for safe and efficient on-line dialogue policy learning. In: Proceedings of the 2017 Conference on Empirical Methods in Natural Language Processing, pp. 2454–2464 (2017)
5. Gašić, M., Mrkšić, N., Su, P.H., Vandyke, D., Wen, T.H., Young, S.: Policy committee for adaptation in multi-domain spoken dialogue systems. In: 2015 IEEE Workshop on Automatic Speech Recognition and Understanding (ASRU), pp. 806–812. IEEE (2015)
6. Gašić, M., et al.: Dialogue manager domain adaptation using gaussian process reinforcement learning. Comput. Speech Lang. **45**, 552–569 (2017)
7. Gašić, M., Young, S.: Gaussian processes for POMDP-based dialogue manager optimization. IEEE/ACM Trans. Audio Speech Lang. Process. **22**(1), 28–40 (2013)
8. Li, X., Chen, Y.N., Li, L., Gao, J., Celikyilmaz, A.: End-to-end task-completion neural dialogue systems. arXiv preprint arXiv:1703.01008 (2017)
9. Liu, B., Lane, I.: Iterative policy learning in end-to-end trainable task-oriented neural dialog models. In: 2017 IEEE Automatic Speech Recognition and Understanding Workshop (ASRU), pp. 482–489. IEEE (2017)
10. Peng, B., Li, X., Gao, J., Liu, J., Chen, Y.N., Wong, K.F.: Adversarial advantage actor-critic model for task-completion dialogue policy learning. In: 2018 IEEE International Conference on Acoustics, Speech and Signal Processing (ICASSP), pp. 6149–6153. IEEE (2018)
11. Peng, B., et al.: Composite task-completion dialogue policy learning via hierarchical deep reinforcement learning. arXiv preprint arXiv:1704.03084 (2017)
12. Scarselli, F., Gori, M., Tsoi, A.C., Hagenbuchner, M., Monfardini, G.: The graph neural network model. IEEE Trans. Neural Netw. **20**(1), 61–80 (2009)
13. Su, P.H., Budzianowski, P., Ultes, S., Gasic, M., Young, S.: Sample-efficient actor-critic reinforcement learning with supervised data for dialogue management. In: Proceedings of the 18th Annual SIGdial Meeting on Discourse and Dialogue, pp. 147–157 (2017)
14. Sutton, R.S., Precup, D., Singh, S.P.: Intra-option learning about temporally abstract actions. In: ICML, vol. 98, pp. 556–564 (1998)
15. Tang, D., Li, X., Gao, J., Wang, C., Li, L., Jebara, T.: Subgoal discovery for hierarchical dialogue policy learning. In: Proceedings of the 2018 Conference on Empirical Methods in Natural Language Processing, pp. 2298–2309 (2018)
16. Thomson, B., Young, S.: Bayesian update of dialogue state: a POMDP framework for spoken dialogue systems. Comput. Speech Lang. **24**(4), 562–588 (2010)
17. Ultes, S., et al.: PyDial: a multi-domain statistical dialogue system toolkit. In: Proceedings of ACL 2017, System Demonstrations, pp. 73–78 (2017)
18. Wang, T., Liao, R., Ba, J., Fidler, S.: NerveNet: learning structured policy with graph neural networks (2018)
19. Williams, J.D., Asadi, K., Zweig, G.: Hybrid code networks: practical and efficient end-to-end dialog control with supervised and reinforcement learning. In: Proceedings of the 55th Annual Meeting of the Association for Computational Linguistics (Volume 1: Long Papers), pp. 665–677 (2017)
20. Williams, J.D., Zweig, G.: End-to-end LSTM-based dialog control optimized with supervised and reinforcement learning. arXiv preprint arXiv:1606.01269 (2016)
21. Young, S., Gašić, M., Thomson, B., Williams, J.D.: POMDP-based statistical spoken dialog systems: a review. Proc. IEEE **101**(5), 1160–1179 (2013)

# Deep Reinforcement Learning for On-line Dialogue State Tracking

Zhi Chen, Lu Chen[✉], Xiang Zhou, and Kai Yu[✉]

X-LANCE Lab, Department of Computer Science and Engineering, MoE Key Lab of Artificial Intelligence, AI Institute, Shanghai Jiao Tong University, Shanghai, China
{zhenchi713,chenlusz,kai.yu}@sjtu.edu.cn

**Abstract.** Dialogue state tracking (DST) is a crucial module in dialogue management. It is usually cast as a supervised training problem, which is not convenient for on-line optimization. In this paper, a novel companion teaching based deep reinforcement learning (DRL) framework for on-line DST optimization is proposed. To the best of our knowledge, this is the first effort to optimize the DST module within DRL framework for on-line task-oriented spoken dialogue systems. In addition, dialogue policy can be further jointly updated. Experiments show that on-line DST optimization can effectively improve the dialogue manager performance while keeping the flexibility of using predefined policy. Joint training of both DST and policy can further improve the performance.

**Keywords:** Task-oriented Dialogue System · Joint Training · Reinforcement Learning

## 1 Introduction

A task-oriented spoken dialogue system usually consists of three modules: input, output and control, shown in Fig. 1. The input module which consists of automatic speech recognition (ASR) and spoken language understanding (SLU) extracts semantic-level user dialogue actions from user speech signal. The control module (referred to as dialogue management) has two missions. One is to maintain dialogue state, an encoding of the machine's understanding about the conversation. Once the information from the input module is received, the dialogue state is updated by dialogue state tracking (DST). The other is to choose a semantic-level machine dialogue action to response the user, which is called dialogue decision policy. The output consists of natural language generation (NLG) and text-to-speech (TTS) synthesis, which convert dialogue action to audio. Dialogue management is an important part of a dialogue system. Nevertheless, there are inevitable ASR and SLU errors which make it hard to track true dialogue state and make decision. In recent statistical dialogue system, the distribution of dialogue state, i.e. *belief state*, is tracked. A well-founded theory for belief tracking and decision making is offered by partially observable Markov Decision Process (POMDP) [14] framework.

L. Zhenhua et al. (Eds.): NCMMSC 2022, CCIS 1765, pp. 278–292, 2023.
https://doi.org/10.1007/978-981-99-2401-1_25

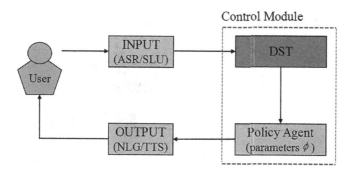

**Fig. 1.** Spoken dialogue system.

Previous DST algorithms can be divided into three families: hand-crafted rules [24, 27], generative models [1,31], and discriminative models [12,25]. Recently, since the Dialog State Tracking Challenges (DSTCs) have provided labelled dialog state tracking data and a common evaluation framework and test-bed, a variety of machine learning methods for DST have been proposed. These methods rely strictly on set of labelled off-line data. Since the labelled data are off-line, the learning process of these supervised learning methods is independent on the dialogue policy module. The key issues of these supervised learning methods are poor generalization and over-tuning. Due to the lack of labels, these approaches can not be easily used for on-line update of DST.

This work marks first step towards employing the deep deterministic reinforcement learning method into dialogue state tracking (DST) module. The performance of the DST module is optimized during the conversation between the user and the dialogue system. We call the DRL-based DST module as the tracking agent. In order to bound the search space of the tracking agent, we propose a companion teaching framework [3]. Furthermore, under this framework, we can train tracking agent and dialogue policy agent jointly with respective deep reinforcement learning (DRL) algorithms in order to make these two agents adaptive to each other. The paper has two main contributions:

- The paper provides a flexible companion teaching framework which makes the DST be able to be optimized in the on-line dialogue system.
- We can jointly train DST agent and dialogue policy agent with different reinforcement learning algorithms.

The rest of the paper is organized as follows. Section 2 gives an overview of related work. In Sect. 3, the framework of on-line DST are presented. The implementation detail is represented in Sect. 4. In Sect. 5, the joint training process is introduced. Section 6 presents experiments conducted to evaluate the proposed framework, followed by the conclusion in Sect. 7.

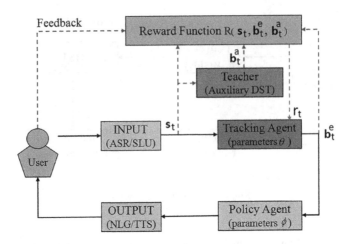

**Fig. 2.** The companion teaching framework for the on-line DST.

## 2  Related Work

Recent mainstream studies on dialogue state tracking are discriminative statistical methods. Some of the approaches encode dialogue history in features to learn a simple classifier. [11] applies a deep neural network as a classifier. [29] proposed a ranking algorithm to construct conjunctions of features. The others of the approaches model dialogue as a sequential process [28], such as conditional random field (CRF) [15] and recurrent neural network (RNN) [12]. All of these approaches need massive labelled in-domain data for training, they are belong to off-line and static methods.

In contrast to dialogue state tracking, the dialogue policy in task-oriented SDS has long been trained using deep reinforcement learning (DRL) which includes value function approximation methods, like deep Q-network (DQN) [2–4, 19, 33], policy gradient methods, e.g. REINFORCE [23, 30], advantage actor-critic (A2C) [5], and inverse reinforcement learning [26]. In our experiments, the dialogue policies of our provided systems are optimized by DQN [18] under POMDP framework. Our proposed framework is also inspired by the success of the companion teaching methods [3] in the dialogue policy.

In this work, we propose a companion teaching framework to generate the tracker from the on-line dialogue system. But the space of the belief state is continuous, it is difficult to be optimized by normal RL algorithms. [8] provided an efficient method to extend deep reinforcement learning to the class of parameterized action space MDPs and extend the deep deterministic policy gradient (DDPG) algorithm [16] for bounding the action space gradients suggested by the critic. This method greatly reduces the difficulty of the exploration.

The closest method is the Natural Actor and Belief Critic (NABC) algorithm [13] which jointly optimizes both the tracker and the policy parameters. However, the tracker in [13] uses a dynamic Bayesian Network to represent the dialogue state. In our work, the dialogue management system which includes the DST and the dialogue policy is purely statistical. Recently, [17] proposed an end-to-end dialogue management system, which directly connects the dialogue state tracking module and the dialogue decision policy module with the reinforcement learning method. There are many problems with this approach in the practical dialogue system, where end-to-end dialogue systems are not as flexible and scalable as the modular dialogue system.

## 3   On-line DST via Interaction

Discriminative machine learning approaches are now the state-of-the-art in DST. However, these methods have several limitations. Firstly, they are supervised learning (SL) approaches which require massive off-line data annotation. This is not only expensive but also infeasible for on-line learning. Secondly, given limited labelled data, over-tuning may easily happen for SL approaches, which leads to poor generalization. Thirdly, since SL-based DST approaches are independent of dialogue policy, the DST module can't dynamically adapt to the habit of the user. These limitations prohibit DST module from on-line update. To address this problem, we propose a deep reinforcement learning (DRL) framework for DST optimization via on-line interaction.

Reinforcement learning (RL) has been popular for updating the dialogue policy module in a task-oriented dialogue system for a long time. However, except for a few joint learning model for both DST and policy, RL has not been used specifically for the DST module. In this paper, under the RL framework, we regard the DST as an agent, referred to as *tracking agent*, and the other parts of the dialogue system as the environment. To our best knowledge, this is the first attempt to employ the reinforcement learning framework specifically for on-line DST optimization.

Different from the policy agent, the decision (belief state) made by the tracking agent is continuous. Hence, in this paper, DST is cast as a continuous control problem which has a similar challenge as robot control. There are several advanced algorithms to tackle the continuous control problems, e.g. DDPG algorithm. However, since the continuous belief state is both continuous and high-dimensional, the straightforward application of existing RL algorithms do not work well. In this paper, we borrow the *companion teaching* idea [3] to construct a novel RL framework for DST. Here, an auxiliary well-trained tracker, e.g. a traditional tracker trained off-line, is used as the *teacher* to guide the optimizing process of the actual DST agent (the *student*) to avoid over-tuning and achieve robust and fast convergence.

The companion teaching RL-DST framework is shown in Fig. 2, where $b^a$ is the auxiliary belief state produced by the auxiliary DST model and $b^e$ is the exploration belief state produced by the tracking agent. The difference between $b^a$ and $b^e$ will be fed into the reward signal to significantly reduce the search space of the tracking agent.

It is also worth comparing the proposed framework with end-to-end dialogue systems which can also support on-line update. Firstly, the modular structure of the RL-DST framework allows more flexible and interpretable dialogue management models to be used. For example, interpretable dialogue policy, such as rule-based policy, can be easily used with arbitrary DST models. This flexibility is practically very useful. Secondly, due to the use of a teacher DST model, the optimizing process of the tracking agent requires few dialogue data and the training is more robust.

## 3.1   Input and Output

To avoid confusion with the concepts of the policy agent, we replace *input* and *output* for *state* and *action* of the tracking agent respectively. In this work, only semantic-level dialogue manager is considered. Thus, the input is semantic features of each slot, which are extracted from system action, spoken language understanding (SLU) output and context from the previous turn. The output of the tracking agent is belief state of the corresponding slot at the current turn. In contrast to the system action of the policy agent, the output of the tracking agent, i.e. belief state, is continuous. In this paper, the input of the tracking agent is represented as $\mathbf{s}$ and the output as $\mathbf{b}^e$.

## 3.2   Tracking Policy

The *tracking policy* denotes a mapping function between $\mathbf{s}$ and $\mathbf{b}^e$ which aims to maximize the expected accumulated reward. Since the search space of the tracking agent is continuous, deterministic reinforcement learning algorithms, such as DDPG algorithm, is used to optimize the tracking policy as in the robotic control problem [6,7,16].

## 3.3   Reward Signal

Dialogue system reward is usually defined as a combination of **turn penalty** and **success reward** [4,33]. The policy agent can be effectively optimized using the two reward signals. However, for the tracking agent, due to the large search space caused by the continuous output, the two signals are not sufficient to achieve fast and robust convergence. To address this problem, we design another **basic score** reward signal to constrain the search space of the tracking agent. Therefore, the overall reward of the tracking agent consists of three kinds of signals:

**Turn Penalty**, denoted as $r^{tp}$, is a negative constant value to penalize long dialogues. The assumption here is that shorter dialogue is better.

**Success Reward**, denoted as $r^{sr}$, is a delayed reward for the whole dialogue at the last turn. When the conversation between the user and the machine is over, the user gives an evaluation value to judge the performance of the dialogue system. If the whole conversation has not achieved the user's goal, the success reward will be zero. Otherwise, success reward will be a positive constant value.

**Basic Score**, denoted as $r^{bs}$, is used to reduce the search space for tracking agent. As shown in Fig. 2, an auxiliary DST is used. We use the auxiliary belief state $\mathbf{b}^a$ to guide the exploration of the tracking agent. If the exploration belief state $\mathbf{b}^e$ is far away from the auxiliary belief state, a penalty is given as in Eq. (1). Thus, basic score is inversely proportional to the distance between auxiliary belief state and exploration belief state.

$$r^{bs} = -\alpha||\mathbf{b}^e - \mathbf{b}^a||_2$$

where $||\cdot||_2$ is L2 distance and $\alpha \geq 0$ is referred to as **trust factor**. With larger $\alpha$, performance of the tracking agent is closer to the auxiliary DST model.

In the middle of a conversation, the immediate reward of the exploration belief state is $r^{tp} + r^{bs}$, the immediate reward of the last turn is $r^{sr}$.

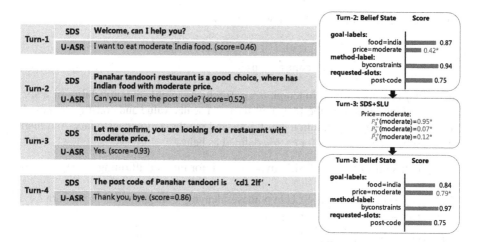

**Fig. 3.** Example with polynomial dialogue state tracking.

## 4    Implementation Detail

In the companion teaching RL-DST framework, the auxiliary DST can make use of arbitrary well trained DST model and the tracking agent can be optimized by any deterministic reinforcement learning algorithm. In this section, we will introduce the dialogue tasks as well as the specific algorithm implementations, though the actual algorithms are not constrained to the below choices. Note that, the tracking agent, i.e. the DST to be optimized, takes a form of deep neural network in this paper.

In this work, we evaluate the proposed framework on the task-oriented dialogue systems in the restaurant/tourism domain in DSTC2/3 [9, 10]. These systems are *slot − based* dialogue systems. There are three slot types: **goal constraint, request slots** and **search method**. The **goal constraint**s are constraints of the information/restaurant which the user is looking for. The **search methods** describe the way the user is trying to interact with the system. The **request slots** are demands which the user has requested. The three different types of slots have different influences on the dialogue performance. Therefore, we use multiple tracking agents, each agent per type, to represent dialogue tracking policy instead of only one overall tracking agent. Each agent has its own input and output. The final overall output is simply the concatenation of the outputs from all agents.

### 4.1    Auxiliary Polynomial Tracker

In this paper, a polynomial tracker is used as the auxiliary DST. It is also referred to as *Constrained Markov Bayesian Polynomial* (CMBP) [32] which is a hybrid model combining both data-driven and rule-based models. CMBP has small number of parameters and good generalization ability. In CMBP, the belief state at current turn is assumed to be dependent on the observations of the current turn and the belief state of the previous turn. A general form of CMBP is shown as below:

$$b_{t+1}(v) = \mathcal{P}(P_{t+1}^+(v), P_{t+1}^-(v), \tilde{P}_{t+1}^+(v), \tilde{P}_{t+1}^-(v), b_t^r, b_t(v)),$$

$$\text{s.t.constraints,} \tag{1}$$

where $\mathcal{P}(\cdot)$ is a polynomial function, $b_{t+1}(v)$ which denotes the probability of a specific slot taking value $v$ at the $(t+1)^{th}$ turn is a scalar value and constraints include probabilistic constraints, intuition constraints and regularization constraints. And there are six probabilistic features for each $v$ defined as below

- $P_{t+1}^+(v)$: sum of scores of SLU hypotheses informing or affirming value $v$ at turn $t+1$
- $P_{t+1}^-(v)$: sum of scores of SLU hypotheses denying or negating value $v$ at turn $t+1$
- $\tilde{P}_{t+1}^+(v) = \sum_{v' \notin \{v, None\}} P_{t+1}^+(v')$
- $\tilde{P}_{t+1}^-(v) = \sum_{v' \notin \{v, None\}} P_{t+1}^-(v')$
- $b_t^r$: probability of the value being 'None' (the value not mentioned) at turn $t$
- $b_t(v)$: belief of "the value being $v$ at turn $t$"

In this paper, polynomial order is 3. The coefficients of polynomial $\mathcal{P}(\cdot)$ are optimized by the off-line pre-collected training data. Each slot type in DSTC2/3 (goal, request, method) has its own polynomial model, represented by $\mathcal{P}_g(\cdot)$, $\mathcal{P}_r(\cdot)$ and $\mathcal{P}_m(\cdot)$ respectively. The belief state of different slot-value pairs within the same slot type is updated by the same polynomial. For example, in our work, we set $\mathcal{P}_g(\cdot)$ as:

$$b_{t+1}(v) = (b_t(v) + P_{t+1}^+(v) * (1 - b_t(v))) * (1 - P_{t+1}^-(v) - \tilde{P}_{t+1}^+(v)).$$

An example of updating belief state of the slot *pricerange* using polynomial tracker is shown in Fig. 3.

## 4.2 Tracking Agents

Three types of the slots (goal, request, method) in DSTC2/3 are not to affect each other. Therefore, the DST tracking agent can be decomposed into three independent tracking agents for DSTC2/3 tasks, represented by **TA_G**, **TA_R** and **TA_M** in Fig. 4. These tracking agents have individual RL components as described in Sect. 3. The three tracking agents correspond to the auxiliary DST trackers $\mathcal{P}_g(\cdot)$, $\mathcal{P}_r(\cdot)$ and $\mathcal{P}_m(\cdot)$ respectively. Note that the forms of the DST tracking agents are deep neural networks instead of polynomials.

In this paper, the input of each tracking agent which represents as $s^g$, $s^r$ and $s^m$ is consistent with the input of polynomial tracker represented in Eq. (1) where each slot is represented by six probabilistic features. The output of each tracking agent represented by $b^{e_g}$, $b^{e_r}$ and $b^{e_m}$ in Fig. 4 is belief state of corresponding slots at the next turn. In this work, we adopt deep deterministic policy gradient (DDPG) algorithm to optimize these three tracking agents. The flexibility of our framework is that we can optimize the selective parts of DST module, the other parts of belief state can still be produced by the auxiliary polynomial DST. We can also figure out which parts of DST module have the bigger effect on the dialogue performance.

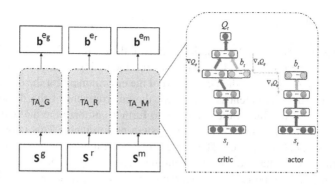

**Fig. 4.** Multi-agent Tracking in DSTC2/3.

### 4.3 DDPG for Tracking Policy

In order to optimize three tracking agents in Fig. 4 which have continuous and high dimensional output spaces, we use DDPG algorithm [16] which is an actor-critic, model-free algorithm based on the deterministic policy gradient that can operate over continuous action spaces. This algorithm combines the actor-critic approach with insights from the DQN algorithm which has a replay buffer and adopts the soft-update strategy.

During training of the three agents, there are three experience memories for each tracking agent respectively. The format of the data in memories is $(\mathbf{s}_t, \mathbf{b}_t^e, r_t, \mathbf{s}_{t+1})$. $\mathbf{s}_t$ is slot feature vector and $\mathbf{b}_t^e$ is the exploration belief state of corresponding slots. The immediate reward $r_t$ is produced by reward function $R(\mathbf{s}_t, \mathbf{b}_t^e, \mathbf{b}_t^a)$ at each turn, presented at Sect. 3.3.

The DDPG algorithm uses the deterministic policy gradient (DPG) method to update the deep neural networks. There are two functions in DDPG algorithm: the actor policy function $\pi(\mathbf{s}_t|\theta)$ deterministically maps the input to output, and the critic function $Q(\mathbf{s}_t, \mathbf{b}_t^e|\beta)$ is learned using the Bellman equation as in Q-learning which aims to minimize the following loss function,

$$L(\beta) = \mathbb{E}_{\mathbf{s}_t, \mathbf{b}_t^e, r_t}[(Q(\mathbf{s}_t, \mathbf{b}_t^e|\beta) - y_t)^2], \tag{2}$$

where $y_t = r_t + \lambda Q(\mathbf{s}_{t+1}, \pi(\mathbf{s}_{t+1}|\beta)), r_t)$ is immediate reward at $t^{th}$ turn and $\lambda \in [0,1]$ is discount factor.

The target of the actor policy is to maximize the cumulative discounted reward from the start state, denoted by the performance objective $J(\pi) = \mathbb{E}[\sum_{t=1}^{T} \gamma^{t-1} r_t | \pi]$. [22] proved that the following equation is the actor policy gradient, the gradient of the policy' s performance:

$$\nabla_\theta J \approx \mathbb{E}_{\mathbf{s}_t}[\nabla_{\mathbf{b}^e} Q(\mathbf{s}, \mathbf{b}^e|\beta)|_{\mathbf{s}=\mathbf{s}_t, \mathbf{b}^e=\pi(\mathbf{s}_t)} \nabla_\theta \pi(\mathbf{s}|\theta)|_{\mathbf{s}=\mathbf{s}_t}], \tag{3}$$

where $\nabla_{\mathbf{b}^e} Q(\mathbf{s}, \mathbf{b}^e|\beta)$ denotes the gradient of the critic with respect to actions and $\nabla_\theta \pi(\mathbf{s}|\theta)$ is a Jacobian matrix such that each column is the gradient $\nabla_\theta[\pi(\mathbf{s}|\theta)]_d$ of the $d^{th}$ action dimension of the policy with respect to the policy parameters $\theta$. The implementation details of the DDPG algorithm are provided in [16].

## 5 Joint Training Process

In Sect. 4, we discuss the implementation details of the on-line DST in DSTC2/3 cases. During the learning process of the tracking agent, the dialogue policy is fixed and the tracker keeps changing. Since the DST is part of the environment for the dialogue policy agent, when the tracking agent is optimized, the environment of the dialogue policy agent is also changed. Thus, we can choose to further optimize the dialogue policy in order to get even more improved dialogue system performance. This is referred to as *joint training* of DST and policy. The process of joint training consists of four phases: the pre-training of the dialogue policy agent, the pre-training of the tracking agent, the training of the dialogue policy agent and the training of the tracking agent. The details of the joint training shows in Algorithm 1.

---

**Algorithm 1:** The process of joint training

---

1: Initialize dialogue policy $Q(\phi)$, **TA_G** tracking agent $Q(\beta^g), \pi(\theta^g)$, **TA_R** tracking agent $Q(\beta^r), \pi(\theta^r)$, **TA_M** tracking agent $Q(\beta^m), \pi(\theta^m)$
   // *pre-train dialogue policy agent*
2: Set polynomial method as the tracker of the system
3: **for** *episode* $= 1 : N_1$ **do**
4:    Update dialogue policy using DQN algorithm
5: **end for**
   // *pre-train tracking agents*
6: **for** *episode* $= 1 : N_2$ **do**
7:    Update the actors $\pi(\theta^g), \pi(\theta^r), \pi(\theta^m)$ of the tracking agents by minimizing mean squared error with the output of polynomial tracker $\mathbf{b}^{ag}, \mathbf{b}^{ar}$ and $\mathbf{b}^{am}$.
8: **end for**
   // *optimize tracking agents*
9: Set multi-tracking agent as the tracker of the system
10: **for** *episode* $= 1 : N_3$ **do**
11:    Update the critics $Q(\beta^g), Q(\beta^r), Q(\beta^m)$ of the multi-tracking agent by minimizing equation (2)
12:    Update the actors $\pi(\theta^g), \pi(\theta^r), \pi(\theta^m)$ of the multi-tracking agent by equation (3)
13: **end for**
   // *optimize dialogue policy agent*
14: **for** *episode* $= 1 : N_4$ **do**
15:    Update dialogue policy using DQN algorithm
16: **end for**

---

## 6 Experiments

Two objectives are set for the experiments: (1) Verifying the performance of the SDS with the optimized on-line DST. (2) Verifying the performance of the dialogue system which jointly train the DST and the dialogue policy.

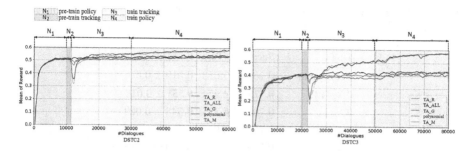

**Fig. 5.** The learning curves of joint training dialogue systems and baseline system in DSTC2 (left) and DSTC3 (right).

## 6.1   Dataset

The proposed framework is evaluated on domains of DSTC2/3 [9, 10]. In DSTC2, there are 8 requestable slots and 4 informable slots. In DSTC3, there are 12 requestable slot and 8 informable slots. Therefore the task in DSTC3 is more complex. Furthermore, the semantic error rate in DSTC3 is higher than the semantic error rate in DSTC2.

Based on the datasets in DSTC2/3, an agenda-based user simulator [20] with error model [21] was implemented to emulate the behaviour of the human user and errors from the input module.

## 6.2   Systems

In our experiments, six spoken dialogue systems with different DST models were compared:

- **Polynomial** is the baseline system. The *polynomial* DST as described in Sect. 4.1 is used. The corresponding policy agent is a two-layer DQN network with 128 nodes per layer.
- **TA_G** is a DST tracking agent as in Fig. 4. It only estimates the belief state of **goal constraint** and the other two parts of belief state are produced by the polynomial tracker.
- **TA_R** is a DST tracking agent as in Fig. 4. It only estimates the belief state of **request slots** and the other two parts of belief state are produced by the polynomial tracker.
- **TA_M** is a DST tracking agent as in Fig. 4. It only estimates the belief state of **search method** and the other two parts of belief state are produced by the polynomial tracker.
- **TA_ALL** is a DST tracking agent as in Fig. 4. Here, the whole belief state is directly produced by the above three tracking agents.
- **TA_noteaching** is similar to **TA_ALL** except the **basic score** reward signal is not used. This is equivalent to directly on-line directly train a neural network DST tracker.

In traditional supervised-learning based DST approaches, metrics such as accuracy or L2 norm are used for evaluation. However, on-line DST optimization does not require

**Table 1.** The performances of tracking agents in DSTC2. The symbol '-' means the dialogue system crashed.

| DST | Success | #Turn | Reward |
|---|---|---|---|
| Polynomial | 0.769 | 5.013 | $0.519 \pm 0.016$ |
| TA_ALL | **0.775** | 4.474 | **0.551** $\pm 0.018$ |
| TA_G | 0.767 | **4.375** | $0.548 \pm 0.020$ |
| TA_R | 0.763 | 5.057 | $0.510 \pm 0.022$ |
| TA_M | 0.765 | 5.121 | $0.509 \pm 0.018$ |
| TA_noteaching | – | – | – |

**Table 2.** The performances of tracking agents in DSTC3. The symbol '-' means the dialogue system crashed.

| DST | Success | #Turn | Reward |
|---|---|---|---|
| Polynomial | **0.744** | 6.566 | $0.415 \pm 0.077$ |
| TA_ALL | 0.713 | **4.117** | **0.507** $\pm 0.083$ |
| TA_G | 0.719 | 4.290 | **0.505** $\pm 0.075$ |
| TA_R | 0.701 | 6.438 | $0.379 \pm 0.028$ |
| TA_M | 0.731 | 6.540 | $0.404 \pm 0.021$ |
| TA_noteaching | – | – | – |

semantic annotation and the optimization objective is to improve dialogue performance. Hence, in this paper, metrics for dialogue performance are employed to evaluate on-line DST performances. There are two metrics used for evaluating the dialogue system performance: average length and success rate. For the reward in Sect. 3.3, at each turn, the turn penalty is $-0.05$ and the dialogue success reward is 1. The summation of the two rewards are used for evaluation, hence, the reward in below experiment tables are between 0 and 1. The trust factors of the basic score in **TA_G**, **TA_R** and **TA_M** tracking agents are 0.2, 0.2, 4 in DSTC2 and 0.07, 0.07, 4 in DSTC3. For each set-up, the moving reward and dialogue success rate are recorded with a window size of 1000. The final results are the average of 25 runs.

### 6.3  DRL-based DST Evaluation

In this subsection, we evaluate the performances of the systems with five different on-line DST models (**TA_G**, **TA_R**, **TA_M**, **TA_ALL** and **TA_noteaching**). The dialogue policy agents of these five systems are optimized by DQN for $N_1$ (10000/20000 in DSTC2/3) episodes/dialogues with the same polynomial trackers. Next, these five systems start to train tracking agents. In the first $N_2$ (1000 in DSTC2/3) episodes, we pre-train *actor* part of DDPG with the output of polynomial tracker using mean squared error (MSE) in all tracking agents. After pre-training, tracking agents of these five systems are optimized by DDPG for $N_3$ (19000/29000 in DSTC2/3) episodes. In the *polynomial* SDS, the dialogue policy agent is optimized for $N_1 + N_2 + N_3$ episodes.

Table 3. The performances of joint training in DSTC2.

| DST | Success | #Turn | Reward |
|-----|---------|-------|--------|
| Polynomial | 0.784 | 4.995 | 0.535 ± 0.015 |
| TA_ALL | **0.810** | 4.566 | **0.581 ± 0.022** |
| TA_G | 0.805 | **4.497** | **0.580 ± 0.015** |
| TA_R | 0.782 | 5.052 | 0.530 ± 0.014 |
| TA_M | 0.782 | 5.051 | 0.530 ± 0.020 |

Table 4. The performances of joint training in DSTC3.

| DST | Success | #Turn | Reward |
|-----|---------|-------|--------|
| Polynomial | 0.754 | 6.580 | 0.425 ± 0.071 |
| TA_ALL | 0.795 | **4.317** | **0.578 ± 0.064** |
| TA_G | **0.800** | 4.579 | 0.571 ± 0.068 |
| TA_R | 0.747 | 6.654 | 0.414 ± 0.069 |
| TA_M | 0.759 | 6.605 | 0.429 ± 0.022 |

In Fig. 5, after the tracking agents are optimized for almost 10000 episodes, the tracking agents in these four on-line DST systems achieve the convergence nearly in DSTC2/3. It demonstrates that the companion teaching framework for the on-line DST is efficient. In Table 1 and Table 2, the tracking agents in the **TA_ALL** system and the **TA_G** system improve the performances of the SDS significantly in DSTC2 and DSTC3. The tracking agents in the **TA_ALL** and the **TA_G** learned a tracking policy which can track the goal of the user accurately. Thus, compared with the *polynomial* system, the length of dialogue in these two systems decrease sharply. The rewards in these two systems increase significantly. The performances of the **TA_R** system and the **TA_M** system are similar with *polynomial* system. We can conclude that **goal constraint** plays a more important role in dialogue state than **request slots** and **search method**. The **TA_noteaching** system crashed during the optimizing process of the tracking agents. It reflects the effectiveness of our proposed companion teaching framework.

### 6.4 Joint Training Evaluation

In this subsection, we evaluate the performances of the systems (except for the **TA_noteaching** system) which jointly train dialogue policy agent and tracking agent. In the first $N_1 + N_2 + N_3$ episodes, training processes of five models have been mentioned in Sect. 6.3. As shown in Fig. 5, in the latter $N_4$ (30000 in DSTC2/3) episodes, four models which contain tracking agents stop optimizing corresponding tracking agents and start to optimize dialogue policy agent and the baseline system continues to train dialogue policy agent. In Fig. 5, we compare the above five systems and the final performances show in Table 3 and Table 4. Compared with the results of the optimized tracking agents in Table 1 and Table 2, the success rates in the **TA_ALL** system and the **TA_G** system increase significantly. It demonstrates that the dialogue policies in the **TA_ALL** and the **TA_G** have adapted the optimized tracking agents respectively.

Compared the results in DSTC3 with the results in DSTC2, we can find that the boost of performance in DSTC3 is larger than that in DSTC2. The reason is that the semantic error rate of SLU in DSTC3 is higher than that in DSTC2, therefore the belief state tracker plays a more important role in DSTC3. These results also indicate that our proposed DRL-based tracker is robust to the input errors of SDS.

## 7    Conclusion

This paper provides a DRL-based companion teaching framework to optimize the DST module of the dialogue system. Under this framework, the tracker can be learned during the conversations between the user and the SDS rather than produced by the off-line methods. We can also choose to jointly train dialogue policy agent and the tracking agent under this framework. The experiments showed that the proposed companion teaching framework for the on-line DST system achieved promising performances in DSTC2 and DSTC3.

**Acknowledgments.** We would like to thank all anonymous reviewers for their insightful comments. This work has been supported by the China NSFC Projects (No. 62120106006 and No. 621061 42), Shanghai Municipal Science and Technology Major Project (2021SHZDZX0102), CCF-Tencent Open Fund and Startup Fund for Youngman Research at SJTU (SFYR at SJTU).

## References

1. Bui, T.H., Poel, M., Nijholt, A., Zwiers, J.: A tractable hybrid DDN-POMDP approach to affective dialogue modeling for probabilistic frame-based dialogue systems. Nat. Lang. Eng. **15**(2), 273–307 (2009)
2. Chang, C., Yang, R., Chen, L., Zhou, X., Yu, K.: Affordable on-line dialogue policy learning. In: Proceedings of EMNLP, pp. 2200–2209 (2017)
3. Chen, L., Yang, R., Chang, C., Ye, Z., Zhou, X., Yu, K.: On-line dialogue policy learning with companion teaching. In: Proceedings of EACL, p. 198 (2017)
4. Cuayáhuitl, H., Keizer, S., Lemon, O.: Strategic dialogue management via deep reinforcement learning. In: NIPS DRL Workshop (2015)
5. Fatemi, M., El Asri, L., Schulz, H., He, J., Suleman, K.: Policy networks with two-stage training for dialogue systems. In: Proceedings of SIGDIAL, pp. 101–110 (2016)
6. Grigorescu, S., Trasnea, B., Cocias, T., Macesanu, G.: A survey of deep learning techniques for autonomous driving. J. Field Rob. **37**(3), 362–386 (2020)
7. Gu, S., Holly, E., Lillicrap, T., Levine, S.: Deep reinforcement learning for robotic manipulation with asynchronous off-policy updates. In: Proceedings of ICRA, pp. 3389–3396. IEEE (2017)
8. Hausknecht, M., Stone, P.: Deep reinforcement learning in parameterized action space. In: Proceedings ICLR, San Juan, Puerto Rico, May 2016. http://www.cs.utexas.edu/users/ai-lab/?hausknecht:iclr16
9. Henderson, M., Thomson, B., Williams, J.D.: The second dialog state tracking challenge. In: Proceedings of SIGDIAL, pp. 263–272 (2014)

10. Henderson, M., Thomson, B., Williams, J.D.: The third dialog state tracking challenge. In: SLT Workshop, pp. 324–329. IEEE (2014)
11. Henderson, M., Thomson, B., Young, S.: Deep neural network approach for the dialog state tracking challenge. In: Proceedings of SIGDIAL, pp. 467–471 (2013)
12. Henderson, M., Thomson, B., Young, S.: Word-based dialog state tracking with recurrent neural networks. In: Proceedings of SIGDIAL, pp. 292–299 (2014)
13. Jurčíček, F., Thomson, B., Young, S.: Reinforcement learning for parameter estimation in statistical spoken dialogue systems. Comput. Speech Lang. **26**(3), 168–192 (2012)
14. Kaelbling, L.P., Littman, M.L., Cassandra, A.R.: Planning and acting in partially observable stochastic domains. Artif. Intell. **101**(1–2), 99–134 (1998)
15. Lee, S.: Structured discriminative model for dialog state tracking. In: Proceedings of SIGDIAL, pp. 442–451 (2013)
16. Lillicrap, T.P., et al.: Continuous control with deep reinforcement learning. arXiv preprint arXiv:1509.02971 (2015)
17. Liu, B., Lane, I.: An end-to-end trainable neural network model with belief tracking for task-oriented dialog. arXiv preprint arXiv:1708.05956 (2017)
18. Mnih, V., et al.: Human-level control through deep reinforcement learning. Nature **518**(7540), 529–533 (2015)
19. Peng, B., Li, X., Gao, J., Liu, J., Wong, K.F.: Deep Dyna-Q: integrating planning for task-completion dialogue policy learning. In: Proceedings of the 56th Annual Meeting of the Association for Computational Linguistics (Volume 1: Long Papers), pp. 2182–2192 (2018)
20. Schatzmann, J., Thomson, B., Weilhammer, K., Ye, H., Young, S.: Agenda-based user simulation for bootstrapping a POMDP dialogue system. In: Proceedings of NAACL, pp. 149–152. ACL, Morristown (2007)
21. Schatzmann, J., Thomson, B., Young, S.: Error simulation for training statistical dialogue systems. In: ASRU, pp. 526–531. IEEE (2007)
22. Silver, D., Lever, G., Heess, N., Degris, T., Wierstra, D., Riedmiller, M.: Deterministic policy gradient algorithms. In: Proceedings of ICML, pp. 387–395 (2014)
23. Su, P.H., Budzianowski, P., Ultes, S., Gasic, M., Young, S.: Sample-efficient actor-critic reinforcement learning with supervised data for dialogue management. In: Proceedings of SIGDIAL, pp. 147–157 (2017)
24. Sun, K., Chen, L., Zhu, S., Yu, K.: A generalized rule based tracker for dialogue state tracking. In: SLT Workshop, pp. 330–335. IEEE (2014)
25. Sun, K., Chen, L., Zhu, S., Yu, K.: The SJTU system for dialog state tracking challenge 2. In: Proceedings of SIGDIAL, pp. 318–326 (2014)
26. Takanobu, R., Zhu, H., Huang, M.: Guided dialog policy learning: reward estimation for multi-domain task-oriented dialog. In: Proceedings of the 2019 Conference on Empirical Methods in Natural Language Processing and the 9th International Joint Conference on Natural Language Processing (EMNLP-IJCNLP), pp. 100–110 (2019)
27. Wang, Z., Lemon, O.: A simple and generic belief tracking mechanism for the dialog state tracking challenge: on the believability of observed information. In: Proceedings of SIGDIAL, pp. 423–432 (2013)
28. Williams, J., Raux, A., Henderson, M.: The dialog state tracking challenge series: a review. Dialogue Discourse **7**(3), 4–33 (2016)
29. Williams, J.D.: Web-style ranking and SLU combination for dialog state tracking. In: Proceedings of SIGDIAL, pp. 282–291 (2014)
30. Williams, J.D., Asadi, K., Zweig, G.: Hybrid code networks: practical and efficient end-to-end dialog control with supervised and reinforcement learning. In: Proceedings of ACL, vol. 1, pp. 665–677 (2017)
31. Young, S., et al.: The hidden information state model: a practical framework for POMDP-based spoken dialogue management. Comput. Speech Lang. **24**(2), 150–174 (2010)

32. Yu, K., Sun, K., Chen, L., Zhu, S.: Constrained Markov Bayesian polynomial for efficient dialogue state tracking. IEEE/ACM Trans. Audio Speech Lang. Process. **23**(12), 2177–2188 (2015)
33. Zhao, T., Eskenazi, M.: Towards end-to-end learning for dialog state tracking and management using deep reinforcement learning. In: Proceedings of SIGDIAL, pp. 1–10 (2016)

# Dual Learning for Dialogue State Tracking

Zhi Chen[1], Lu Chen[1(✉)], Yanbin Zhao[1], Su Zhu[2], and Kai Yu[1(✉)]

[1] X-LANCE Lab, Department of Computer Science and Engineering, MoE Key Lab of
Artificial Intelligence, AI Institute, Shanghai Jiao Tong University, Shanghai, China
{zhenchi713,chenlusz,kai.yu}@sjtu.edu.cn
[2] AISpeech, Jiangshan, China

**Abstract.** In task-oriented multi-turn dialogue systems, dialogue state refers to
a compact representation of the user goal in the context of dialogue history. Dialogue state tracking (DST) is to estimate the dialogue state at each turn. Due to
the dependency on complicated dialogue history contexts, DST data annotation
is more expensive than single-sentence language understanding, which makes
the task more challenging. In this work, we formulate DST as a sequence generation problem and propose a novel dual-learning framework to make full use of
unlabeled data. In the dual-learning framework, there are two agents: the primal tracker agent (utterance-to-state generator) and the dual utterance generator agent (state-to-utterance generator). Compared with traditional supervised
learning framework, dual learning can iteratively update both agents through the
reconstruction error and reward signal respectively without labeled data. Reward
sparsity problem is hard to solve in previous DST methods. In this work, the
reformulation of DST as a sequence generation model effectively alleviates this
problem. We call this primal tracker agent dual-DST. Experimental results on
MultiWOZ2.1 dataset show that the proposed dual-DST works very well, especially when labelled data is limited. It achieves comparable performance to the
system where labeled data is fully used.

**Keywords:** Dual Learning · Dialogue State Tracking · Reinforcement Learning

## 1 Introduction

Dialogue state tracker is a core part of the task-oriented dialogue system, which records
the dialogue state. The dialogue state consists of a set of *domain-slot-value* triples,
where the specific value represents the user goal, e.g., $hotel(price = cheap)$. The
dialogue system responds to the user just based on the dialogue state. Thus, in order
to make the dialogue process natural and fluent, it is essential to extract the dialogue
state from the dialogue context accurately. However, the paucity of annotated data is
the main challenge in this field. In this work, we solve a key problem that how to
learn from the unlabeled data in DST task. We design a dual learning framework for
DST task, where the dialogue state tracker is the primal agent and the dual agent is the
utterance generator. Within the dual learning framework, these two primal-dual agents
help to update each other through external reward signals and reconstruction errors by

using unlabeled data. It only needs a few of labeled dialogue data to warm up these two primal-dual agents.

However, there are two main challenges when combining dual learning framework with previous dialogue state tracking (DST) methods:

**How to represent dialogue state under dual learning framework?** Dual learning method is first proposed in the neural machine translation (NMT) task. The outputs of the primal-dual agents in NMT task are both sequential natural languages. However, in DST task, the output of the dialogue state tracker consists of isolated domain-slot-value triples. The traditional DST task is formulated as a classification problem with the given ontology, where all the possible values of the corresponding slot are listed. Under this problem definition, the previous classification methods just choose the right value for each slot. The recent innovated tracker TRADE [25] directly generates the values slot by slot using copy mechanism from dialogue context. However, these tracker methods get slot values independently. During the dual learning loop, it is hard to get reward signal from these independent slot values. The reward signal from dual utterance generator is also hard to allocate to these isolated value generation processes. Since the relations of the predicted values are not modeled and they are assumed to be independent with each other, it would face serious reward sparse problem. In this work, we reformulate the dialogue state tracking task as a sequential generation task. The whole dialogue state is represented by a sequence with structured information. For example, the state $hotel(price = cheap, area = centre), taxi(destination = cambridge)$ can be represented as "$<hotel>$ $<price>$ cheap $<area>$ centre $</hotel>$ $<taxi>$ $<destination>$ cambridge $</taxi>$".

**Is it reasonable that generating the whole dialogue context from dialogue state?** The intuitive dual task of the state tracker is dialogue context generation. However, in MultiWOZ 2.1 [7] dataset, the dialogue context has more than 10 turns on average and the average length of each sentence is over 10 tokens. It is very difficult in generating accurately a dialogue context with a dialogue state. Because the dialogue context is too long, it is hard to guarantee that the generated dialogue context contains the same semantics with the given state. In this work, we simplify the dual task into a user utterance generation task which ignores the specific values of the given state. The input of the dual task is composed of two parts (i.e., the delexicalized system utterance and the turn state), and its output is the delexicalized user utterance. The delexicalized script is copied from the released code[1]. The system utterance and user utterance can be lexicalized respectively according to the given turn state. We get a new pseudo-labeled dialogue turn. In order to produce multi-turn pseudo-labeled data, we sample a labeled dialogue data and combine it with the pseudo-labeled dialogue turn, where the dialogue turn directly adds to the end of the sampled dialogue context and the turn state covers into the label of the sampled state. Finally, we get a new dialogue context and pseudo label of the state, as the intuitive dual-task does.

The main contributions of this paper are summarized as follows:

– An innovative dialogue state tracking framework based on dual learning is proposed, which can make full use of the unlabeled dialogue data for DST task.

---

[1] https://github.com/ConvLab/ConvLab.

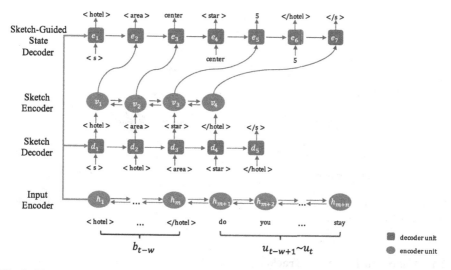

**Fig. 1.** The coarse-to-fine tracker model, which consists of four parts: context encoder, state sketch decoder, sketch encoder and sketch-guided state decoder.

- In this paper, we reformulate the dialogue state tracking as a sequence generation task and propose an efficient state generation model.
- In MultiWOZ 2.1 dataset, our proposed tracker achieves an encouraging joint accuracy. Under dual learning framework, when the labeled dialogue data is limited, the dual-DST works very well.

## 2  Tracker and Dual Task

In this section, we introduce the primal-dual models for DST task under dual learning framework.

Different from previous DST approaches, we formulate the dialogue state tracking task as a sequence generation task. We represent the dialogue state as a structured sequence, rather than a set of isolated state triples. There are two important benefits: (1) The structured state representation keeps the relation information among the different slot values. The relation of these values contains some useful information, for example, the value of the slot *departure* is different from the value of the *destination* in flight ticket booking task. (2) Compared with isolated state representation, the state sequence is more applicable to the dual learning. It is easy to measure using BLEU score and evaluate using normal language model (LM) [17].

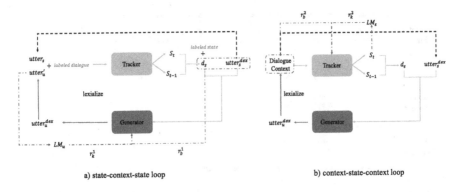

a) state-context-state loop          b) context-state-context loop

**Fig. 2.** Abstraction of the dual learning framework. The dotted box means the start input content of the dual learning game.

## 2.1 Coarse-to-Fine State Tracker

In this work, we adopt coarse-to-fine decoding method [6] to generate the sequential dialogue state. If specific values in sequential dialogue state are removed, we denote the rest representation as state sketch, which only contains domain-slot information, e.g., "$<hotel>$ $<price>$ $<area>$ $</hotel>$ $<taxi>$ $<destination>$ $</taxi>$". In order to simplify state generation, the coarse-to-fine method first generates the state sketch and then produces the final state guided by the state sketch. The coarse-to-fine state generation model consists of four parts: dialogue context encoder, state sketch decoder, sketch encoder and sketch-guided state decoder, as shown in Fig. 1.

**Context Encoder:** The input $x$ of coarse-to-fine tracker is composed of two components: current $w$ dialogue turns $u_{t-w+1} \sim u_t$ and $(t-w)$-th dialogue state $b_{t-w}$, where $w$ is window size of the dialogue context and earlier dialogue utterances is replaced by $(t-w)$-th dialogue state. In this work, we directly concatenate them together and use bi-directional gated recurrent units (GRU) to encode the input as:

$$\overrightarrow{\mathbf{h}}_i = f_{GRU}^x(\mathbf{h}_{i-1}, \mathbf{x}_i), i = 1, \ldots, |x|, \tag{1}$$

$$\overleftarrow{\mathbf{h}}_i = f_{GRU}^x(\mathbf{h}_{i+1}, \mathbf{x}_i), i = |x|, \ldots, 1, \tag{2}$$

$$\mathbf{h}_i = [\overrightarrow{\mathbf{h}}_i, \overleftarrow{\mathbf{h}}_i] \tag{3}$$

where $\mathbf{x}_i$ is embedding of $i$-th token of the input $x$, $[\cdot, \cdot]$ means the concatenation of two vectors and $f_{GRU}^x$ is the input GRU function.

**State Sketch Decoder:** The sketch decoder generates a state sketch $a$ conditioned on the encoded context. We use a unidirectional GRU to decode the state sketch with the attention mechanism [16]. At $t$-th time step of sketch decoding, the hidden vector is computed by $\mathbf{d}_t = f_{GRU}^a(\mathbf{d}_{t-1}, \mathbf{a}_{t-1})$, where $f_{GRU}^a$ is the GRU function and $\mathbf{a}_{t-1}$ is the embedding of previously predicted token. The initial hidden state $\mathbf{d}_0$ is $\overleftarrow{\mathbf{h}}_1$. The attention weight from $t$-th decoding vector to $i$-th vector in encoder is $s_i^t = \frac{\exp(u_i^t)}{\sum_{j=1}^{|x|} \exp(u_j^t)}$.

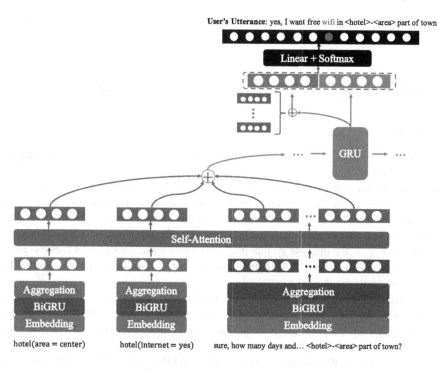

**Fig. 3.** Utterance generation model.

The attention score $u_i^t$ is computed by

$$u_i^t = \mathbf{v}^T \tanh(\mathbf{W}_1 \mathbf{d}_t + \mathbf{W}_2 \mathbf{h}_i + \mathbf{b}), \tag{4}$$

where $\mathbf{v}$, $\mathbf{W}_1$, $\mathbf{W}_2$ and $\mathbf{b}$ are parameters. Then we calculate the distribution of the $t$-th sketch token $p(a_t|a_{<t})$ using

$$p(a_t|a_{<t}) = \text{softmax}(\mathbf{W}_a[\mathbf{d}_t, \mathbf{s}_t] + \mathbf{b}_a), \tag{5}$$

$$\mathbf{s}_t = \sum_{i=1}^{|x|} u_i^t \mathbf{h}_i, \tag{6}$$

where $\mathbf{W}_a$ and $\mathbf{b}_a$ are trainable parameters. Generation terminates until the end token of sequence "$<$EOB$>$" is emitted.

**Sketch Encoder:** We use another bidirectional GRU to map the sketch state into a sequence of sketch vectors $\{\mathbf{v}_i\}_{i=1}^{|a|}$, as context encoder does.

**Sketch-Guided State Decoder:** The final state generation is similar to sketch generation. The difference comes from that the state generation tries to use the generated sketch state. In sketch generation process, the input of the sketch decoder is always previously predicted token. However, during state generation, the input of state decoder at

$t$-th time step $\mathbf{z}_t$ is

$$\mathbf{z}_t = \begin{cases} \mathbf{v}_k, & y_{t-1} \text{ is equal to } a_k \\ \mathbf{y}_{t-1}, & \text{otherwise,} \end{cases} \tag{7}$$

where $\mathbf{y}_{t-1}$ is the embedding of the predicted token at $(t-1)$-th time step.

## 2.2   Dual Task

As introduced in Sect. 1, the dual task of dialogue state tracker is simplified into a user utterance simulation task.

**Encoder:** The input of utterance generation model is composed of two parts: turn state and system utterance (or wizard utterance). The turn state means the dialogue state mentioned by current dialogue turn, which consists of several domain-slot-value triples. We use a bidirectional GRU to encode each triple into a state vector respectively, as shown in Fig. 3. We map the system utterance into a sequence of token vectors. Then we use a self-attention layer [24] to encode state vectors and token vectors together to get final encoded vector.

**Decoder:** The utterance decoder generates the user utterance conditioned on the designed turn state and system utterance. We use a unidirectional GRU to generate the user utterance with attention mechanism. The initial hidden state of the decoder is sum pooling of final encoded vector.

In the dual task, the given system utterance and the generated user utterance are delexicalized, which means that specific values of the dialogue state in two utterances are removed and replaced by common *domain-slot* flags. For example, if the turn state is $hotel(star = 5)$, the system utterance could be "Do you want to reserve <hotel>-<star> star hotel?" and the user utterance could be "Yes, I need <hotel>-<star> star.". Inversely, when the delexicalized utterance is given, we can use the corresponding turn state to get lexicalized utterance. Because the delexicalized system utterance is easy to collect, the function of the dual model can be regarded as to generate a lexicalized dialogue turn given a turn dialogue state.

## 3   Dual Learning for DST

In this section, we present the dual learning mechanism for dialogue state tracking. Before introduce the dual learning method for DST, we define the state tracking model and dual generation model as $P(\cdot|\Theta_{u2s})$ and $P(\cdot|\Theta_{s2u})$, respectively. Similar to dual-NMT [11], we have also two pretrained language models to evaluate the generated state and user utterance, which are indicated as $LM_s$ and $LM_u$. Noticing that we pretrain language model for the sketch of the dialogue state, where the slot values are removed. We regard two language models as two kinds of external knowledge. The dual game of DST task consists of two sub-games: state reconstruction and utterance reconstruction. In other words, the dual learning method contains two kinds of training loop. The abstract of the dual learning method shows in Fig. 2.

The first training loop for **state reconstruction** starts from a turn state. The utterance generator $P(\cdot|\Theta_{s2u})$ generates the delexicalized user utterance with a sampled

---

**Algorithm 1:** Dual learning method for dialogue state tracking

**Input**: unlabeled dialogue data $D_u$, unlabeled turn state $D_s$, labeled dialogue-state pairs $(\hat{D}_u, \hat{D}_s)$, corresponding delexicalized dialogue context $\hat{D}_u^{\text{dex}}$, the language model of user utterance $LM_u$, the language model of coarse state $LM_s$, state tracker $P(\cdot|\Theta_{u2s})$, utterance generator $P(\cdot|\Theta_{s2u})$

**repeat**

      ◁ State-Context-State Loop;

 (1) Sample an unlabeled turn state $d_s$ from $D_s$ and a related delexicalized system utterance $uttr_s^{\text{dex}}$;

 (2) Generate delexicalized user utterance $uttr_u^{\text{dex}}$ using generator $P(\cdot|\Theta_{s2u})$;

 (3) Lexicalize $uttr_s^{\text{dex}}$ and $uttr_u^{\text{dex}}$ using turn state and get a dialogue turn $d_u = (uttr_s, uttr_u)$;

 (4) Evaluate the user utterance $uttr_u$ using $LM_u$ and get external-knowledge reward $r_k^1$ ;

 (5) Sample a labeled dialogue-state pair $(\hat{d}_u, \hat{d}_s)$ and combine this pair with $(d_u, d_s)$ to get a new dialogue-state pair $(\bar{d}_u, \bar{d}_s)$ ;

 (6) Update tracker $P(\cdot|\Theta_{u2s})$ using $(\bar{d}_u, \bar{d}_s)$;

 (7) Generate the dialogue state $\bar{d}_s'$ using tracker $P(\cdot|\Theta_{u2s})$ and get BLEU score reward $r_b^1$ with $\bar{d}_s$ ;

 (8) Update generator $P(\cdot|\Theta_{s2u})$ by policy gradient loss with reward $r^1 = \alpha r_k^1 + (1 - \alpha) r_b^1$;

      ◁ Context-State-Context Loop;

 (9) Sample a unlabeled dialogue context $d_u$ from $D_u$;

 (10) Generate the dialogue state $s_t$ and previous dialogue state $s_{t-1}$ and get $t$-th turn state $d_s$;

 (11) Evaluate the sketch of state $s_t$ using $LM_s$ and get external-knowledge reward $r_k^2$;

 (12) Get delexicalized utterances $(uttr_s^{\text{dex}}, uttr_u^{\text{dex}})$ of $t$-th turn in $d_u$ with turn state $d_s$;

 (13) Update generator $P(\cdot|\Theta_{s2u})$ by cross-entropy loss with $uttr_s^{\text{dex}}$, $d_s$ and $uttr_u^{\text{dex}}$;

 (14) Generate the user utterance using $P(\cdot|\Theta_{s2u})$ with $uttr_s^{\text{dex}}$ and $d_s$ and lexicalize it into $uttr_u'$;

 (15) Calculate the BLEU score of $uttr_u'$ as the reward $r_b^2$ and update tracker $P(\cdot|\Theta_{u2s})$ by policy gradient loss with reward $r^2 = \alpha r_k^2 + (1 - \alpha) r_b^2$;

**until** *Convergence*;

---

delexicalized system utterance. Noticing that the sampled utterance normally contains some domain-slots. The generated utterance can be evaluated by $LM_u$. We use logarithmic of the utterance probability calculated by language model as external-knowledge reward $r_k^1$. Then we pair the given state and the generated utterances as pseudo labeled data to update the tracker $P(\cdot|\Theta_{u2s})$. Because this pair data contains only one turn, we sample from labeled multi-turn data and combine them together to get new multi-turn data. The tracker $P(\cdot|\Theta_{u2s})$ can further predict the state of concatenated utterances in the new multi-turn data. Then we can get BLEU score of the predicted state with combined state. The BLEU score can be regarded as another reward $r_b^1$ to indicate the quality of the generated utterance. At the end of this loop, the generator $P(\cdot|\Theta_{s2u})$ can be updated using weight-sum reward $r^1 = \alpha r_k^1 + (1 - \alpha)r_b^1$ by policy gradient loss [21], where $\alpha$ is the hyper-parameter. The data flow of state reconstruction game is state-context-state.

The second training loop for **utterance reconstruction** starts from dialogue context with $t$ dialogue turns. The tracker $P(\cdot|\Theta_{u2s})$ predicts the $t$-th dialogue state $s_t$ and the previous state $s_{t-1}$. The sketch of the predicted state $s_t$ can be evaluated by $LM_s$. The external-knowledge reward $r_k^2$ is still logarithmic of the probability of the generated state sketch. Then we can get the $t$-th turn state $d_s$. We can further get $t$-th delexical-ized system utterance $utter_s^{\text{dex}}$ using the $d_s$. The generator generates the user utterance with turn state $d_s$ and system utterance $utter_s^{\text{dex}}$. Then we calculate the BLEU score of the user utterance $utter_u'$, which is lexicalized from the generated user utterance. The BLEU score is an implicit reward $r_b^2$ to measure the generated state. Similarly, the tracker $P(\cdot|\Theta_{u2s})$ can be updated using weight-sum reward $r^2 = \alpha r_k^2 + (1 - \alpha)r_b^2$ by policy gradient loss. The data flow of utterance reconstruction game is context-state-context.

The specific process of the dual learning for DST is shown in Algorithm 1, where state-context-state loop means state reconstruction process and context-state-context loop indicates utterance reconstruction process.

## 4    Experiments

### 4.1    Dataset

We evaluate our methods in MultiWOZ 2.1 dataset, which is the largest task-oriented dialogue dataset for multi-domain dialogue state tracking task. MultiWOZ 2.1 dataset contains 8438 multi-domain dialogues and spans 7 dialogue domains. For dialogue state tracking task, there are only 5 domains (*restaurant, hotel, attraction, taxi, train*) in validation and test set. The domains *hospital, bus* only exist in training set. Around 70% dialogues have more than 10 turns and the average length of the utterances in the dialogue is over 10.

### 4.2    Training Details

Similar to TRADE, we initialize all the embeddings using the concatenation of Glove embeddings [18] and character embeddings [10]. We set the window size as 10 turns. The hidden size of all GRUs is 500. Under the dual learning framework, there are two training phases: pretraining phase and dual learning phase. The pretraining phase aims to warm up the state tracker and the utterance generator with labeled data. We adopt Adam [14] optimizer with learning rate 1e−4. During the dual learning phase, the learning rate is 1e−5. In order to stabilize the dual learning, we still use the cross-entropy loss to update the above two models with labeled data. The reward weight $\alpha$ is 0.5.

**Table 1.** The results of baseline models and our proposed coarse-to-fine tracker in MultiWOZ 2.1 dataset. +BERT means that the tracking model encodes the utterances using pretrained BERT. ITC means the inference time complexity, which measures the calculation time of evaluating state. In ITC column, $M$ is the number of slots and $N$ is the number of values. Joint Acc. means the joint goal accuracy.

| Model | +BERT | Joint Acc. MultiWOZ 2.1 | ITC |
|---|---|---|---|
| DS-DST | Y | 51.21% | $O(M)$ |
| SOM-DST | Y | 52.57% | $O(1)$ |
| DST-picklist | Y | **53.30%** | $O(MN)$ |
| HJST | N | 35.55% | $O(M)$ |
| DST Reader | N | 36.40% | $O(M)$ |
| FJST | N | 38.00% | $O(M)$ |
| HyST | N | 38.10% | $O(M)$ |
| TRADE | N | 45.60% | $O(M)$ |
| Coarse2Fine DST(ours) | N | 48.79% | $O(1)$ |
| dual-DST(ours) | N | **49.88%** | $O(1)$ |

## 4.3 Baseline Methods

We first compare our proposed coarse-to-fine state tracker with previous state tracking methods, when all the labeled training data is used.

- **FJST** [7] and **HJST** [7] are two straightforward methods, which directly predict all the slot values based on the encoded dialogue history. Instead of directly concatenating the whole dialogue history as input in FJST, HJST takes the hierarchical model as the encoder.
- **HyST** [9] is a hybrid method that improves HJST by adding the value-copy mechanism.
- **TRADE** [25] directly generates the slot value from the dialogue history.
- **DS-DST** [28] and **DST-picklist** [28] divide the slots as uncountable type and countable type and generate the slot value in a hybrid method like HyST. Compared with DS-DST, DST-picklist knows all the candidate values of the slots, including uncountable slots.
- **SOM-DST** [13] feeds dialogue history and previous state as the input and modifies the state with dialogue history into the current state.
- **DST Reader** [8] formulates DST task as a machine reading task and leverages the corresponding method to solve the multi-domain task.

The second experiment is to invalid the dual learning framework for DST task. In this experiment setup, we randomly sample 20%, 40%, 60% and 80% labeled data in training data. The rest data is used as unlabeled data. We compare dual learning method with **pseudo labeling method**, which is an important approach to use the unlabeled data. The pseudo labeling method first uses the sampled labeled data to pretrain our proposed tracker. During the training of pseudo labeling method, the pretrained tracker is used to generate the state of the unlabeled dialogue context. Then, the dialogue context and the generated state are paired together as the pseudo-labeled data to retrain the tracker. In order to stabilize the training process of the pseudo labeling method, we also mixture the pseudo labeled data and labeled data as a batch to update the pretrained tracker.

## 4.4   Results

*The performance of our trackers:* As shown in Table 1, our proposed coarse-to-fine tracker achieves the highest joint goal accuracy in the BERT-free models. Our proposed tracker directly generates all the slot values, which is represented as a structured sequence. Compared with the methods that predict the values slot by slot, the inference time complexity (ITC) is O(1). This property is important for the dialogue system. The response time of the dialogue system effects user experience seriously. Compared with SOM-DST, our proposed tracker does not rely on the pretrained BERT [5], whose model size is more than 110M. This is another challenge for memory-starve devices. Compared with recently proposed TRADE, our proposed coarse-to-fine tracker not only reduces the inference time, but also gets the absolute 3.19% joint goal accuracy improvement.

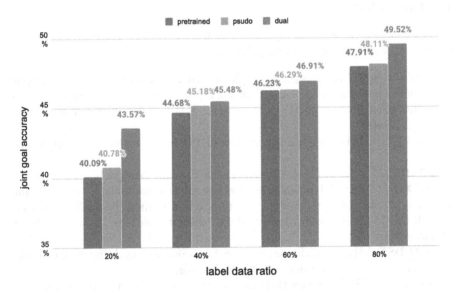

**Fig. 4.** The joint goal accuracy with unlabeled data.

*The performance of dual learning:* In order to validate the effectiveness of our proposed dual learning framework, we randomly sample parts of training dataset as the small training set. The rest data is regarded as unlabeled data. In this experiment, we randomly sample 20%, 40%, 60% and 80% labeled data. As shown in Fig. 4, we can see that the joint goal accuracy improves as the labeled data increases. It indicates that the scale of the annotated data is a big challenge for the multi-domain DST task. In this work, we propose a dual learning framework for DST to help to improve the performance of the tracker with the unlabeled data.

When the training data is starved, the dual learning method can improve the performance of the pretrained tracker by efficiently using the unlabeled data. Compared with pseudo labeling method, our proposed dual learning method is able to treat the external knowledge (two kinds of language models: coarse state language model and user utterance language model) as reward function to feedback to the tracker and improve the performance. Especially, when the labeled data is extremely limited that only has 20% sampled data, the dual learning method achieves a larger performance gain than the pseudo labeling method. As shown in Fig. 4, we can see that the pseudo labeling method only gets less improvement. As we introduce in Sect. 4.1, the multi-domain DST task in MultiWOZ 2.1 dataset is more complex than single-domain DST task. The positive influence of the pseudo labeled data for pretrained tracker is limited.

When the training data is fully used, the dual learning method can be still used to fine-tune the pretrained tracker. During the dual learning process, all the training data can be regarded as the unlabeled data. As shown in Table 1, the dual-DST can get further improvement from the fully pretrained tracker.

# 5    Related Work

*Multi-Domain DST:* With the release of MultiWOZ dataset [1], one of the largest task-oriented dialogue datasets, many advanced dialogue state tracking methods for multi-domain task have been proposed. The previously proposed multi-domain state tracking approaches can be divided into two categories: classification [7,26] and generation [3,12,15,19,25]. The classification methods usually require that all the possible slot values are given by ontology. However, in real dialogue scenarios, some slot values cannot be enumerated. To alleviate this problem, the generative methods have been proposed, where the slot values are directly generated from the dialogue history. Like the classification methods, most of the generative methods generate slot value one by one, until all the slots on different domains have been visited. The methods that predict the slot values independently can not be used in dual learning framework. In this work, we redefine the dialogue state as a structured representation. We further propose a coarse-to-fine tracking method to directly generate the structured dialogue state.

*Dual learning:* Dual learning method is first proposed to improve neural machine translation (NMT) [11]. In NMT task, the primal task and the dual task are symmetric, while not in DST task. We design a state tracking model and an utterance generation model under the dual learning framework of DST. The idea of dual learning has been applied into various tasks, such as Question Answer [22]/ Generation [4,23], Image-to-Image Translation [27], Open-domain Information Extraction/Narration [20] and Semantic Parsing [2]. To the best of our knowledge, we are the first to introduce the dual learning in dialogue state tracking.

## 6 Conclusion

In this work, we first reformulate the dialogue state tracking task as a sequence generation task. Then we adopt a coarse-to-fine decoding method to directly generate the structured state sequence. The proposed coarse-to-fine tracker achieves the best performance among BERT-free methods. The main contribution of this work lies on building a dual learning framework for multi-domain DST task. The experimental results indicate that our proposed dual learning method can efficiently improve the pretrained tracker with unlabeled data. In future work, we will further improve the state tracking model and dual utterance generation model using pretrained models, e.g. BERT.

**Acknowledgments.** We would like to thank all anonymous reviewers for their insightful comments. This work has been supported by the China NSFC Projects (No. 62120106006 and No. 621061 42), Shanghai Municipal Science and Technology Major Project (2021SHZDZX0102), CCF-Tencent Open Fund and Startup Fund for Youngman Research at SJTU (SFYR at SJTU).

## References

1. Budzianowski, P., et al.: MultiWOZ-a large-scale multi-domain wizard-of-OZ dataset for task-oriented dialogue modelling. In: Proceedings of the 2018 Conference on Empirical Methods in Natural Language Processing, pp. 5016–5026 (2018)
2. Cao, R., Zhu, S., Liu, C., Li, J., Yu, K.: Semantic parsing with dual learning. In: Proceedings of the 57th Annual Meeting of the Association for Computational Linguistics, pp. 51–64 (2019)
3. Chen, Z., et al.: UniDU: towards a unified generative dialogue understanding framework. In: Proceedings of the 23rd Annual Meeting of the Special Interest Group on Discourse and Dialogue, pp. 442–455. Association for Computational Linguistics, Edinburgh, September 2022. https://aclanthology.org/2022.sigdial-1.43
4. Chen, Z., et al.: Decoupled dialogue modeling and semantic parsing for multi-turn text-to-SQL. In: Findings of the Association for Computational Linguistics: ACL-IJCNLP 2021, pp. 3063–3074 (2021)
5. Devlin, J., Chang, M.W., Lee, K., Toutanova, K.: BERT: pre-training of deep bidirectional transformers for language understanding. arXiv preprint arXiv:1810.04805 (2018)
6. Dong, L., Lapata, M.: Coarse-to-fine decoding for neural semantic parsing. In: Proceedings of the 56th Annual Meeting of the Association for Computational Linguistics (Volume 1: Long Papers), pp. 731–742 (2018)
7. Eric, M., et al.: MultiWOZ 2.1: multi-domain dialogue state corrections and state tracking baselines. arXiv preprint arXiv:1907.01669 (2019)
8. Gao, S., Sethi, A., Agarwal, S., Chung, T., Hakkani-Tur, D., AI, A.A.: Dialog state tracking: a neural reading comprehension approach. In: 20th Annual Meeting of the Special Interest Group on Discourse and Dialogue, p. 264 (2019)
9. Goel, R., Paul, S., Hakkani-Tür, D.: HyST: a hybrid approach for flexible and accurate dialogue state tracking. Proc. Interspeech **2019**, 1458–1462 (2019)
10. Hashimoto, K., Tsuruoka, Y., Socher, R., et al.: A joint many-task model: growing a neural network for multiple NLP tasks. In: Proceedings of the 2017 Conference on Empirical Methods in Natural Language Processing, pp. 1923–1933 (2017)
11. He, D., et al.: Dual learning for machine translation. In: Advances in Neural Information Processing Systems, pp. 820–828 (2016)

12. Hosseini-Asl, E., McCann, B., Wu, C.S., Yavuz, S., Socher, R.: A simple language model for task-oriented dialogue. arXiv preprint arXiv:2005.00796 (2020)

13. Kim, S., Yang, S., Kim, G., Lee, S.W.: Efficient dialogue state tracking by selectively over-writing memory. arXiv preprint arXiv:1911.03906 (2019)

14. Kingma, D.P., Ba, J.: Adam: a method for stochastic optimization. arXiv preprint arXiv:1412.6980 (2014)

15. Lee, C.H., Cheng, H., Ostendorf, M.: Dialogue state tracking with a language model using schema-driven prompting. In: Proceedings of the 2021 Conference on Empirical Methods in Natural Language Processing, pp. 4937–4949 (2021)

16. Luong, M.T., Pham, H., Manning, C.D.: Effective approaches to attention-based neural machine translation. In: Proceedings of the 2015 Conference on Empirical Methods in Natural Language Processing, pp. 1412–1421 (2015)

17. Mikolov, T., Karafiát, M., Burget, L., Černocký, J., Khudanpur, S.: Recurrent neural network based language model. In: Eleventh Annual Conference of the International Speech Communication Association (2010)

18. Pennington, J., Socher, R., Manning, C.: Glove: global vectors for word representation. In: Proceedings of the 2014 Conference on Empirical Methods in Natural Language Processing (EMNLP), pp. 1532–1543 (2014)

19. Su, Y., et al.: Multi-task pre-training for plug-and-play task-oriented dialogue system. In: Proceedings of the 60th Annual Meeting of the Association for Computational Linguistics (Volume 1: Long Papers), pp. 4661–4676 (2022)

20. Sun, M., Li, X., Li, P.: Logician and orator: learning from the duality between language and knowledge in open domain. In: Proceedings of the 2018 Conference on Empirical Methods in Natural Language Processing, pp. 2119–2130 (2018)

21. Sutton, R.S., McAllester, D.A., Singh, S.P., Mansour, Y.: Policy gradient methods for reinforcement learning with function approximation. In: Advances in Neural Information Processing Systems, pp. 1057–1063 (2000)

22. Tang, D., Duan, N., Qin, T., Yan, Z., Zhou, M.: Question answering and question generation as dual tasks. arXiv preprint arXiv:1706.02027 (2017)

23. Tang, D., et al.: Learning to collaborate for question answering and asking. In: Proceedings of the 2018 Conference of the North American Chapter of the Association for Computational Linguistics: Human Language Technologies, Volume 1 (Long Papers), pp. 1564–1574 (2018)

24. Vaswani, A., et al.: Attention is all you need. In: Advances in Neural Information Processing Systems, pp. 5998–6008 (2017)

25. Wu, C.S., Madotto, A., Hosseini-Asl, E., Xiong, C., Socher, R., Fung, P.: Transferable multi-domain state generator for task-oriented dialogue systems. In: Proceedings of the 57th Annual Meeting of the Association for Computational Linguistics, pp. 808–819. Association for Computational Linguistics, Florence, July 2019

26. Ye, F., Manotumruksa, J., Zhang, Q., Li, S., Yilmaz, E.: Slot self-attentive dialogue state tracking. In: Proceedings of the Web Conference 2021. pp. 1598–1608 (2021)

27. Yi, Z., Zhang, H., Tan, P., Gong, M.: DualGAN: unsupervised dual learning for image-to-image translation. In: Proceedings of the IEEE International Conference on Computer Vision, pp. 2849–2857 (2017)

28. Zhang, J.G., et al.: Find or classify? Dual strategy for slot-value predictions on multi-domain dialog state tracking. arXiv preprint arXiv:1910.03544 (2019)

# Automatic Stress Annotation and Prediction for Expressive Mandarin TTS

Wendi He[✉], Yiting Lin, Jianhao Ye, Hongbin Zhou, Kaimeng Ren, Tianwei He, Pengfei Tan, and Heng Lu

Ximalaya Inc., No. 799, Dangui Road, Shanghai, China
{cloris.he,yiting.lin,jianhao.ye,hongbin.zhou,irving.ren,
tianwei.he,pengfei.tan,bear.lu}@ximalaya.com

**Abstract.** The current text-to-speech technique has developed to a close-to-human state, and more research interest has been paid to highly expressive and more controllable speech synthesis. Stress detection and modeling in the Mandarin TTS(Text-to-speech) system have been verified to be an efficient and direct way to enhance the rhythm and prosody performance in previous studies. But labeling stress in training data manually needs linguistic knowledge and is also time-consuming. In this paper, an automatic syllable-level stress annotation mechanism is proposed. Then based on the automatically annotated stress labels, a transformer-based ALBERT front-end module is built for stress label prediction from the text. In the experiment part, a DurIAN-based expressive text-to-speech system is built with the proposed automatic stress annotation and prediction module. Experiments show the proposed method can consistently predict stress from linguistic context input, and speech synthesis systems with proposed stress annotation and prediction components outperform baseline systems.

**Keywords:** Speech synthesis · Stress modeling · Expressive speech synthesis · Mandarin Text-to-speech · Rule-based

## 1 Introduction

Natural and expressive mandarin speech synthesis continues to receive increased attention. Mandarin, as a tonal language, performs to emphasize the importance of different tones in distinguishing the meaning of words, ie. "书" (**shu1** means book) and "树" (**shu4** means tree). In natural conversation, both intonation and syllable tone cause changes in the meaning of the sentence. Stressed and unstressed syllables often demonstrate where the felicitous emphasis is placed within a sentence and the improper location of stress always makes the speech flow unnatural and even difficult to understand. Therefore, stress modeling and stress detection mechanism is fundamental research in Text-to-speech system.

Stress is the perceptual prominence within a prosodic word or utterance [5] which could bring about the fluctuation of pitch contour. However, it is still controversy regarding the definition of stress in Mandarin. In this paper, we consider

© The Author(s), under exclusive license to Springer Nature Singapore Pte Ltd. 2023
L. Zhenhua et al. (Eds.): NCMMSC 2022, CCIS 1765, pp. 306–317, 2023.
https://doi.org/10.1007/978-981-99-2401-1_27

stress to be a syllable-level phenomenon with four degrees: Strong Stressed(**SS**), Regular Stressed(**RS**), and Unstressed(**US**). The majority of previous studies show that pitch and duration have been proven to be the two complementary important acoustic features to influence stress perception by researchers [3,20,21]. Further research [13] proved that some specific combinations of tonal patterns indicate whether the syllable is stressed or not.

In existing related work of stress detection and annotation, the Fujisaki model [4] describes the fundamental frequency contour as the superposition of the outputs of phrase and tone (or accent) commands that match the two-level hierarchical stress model we proposed. [7,11] utilize this model for generating stress and controlling prosodic F0. Furthermore, other statistical methods based on a large corpus are of great help in the study of stress annotation, like ToBI [22], C-ToBI [10] etc. Relying entirely on manual annotation of stress is undoubtedly very time-consuming and inaccurate. [24] used Decision Trees and Markov models to determine the type of syllable discontinuity and the presence of stress. [12,16] described the performance of the machine learning-based utilization like SVM, CART, and AdaBoost with CART in both English and Mandarin. [26] introduced the use of continuous lexical embedding in the BLSTM model which gains an F1 score. Inspired by [1,11,13,21,23,26], the main contribution of this paper is that we proposed a heuristic automatically rule-based stress detection mechanism from audio files for labeling stress to deal with the time-consuming and the bias when annotating manually by different annotators. Besides, these labeling data successfully model the stress in both acoustic model DurIAN [25] and front-end model ALBERT [9] for building a controllable stress TTS. The effectiveness of and the performance of stress controllable TTS system is evaluated using comprehensive objective and subjective experiments. The audio samples are available at https://misakamikoto96.github.io/stress-tts.github.io/.

## 2   Methodology

### 2.1   Proposed Method for Stress Detection

**Data Preparation and Statistic Result.** First, we carried out an analysis of a large-scale corpus to investigate prosodic variation in different contexts. The corpus used for the experimental study in this paper is taken from the single-speaker Mandarin novel-style corpus with distinctive style and variable pitch contour to capture more perceptual prominence within words or utterances. The linguistic experts annotate the corpus with prosody boundaries, pinyin, and stress by using Praat [2] toolkit. Then, all the syllable units are classified in to different categories, see in Table 1. From the theoretical and heuristically statistical results of the experiment mentioned above, we conclude that: (1) Stressed/Unstressed classification of syllables is meaningful for improving the naturalness of mandarin speech synthesis. (2)By observing the pitch curve, we can determine to stress or unstress a syllable based on whether the syllable has a significantly higher pitch maximum than its preceding or following syllables, or whether a syllable is significantly performed better than non-stressed syllables in their ability to maintain its 'typical tonal patterns shape' [21]. The study result illustrates that stress is influenced not only by pitch range and duration of the syllables, but also by the neighboring silence, neighboring tone pattern, and neighboring stresses. Therefore, we proposed an automatically rule-based mechanism for syllable-level stress labeling from audio files.

**Table 1.** The part of categories for syllable unit classification

| Categorise | Description |
| --- | --- |
| tone_categorise | Tone categorise |
| stressed_or_not | Whether the syllable is stressed |
| left/right_prosodic_categories | the prosody boundaries categories of its preceding/next syllable |
| left/right_syllable_tone | the tone categories of its preceding/next syllable |
| left/right_f0_max | the $max(f_0)$ of its preceding/next syllable |

**Details.** The code reproduction could be done with the pseudo-code of three steps of the Stress Detection Rule, all the scalar coefficients included are originated from our previous statistical results, control experiments, and empirical values.

- **$ST$** stands for the sequence of Stress labels for the input text, including Strong stressed(SS), Regular stressed(RS), and Unstressed(US), zero-initialized from the pre-annotation step.
- **$P^{syl}$** stands for the discrete semitone value $[[P_1], [P_2], ..., [P_t]]$, $P_i$ contains the set of points of syllable-level semitone contour for the i-th syllable.
- **$P^{phr}$** stands for discrete points value of phrase-level semitone contour (split a sentence into phrases by punctuation and prosody boundary $>=$ IPH))
- **$P$** adds all the elements $P_i$ of an iterable $P^{syl}$ , $P = [P_1, P_2, ..., P_t]$, stand for all the points of the whole sequence.
- **$To$** stands for Tone including $T1$(high), $T2$(rising), $T3$(dipping), $T4$(falling) and $T5$(neutral).
- **$D$** stands for the slicing timeline list of all syllables. $D_i[-1] - D_i[0]$ means the duration list for the i-th syllable, $D = [D_1, D_2, ..., D_t]$
- **$\Delta O$** stands for the $|\Delta timestamps|$ of the highest and lowest points on the pitch curve of a syllable.
- **$k$** stands for the slope between the highest and lowest point on the pitch curve, $k = |max(P_i^{syl}) - min(P_i^{syl})|/\Delta O$

*Initial settings*. Firstly, Kaldi toolkit [18] is utilized to do forced-alignment for getting phoneme-level duration. Based primarily on experience, we pre-set an initial pitch array getting from WORLD [15] and perform linear interpolation, then align it to be the syllable-level F0 sequence, which is defined as $P_{init}$. As humans perceive differences in pitch levels as approximate logarithmic levels, rather than linear levels, all pitch values were normalized in this study from the hertz scale to the semitone scale. Each current syllable-level semitone value $S_i = 12 \times \log_2(F0_i/F0^{ref})$, where $F0$ stands for syllable-level $F0$ value sequence with valid hop length, and $F0^{ref}$ usually represent the lower limit of the speaker's vocal range.

*Pre-annotation*. The rule for syllable-level pre-annotation of stress label is firstly done by detecting the convex pitch contour in the whole sentence. We set a zero-initialized stress label array $St = \{St_1, St_2, St_3, ..., St_t\}$ which length is equal to the syllable-level linguistic feature. In syllable-level semitone contour $S$, we detect those syllables where the peak of the contour is highest than its previous one and its next one. We detect a syllable-level semitone value $S_i$ whose highest point is the maximum of $[\max(S_{i-1}), \max(S_i), \max(S_{i+1})]$, and then, we calculate the stress level on this syllable by the equation below:

$$St_i = \begin{cases} S_{i_{max}} > Med(S) + 1.5 * \delta(S), & \textbf{SS} \to 2 \\ S_{i_{max}} > Med(S) + 0.5 * \delta(S), & \textbf{RS} \to 1 \\ else, & \textbf{US} \to 0 \end{cases}$$

*Forward and Cancel Movement*. In order to maintain the integrity of the tone contour, the movement of the stress label might be moved forward or canceled. For syllables whose pre-labeled stress results are unequal to 0, we initially adjust the position of the stress label based on its pitch curve. The $ST_i$ of the i-th syllable is shifted forward to the nearest j-th syllable($j < i$), with its $T_{i-n} \neq 5$, and the current $ST_i = 0$. See in Algorithm1.

*Supplementary Movement*. For dealing with syllables in which the stress label is unmarked in the pre-annotation phase when its tone is T2 or T3. When considering the falling-rise tone and rising tone, the probability of stress is highly related to the syllable duration to express the feeling of stress in Mandarin speech. *Supplementary* means the action to re-label some syllable with significant duration characteristics.

*Expand Movement*. To detect the syllable has been omit-stressed in the previous steps when a current i-th syllable $Syl_i$ is stressed, there is a comparison among the maximum of syllable-level semitone value of its preceding syllable $Syl_{i-1}$, following syllable $Syl_{i+1}$, and the overall pitch level of the entire sentence when the tone of $Syl_{i-1}$ or $Syl_{i+1}$ is T1 or T4. *Expand* represents extending the stress label to its eligible preceding and following part in a sequence. See in Algorithm 3.

---

**Algorithm 1.** Forward movement of stress label

---

**Ensure:** $S_i \neq 0$
  **for** $i \leftarrow 1$ **to** $Len(ST)$ **do**
    **if** $T_i = 5$ and $T_i = 3$ **then**
      $*MoveForward$
    **else if** $T_i = 2$ **then**
      **if** $max(P_i^{syl}) - min(P_i^{syl}) < 0.5 * \delta(St)$ **then**
        $*MoveForward$
      **else**
        $*Cancel$
      **end if**
    **else if** $T_i = 4$ **then**
      **if** $k < 10$ and $T_{i-1}$ **in**$[2, 3]$ **then**
        $*Move\ Forward$
      **else if** $k > 10$ and $max(P_i^{syl}) < Med(P) + 0.75 * \delta(P)$ **then**
        $*Cancel$
      **else**
        $*Cancel$
      **end if**
    **else if** $T_i = 1$ and k **then**
      **if** $k < 15$ and $max(P_i^{syl}) < Med(P) + 0.5 * \delta(P)$ **then**
        $*Cancel$
      **end if**
    **end if**
  **end for**

---

---

**Algorithm 2.** Supplementary Movement of stress label

---

  **for** $i \leftarrow 1$ **to** $Len(ST)$ **do**
    **if** $T_i = 3$ **then**
      **if** $min(P_{i+1}^{syl}) \leq min(min(P_i^{syl}), max(P_{i+1}^{syl}))$ **then**
        **if** $max(P_i^{syl}) = min(P)$ and $D_i > max(D_{i-1}, D_{i+1})$ and $D_i > avg(Dur)$ **then**
          $*ST_i = 2$
        **end if**
      **else if** $max(P_i^{syl}) <= min(P_i^{phr})$ and $D_i > avg(D)$ **then**
        $*ST_i = 1$
      **end if**
    **else if** $Tone_i = 2$ **then**
      **if** $k > 30$**and**$|\Delta O| > 0.5 * D_i$  **then**
        **if** $D_i > 1.75 * avg(D)$ **then**
          $*ST_i = 2$
        **else if** $D_i > 1.2 * avg(D)$ **then**
          $*ST_i = 1$
        **end if**
      **end if**
    **end if**
  **end for**

---

**Algorithm 3.** Expand Movement of stress label

**Ensure:** $St_i \neq 0$
  **for** $i \leftarrow 1$ **to** $Len(ST)$ **do**
    **if** $max(P_i^{syl}) > Med(P) + 0.5 * \delta(P)$ **then**
      **if** $max(P_i^{syl}) - max(P_{i+1}^{syl}) < 0.45 * \delta(P)$ **or** $max(P_i^{syl}) - max(P_{i-1}^{syl}) < 0.45 * \delta(P)$ **then**
        **if** $T_{i-1}$ **in** $[1, 4]$ **then**
          *Forward Expansion of $ST_i$ until condition false*
        **end if**
        **if** $T_{i+1}$ **in** $[1, 4]$ **then**
          *Backward Expansion of $ST_i$ until condition false*
        **end if**
      **end if**
    **end if**
  **end for**

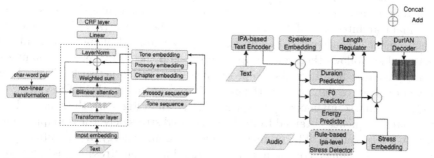

(a) *The pipeline of the textual-level fron-tend stress prediction*

(b) *The main architecture of our proposed stress-modeling TTS system*

**Fig. 1.** The architecture of the models

## 2.2   Textual-Level Stress Prediction

The main architecture of the proposed textual-level stress prediction is shown in Fig. 1(a). A transformer-based pre-trained model ALBERT has been used as the baseline. Compared to ALBERT, multiple stress-related prior information including lexicon character-words pair [14], hierarchical prosody, tone pattern, and chapter-level context information are integrated into the network effectively. Input sentence $T = [t_1, t_2, ..., t_N]$, N is the length of utterance and also the number of stress labels for each syllable $S = [s_1, s_2, ..., s_N]$ with $< eos >$ and $< bos >$ padding in the head and tail.

The char-lexicon pair information $E_{lexicon}$ passed through the Lexicon Adapter which follows the same setup of [14] for the in-depth fusion of lexicon features and **ALBERT** representation. The tone and prosody boundary of each syllable are obtained from the multi-task **ALBERT** based Ximalaya self-developed frontend system, $E_{tone} = e^T(T_t)$ and $E_{prosody} = e^P(P_t)$ where $e^T$ and $e^P$ are trainable randomly initialized embedding lookup tables with size 5 and 3 (as $T_{class} = \{T1, T2, T3, T4, T5\}$ and $P_{class} = \{PW, PPH, IPH\}$) respectively. Also, inspired by [6], from the similar structure setup of its conversational context encoder, we select the sequence of **BERT** sentence embeddings $E_{t-2:t}$ with the length c from $E_{t-2}$ to the current sentence $E_t$ as the inputs to get the output $E_{chapter}$ to directly extract stress position-related contextual features from sentence embeddings. We finetune from the pre-trained Albert-Chinese-base model with randomly initialized dense layers and use the CRF layer for sequence labeling. In the training process, the model task is to minimize the sentence-level negative log-likelihood loss as well as in the decoding stage, the most likely sequence will be obtained by the Viterbi algorithm, as the setting in [9].

### 2.3    Modeling Stress in Acoustic Model

To generate a target Mel-spectrogram sequence including expressive stress-related performance, we tried to ways to do stress modeling in an acoustic model. The main architecture of the acoustic model follows DurIAN, [25] while the enhancement is constructed by adding extra predictors of phoneme-level f0 and energy which follows the same setup as in [19] and the real values are quantized into trainable embeddings. As the prosody bottleneck sub-network [17] has been proven to produce significant performance improvements in the TTS system, we follow the same setup in our architecture. As shown in Fig. 1(b), the syllable-level stress label sequence obtained from the rule-based stress detector is mapped into an IPA-level sequence. The predicted values of f0 and energy are quantized into trainable embeddings, concatenated with 128-dimension stress embedding, and scaled duration output, to condition the decoder for speech synthesis.

## 3    Experiments

For the acoustic model, an internal male audio corpus is used in this experiment containing 13049 sentences in a novel genre including narration and dialogue (about 108,741 prosodic words and 326,225 syllables), which were uttered by a professional male speaker, and saved in WAV files (16-bit mono, 24kHz). 80-dimensional Mel-spectrograms are extracted by using Hanning Window with hop size 240 (frameshift 10ms). Kaldi toolkit [18] is utilized for forced alignment and getting phoneme-level duration and prosody boundaries from silence. F0 and energy are extracted following the same setup in [19]. We obtained the syllable-level f0 by aggregating the slicing-aligned duration of each phoneme. The tone and the prosody boundaries of each syllable are contained in the input IPA sequence, which is used as the input of the rule-based stress extraction tool for generated stress annotation, together with the same level duration and f0.

For the textual-level stress prediction, the corpus we selected are internal parallel datasets of different speakers in the same novel domain and gave the corresponding audio for accent detection and alignment to each word as the target for front-end model training. Moreover, we use several different combinations of linguistic textual features to investigate which have the greatest impact on the accuracy of stress sequence prediction.

## 3.1  Complete Stress-Controllable TTS System

There are two ways to perform stress in the whole TTS system:

- *MANUAL* The stress label sequence is assigned manually by professional linguists based on the semantics of the sentence, from the SSML input.
- *FE* The stress label sequence is predicted from the Albert-based Front-end system as described in para 2.2.
- *BASELINE* The baseline modified DurIAN model without any stress modeling.

## 3.2  Experimental Results

*Performance of rule-based stress extraction.* An initial syllable-level manual check of overall accuracy rates on the annotation result of the stress extraction tool, three linguistic experts are involved in the evaluation stage. We randomly selected 100 audio samples of each speaker distributed in female(data from 2 internal speakers), male(data from 2 internal speakers), child(data from 1 internal child speaker), and male elder(data from 1 internal speaker) respectively with different vocal ranges to verify the accuracy of the rule-based stress extraction tool in detecting stress on Wav. The precision and f1-score of each stress label shown in Table 2 illustrate that the stress detection rule could generally achieve accurate detection of stress for different speakers.

**Table 2.** Precison and F1-score of the generated results of *SS*, *RS* and *US* for each syllable, and the overall Accuracy from the stress extraction tool.

| Speaker Type | SS | | RS | | US | | Acc |
|---|---|---|---|---|---|---|---|
| | prec | f1-score | prec | f1-score | prec | f1-score | |
| Male | 0.9679 | 0.9837 | 0.9760 | 0.9803 | 0.9979 | 0.9953 | 0.9918 |
| Female | 0.9714 | 0.9714 | 0.9382 | 0.9255 | 0.9700 | 0.9738 | 0.9619 |
| Child | 1.0000 | 1.0000 | 0.9796 | 0.9143 | 0.9572 | 0.9755 | 0.9648 |
| Elder | 0.9767 | 0.9882 | 0.9518 | 0.9693 | 0.9959 | 0.9878 | 0.9837 |

*Accuracy of the Frontend model.* As a sequence labeling task, stress prediction should be evaluated with consideration of all the stresses. The trained classifiers are applied to a test corpus to predict the label of each stress. Table 3 shows that the accuracy and the macro-f1 score of *FE model* increased when testing different feature embeddings combination added in the ALBERT model. Tone and prosody boundaries have the greatest impact on the improvement of prediction accuracy.

*Ablation Studies.* Stress, as a high-level feature, is strongly correlated with f0 and energy, which is a refined expression of the prosody feature. There is a consideration that the prediction of the f0 and the energy may perhaps bring about

**Table 3.** Accuracy and macro-f1 score of FE model

| Features used | Accuracy(%) | macro-f1 |
|---|---|---|
| ALBERT baseline | 75.46 | 0.3462 |
| lexicon ALBERT(*LA*) | 76.01 | 0.4548 |
| *LA* + chapter | 76.45 | 0.4532 |
| *LA* + tone | 76.51 | 0.4688 |
| *LA* + prosody | 76.64 | 0.4583 |
| *LA* + prosody + chapter | 76.31 | 0.4776 |
| *LA* + tone + chapter | 76.85 | 0.4697 |
| *LA* + tone + prosody | **76.87** | **0.5076** |
| *LA* + tone + prosody + chapter | 76.75 | 0.4710 |

some conflict or negative impact. So, the ablation study is employed for exploring a better way to maintain more stress information. As displayed in Table 4, we randomly pick up 50 single long sentences and 10 paragraphs in the same genre as the training speaker's data, it shows that the accurate stress annotation generated from our rule-based detector makes that simple 128-dimensional stress embedding to be sufficient to describe the prosody feature, which could replace lower-level features like f0 and energy as the condition of the decoder (Table 5).

**Table 4.** The MOS in different levels, comparison of 3 systems for the **Ablation study**.

| System | MOS of Sentence | MOS of Paragraph |
|---|---|---|
| Stress | **3.86429** | **3.94375** |
| Stress + f0&energy modeling | 3.8619 | 3.825 |
| Baseline | 3.77381 | 3.76875 |

***Similarity.*** Here, Mel-cepstral distortion [8] (MCD) is utilized to calculate the difference between the synthesized results and ground-truth speech, which uses dynamic time warping (DTW) to align these two sequences. We synthesize 100 Wav samples by the manual stress for each character from the ground truth Wav which is annotated by linguists with cross-validation. The smaller result in our experiments (see in Table 4) shows that the synthesized speech is more similar to the ground truth, indicating the great accuracy of the stress annotation rules and the better performance of its modeling in the TTS system.

***Stress modeling and controbility.*** To further verify the efficiency of the stress modeling in the whole TTS system and the importance of stress in contextualized

**Table 5.** The MCD test with synthesized audio file from TTS with stress modeling and TTS Baseline.

| System | MCD |
|---|---|
| Baseline | 7.790668 |
| Stress + f0&energy modeling | 7.645527 |
| Stress | **7.260958** |

scenarios, 50 groups of Q&A test set have been designed and adopted by linguists, each group includes 1 question and its corresponding answer, which is adopted as the input of the speech synthesis system. The examples of Q&A below present the different locations of the stress making various meanings:

- Q1: "张三今天中午吃了【什么】？" (**What** did Zhang San eat for lunch today?)
  A1: "张三今天中午吃了【粽子】。" (Zhang San had **dumplings** for lunch today.)
- Q2: "张三【哪天】中午吃了粽子？" (**Which day** did Zhang San eat dumplings at noon?)
  A2: "张三【今天】中午吃了粽子。" (**Today**, Zhang San had dumplings for lunch.)

The result in Table 6 demonstrates that the *Baseline* model without frontend and backend stress modeling is difficult to perform correct semantics in a Q&A context. Meanwhile, two different ways for stress conditions bring a more accurate semantic representation.

***Audio Naturalness.*** We conduct the MOS (mean opinion score) evaluation on the test set to measure the audio naturalness. We randomly selected 50 sentences

**Table 6.** Score of correctness of the *answer* to its corresponding *question*

| System | Score |
|---|---|
| MANUAL | 4.438 |
| FE | 3.893 |
| BASELINE | 3.301 |

**Table 7.** MOS of naturalness in *sentence-level* test and *chapter-level* test

| System | MOS of Sentence | MOS of Paragraph |
|---|---|---|
| MANUAL | **4.139** | **4.210** |
| FE | 4.050 | 3.981 |
| BASELINE | 3.73 | 3.820 |

with 5 chapters from the test set of our novel domain dataset as an evaluation set for the test phase. The MOS results of the overall naturalness are also shown in Table 7 demonstrating that the stress modeling in TTS outperforms our baseline.

## 4    Conclusion

To conclude, we proposed a generalized rule-based stress detection mechanism for automatic labeling from parallel text-audio files, and the usage of its annotated stress data could successfully do stress modeling in both the acoustic model and the textual front-end model. According to the results of our experiments, a stress-controllable TTS system outperforms the baseline to generate speech with more expressive prosody.

## References

1. Xu, J., Chu, M., He, L., Lu, S., Guan, D., Institute of Acoustics, A.S.B.: The influence of chinese sentence stress on pitch and duration. Chin. J. Acoust. **3**, 7
2. Boersma, P.: Praat, a system for doing phonetics by computer. Glot. Int. **5**(9), 341–345 (2001)
3. Chao, B.: A grammar of spoken chinese. J. Am. Orient. Soc. **92**(1), 136 (1965)
4. Fujisaki, H.: Dynamic characteristics of voice fundamental frequency in speech and singing. In: MacNeilage, P.F. (eds.) The Production of Speech. Springer, New York (1983). https://doi.org/10.1007/978-1-4613-8202-7_3
5. Gordon, M.: Disentangling stress and pitch-accent: a typology of prominence at different prosodic levels. Word Stress: Theoretical and Typological Issues, pp. 83–118 (2014)
6. Guo, H., Zhang, S., Soong, F.K., He, L., Xie, L.: Conversational end-to-end TTS for voice agent (2020). arXiv preprint arXiv:2005.10438, https://arxiv.org/abs/2005.10438
7. Hirose, K., Ochi, K., Minematsu, N.: Corpus-based generation of F0 contours of japanese based on the generation process model and its control for prosodic focus
8. Kubichek, R.: Mel-cepstral distance measure for objective speech quality assessment. In: Proceedings of IEEE Pacific Rim Conference on Communications Computers and Signal Processing. vol. 1, pp. 125–128. IEEE (1993)
9. Lan, Z., Chen, M., Goodman, S., Gimpel, K., Sharma, P., Soricut, R.: ALBERT: a lite BERT for self-supervised learning of language representations (2019). https://doi.org/10.48550/ARXIV.1909.11942, https://arxiv.org/abs/1909.11942
10. Li, A.: Chinese prosody and prosodic labeling of spontaneous speech. In: Speech Prosody 2002, International Conference (2002)
11. Li, Y., Tao, J., Hirose, K., Lai, W., Xu, X.: Hierarchical stress generation with fujisaki model in expressive speech synthesis. In: International Conference on Speech Prosody (2014)
12. Li, Y., Tao, J., Zhang, M., Pan, S., Xu, X.: Text-based unstressed syllable prediction in mandarin. In: INTERSPEECH, pp. 1752–1755 (2010)
13. LIN, Y.: Pitch analysis of Mandarin declarative sentences. Ph.D. thesis, Fudan University (2013)

14. Liu, W., Fu, X., Zhang, Y., Xiao, W.: Lexicon enhanced chinese sequence labeling using BERT adapter (2021). https://doi.org/10.48550/ARXIV.2105.07148, https://arxiv.org/abs/2105.07148
15. Morise, M., Yokomori, F., Ozawa, K.: WORLD: a vocoder-based high-quality speech synthesis system for real-time applications. IEICE Trans. Inf. Syst. **99**(7), 1877–1884 (2016)
16. Ni, C., Liu, W., Xu, B.: Mandarin pitch accent prediction using hierarchical model based ensemble machine learning. In: 2009 IEEE Youth Conference on Information, Computing and Telecommunication, pp. 327–330. IEEE (2009)
17. Pan, S., He, L.: Cross-speaker style transfer with prosody bottleneck in neural speech synthesis (2021). https://doi.org/10.48550/ARXIV.2107.12562, https://arxiv.org/abs/2107.12562
18. Povey, D., et al.: The kaldi speech recognition toolkit. In: IEEE 2011 Workshop on Automatic Speech Recognition and Understanding. No. CONF, IEEE Signal Processing Society (2011)
19. Ren, Y., et al.: FastSpeech 2: fast and high-quality end-to-end text to speech (2020). https://doi.org/10.48550/ARXIV.2006.04558, https://arxiv.org/abs/2006.04558
20. Shao, Y., Han, J., Liu, T., Zhao, Y.: Study on automatic prediction of sentential stress with natural style in chinese. Acta Acustica (2006)
21. Shen, J.: Chinese intonation construction and categorical patterns (1994)
22. Silverman, K.E., et al.: ToBI: a standard for labeling english prosody. In: The Second International Conference on Spoken Language Processing (ICSLP). vol. 2, pp. 867–870 (1992)
23. Tao, J.H., Zhao, S., Cai, L.H.: Automatic stress prediction of chinese speech synthesis. In: International Symposium on Chinese Spoken Language Processing (2002)
24. Wightman, C.W., Ostendorf, M.: Automatic labeling of prosodic patterns. IEEE Trans. Speech Audio Process. **2**(4), 469–481 (1994)
25. Zhang, Z., Tian, Q., Lu, H., Chen, L.H., Liu, S.: AdaDurIAN: few-shot adaptation for neural text-to-speech with durian (2020). https://doi.org/10.48550/ARXIV.2005.05642, https://arxiv.org/abs/2005.05642
26. Zheng, Y., Li, Y., Wen, Z., Liu, B., Tao, J.: Text-based sentential stress prediction using continuous lexical embedding for mandarin speech synthesis. In: International Symposium on Chinese Spoken Language Processing (2017). https://ieeexplore.ieee.org/document/7918425

# MnTTS2: An Open-Source Multi-speaker Mongolian Text-to-Speech Synthesis Dataset

Kailin Liang, Bin Liu, Yifan Hu, Rui Liu$^{(\boxtimes)}$, Feilong Bao, and Guanglai Gao

Inner Mongolia University, Hohhot, China
liurui_imu@163.com, {csfeilong,csggl}@imu.edu.cn

**Abstract.** Text-to-Speech (TTS) synthesis for low-resource languages is an attractive research issue in academia and industry nowadays. Mongolian is the official language of the Inner Mongolia Autonomous Region and a representative low-resource language spoken by over 10 million people worldwide. However, there is a relative lack of open-source datasets for Mongolian TTS. Therefore, we make public an open-source multi-speaker Mongolian TTS dataset, named MnTTS2, for the benefit of related researchers. In this work, we prepare the transcription from various topics and invite three professional Mongolian announcers to form a three-speaker TTS dataset, in which each announcer records 10 h of speeches in Mongolian, resulting 30 h in total. Furthermore, we build the baseline system based on the state-of-the-art FastSpeech2 model and HiFi-GAN vocoder. The experimental results suggest that the constructed MnTTS2 dataset is sufficient to build robust multi-speaker TTS models for real-world applications. The MnTTS2 dataset, training recipe, and pretrained models are released at: https://github.com/ssmlkl/MnTTS2.

**Keywords:** Mongolian · Text-to-Speech (TTS) · Open-Source Dataset · Multi-Speaker

## 1 Introduction

Text-to-Speech (TTS) aims to convert the input text to human-like speech [1]. It is a standard technology in human-computer interaction, such as cell phone voice assistants, car navigation, smart speakers, etc. The field of speech synthesis has developed rapidly in recent years. Different from the traditional methods, which use concatenation [2], statistical modeling [3] based methods to synthesize speech, neural end-to-end TTS models achieve remarkable performance with the help of Encoder-Decoder architecture [4]. Typical models include Tacotron [5],

---

K. Liang, B. Liu and Y. Hu—Equal Contributions. This research was funded by the High-level Talents Introduction Project of Inner Mongolia University (No. 10000-22311201/002) and the Young Scientists Fund of the National Natural Science Foundation of China (No. 62206136).

L. Zhenhua et al. (Eds.): NCMMSC 2022, CCIS 1765, pp. 318–329, 2023.
https://doi.org/10.1007/978-981-99-2401-1_28

Tacotron2 [1], Transformer TTS [6], Deep Voice [7], etc. To further accelerate the inference speed, the non-autoregressive TTS models, such as FastSpeech [8], FastSpeech2(s) [9], are proposed and become the mainstream methods of TTS. Note that armed with the neural network based vocoder, including WaveNet [10], WaveRNN [11], MelGAN [12], HiFi-GAN [13], etc., the TTS model can synthesize speech sounds that are comparable to human sounds.

We note that an important factor in the rapid development of neural TTS mentioned above is the large scale corpus resources. This is especially true for languages such as English and Mandarin, which are widely spoken worldwide. However, low-resource language such as Mongolian [14] have been making slow progress in related research due to the difficulties in corpus collection. Therefore, building a large-scale and high-quality Mongolian TTS dataset is necessary. In addition, our lab have previously open-sourced a single-speaker dataset called MnTTS [15], which was recorded by a young female native Mongolian speaker and received much attention from academia and industry upon its release. This also shows the necessity of continuing to collect and organize Mongolian speech synthesis datasets and opening the baseline model's source code.

Motivated by this, this paper presents a multi-speaker Mongolian TTS dataset, termed as MnTTS2, which increases the number of speakers to three and increases the data size from 8 to 30 h, with an average of 10 h per speaker. The textual content has been further expanded and enriched in the domain. Similar with our MnTTS, MnTTS2 dataset is freely available to academics and industry practitioners.

To demonstrate the reliability of MnTTS2, we combined the state-of-the-art FastSpeech2 [9] model and the HiFi-GAN [13] vocoder to build the accompanied baseline model for MnTTS2. We conduct listening experiments and report the Naturalness Mean Opinion Score (N-MOS) and Speaker Similarity Mean Opinion Score (SS-MOS) results in terms of naturalness and speaker similarity respectively. The experimental results show that our system can achieve satisfactory performance on the MnTTS2, which indicates that the MnTTS2 corpus is practically usable and can be used to build robust multi-speaker TTS system.

The main contributions are summarized as follows. 1) We developed a multi-speaker TTS dataset, termd as MnTTS2, containing three speakers. The total audio duration is about 30 h. The transcribed text covers various domains, such as sports and culture, etc. 2) We used the state-of-the-art non-autoregressive FastSpeech2 model to build the baseline model and validate our MnTTS2. 3) The MnTTS2 dataset, source code, and pre-trained models will be publicly available to academics and industry practitioners.

The rest of the paper is organized as follows. Section 2 revisits the related works about the Mongolian TTS corpus. In Sect. 3, we introduce the details of MnTTS2, including the corpus structure and statistical information. Section 4 explains and discusses the experimental setup and experimental results. Section 5 discusses the challenges faced by Mongolian speech synthesis and the future research directions. Section 6 concludes the paper and summarizes the work and research of this paper.

## 2    Related Work

For mainstream languages such as English and Mandarin, there are many free and publicly available TTS datasets. For example, LJSpeech [16] is a single-speaker dataset for English. To rich the speaker diversity, some multi-speaker TTS dataset are released, such as Libritts [17] for English and Aishell [18] for Chinese.

For the low-resource language such as Mongolian, the available resources are pretty limited. We note that some attempts tried to improve the effect of TTS synthesis under low resource data with unsupervised learning [19], semi-supervised learning [20], and transfer learning [21] methods, etc. However, due to the lack of large-scale training data, all the mentioned methods are difficult to achieve the effect that meets the requirements of practical scenarios.

In order to promote the development of Mongolian TTS, some works built their own Mongolian TTS corpus and designed various models to achieve good results. For example, Huang et al. established the first emotionally controllable Mongolian TTS system and achieved eight emotional embeddings by transfer learning and emotional embedding [22]. Rui Liu et al. introduced a new method to segment Mongolian words into stems and suffixes, which greatly improved the performance of the Mongolian rhyming phrase prediction system [23]. Immediately after that, Rui Liu proposed a DNN-based Mongolian speech synthesis system, which performs better than the traditional HMM [24]. Also, he introduced the Bidirectional Long Term Memory (BiLSTM) model to improve the phrase break prediction step in the traditional speech synthesis system, making it more applicable to Mongolian [25]. Unfortunately, none of the Mongolian TTS dataset from the above works have been released publicly and are not directly available to the public. We also found that some dataset in related fields, such as M2ASR-MONGO [26] for Mongolian speech recognition, are public recently. However, the speech recognition corpus cannot be applied in the TTS filed due to the environment noise and improper speaking style issues etc.

We previously released the single-speaker MnTTS dataset [15], called MnTTS. The total duration of the MnTTS is 8 h, and it was recorded in a studio by a professional female native Mongolian announcer. However, the duration and speaker diversity still needs to be further expanded. In a nutshell, it is necessary to construct a high-quality multi-speaker Mongolian TTS dataset to further promote the Mongolian TTS research, which is the focus of this paper. We will introduce the details of the MnTTS2 at the following subsection.

## 3    MnTTS2 Dataset

In this section, we first revisit the MnTTS dataset briefly and then introduce our MnTTS2 by highlighting the extended content.

## 3.1 MnTTS

In the preliminary work, we presented a high-quality single-speaker Mongolian TTS dataset, called MnTTS [15]. The transcription of the dataset was collected from a wide range of topics, such as policy, sports, culture, etc. The Mongolian script was then converted to Latin sequences to avoid as many miscoding issues as possible. A professional female native Mongolian announcer was invited to record all the audio. A Mongolian volunteer was invited to check and re-align the alignment errors. The audio containing ambient noise and mispronunciation was removed to ensure the overall quality.

MnTTS has received much attention from researchers in the same industry upon its release. Furthermore, the subset was used in the Mongolian Text-to-Speech Challenge under Low-Resource Scenario at NCMMSC2022[1]. The organizers provided two hours of data for all participants to train their models. This competition also promotes the development of intelligent information processing in minority languages within China.

## 3.2 MnTTS2

The construction pipeline of MnTTS2 consists of "Text collection and narration", "Text preprocessing" and "Audio recording and audio-text alignment". We will introduce them in order and then report the corpus structure and statistics.

**Text Collection and Narration.** Similar with MnTTS [15], the first step in building the MnTTS2 dataset is to collect a large amount of transcription. The natural idea for collecting such a text materials is crawl text information from websites and electronic books. The text topics should cover human daily usage scenarios as much as possible. Following this, we crawled 23, 801 sentences, that are rich in content and have a wide range of topics (e.g., politics, culture, economy, sports, etc.), to meet our requirements well. At the same time, we manually filtered and removed some texts with unsuitable content, which may involve sensitive political issues, religious issues or pornographic content. These contents are removed in the hope that our dataset can make a positive contribution to the development of Mongolian language, which is the original intention of our work.

**Text Preprocessing.** Compared to mainstream languages, such as Mandarin and English, traditional Mongolian performs agglutinative characteristic [27]. This makes the Mongolian letters express different styles in different contexts and brings a serious harmonic phenomenon [15]. In order to solve this problem, we transformed the texts into a Latin alphabet, instead of traditional Mongolian representation, for TTS training. The entire pipeline of converting Mongolian texts into Latin sequences is divided into three steps: encoding correction, Latin conversion and text regularization. The detailed description can be found in our previous work MnTTS [15].

---

[1] http://mglip.com/challenge/NCMMSC2022-MTTSC/index.html.

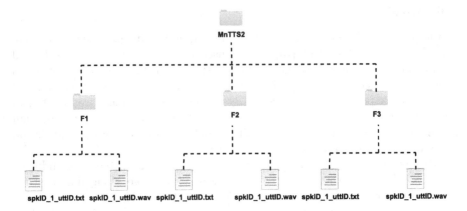

**Fig. 1.** The folder structure of the MnTTS2 corpus.

**Audio Recording and Audio-Text Alignment.** Different with the MnTTS [15], we invited three native Mongolian-speaking announcers to record the audio. Each announcer volunteered to participate and signed an informed consent form to be informed of the data collection and use protocol. F1, F2 and F3 are three native Mongolian speaking females, with F2 being a little girl and F1 and F3 being slightly older in grade. All recordings were done in a standard recording studio at Inner Mongolia University. We choose Adobe Audition[2] as the recording software.

During the recording process, we asked the announcer to keeps a 0.3 s pause at the beginning and end of each audio segment, keeps a constant distance between the lips and the microphone, performs a slight pause at the comma position, and performs an appropriate pitch boost at the question mark position.

To ensure the quality of the recording data, we rechecked the recording data after completing the recording work. Specifically, we invited three volunteers to check each text against its corresponding natural audio. These volunteers are responsible for splitting the recorded audio file into sentences and aligning the split sentences with the text. The Mongolian text is represented by a Latin sequence, where each Latin word in the sequence becomes a word and each letter that makes up the word is called a character. Characters also include punctuation marks, such as commas (','), periods ('.'), question mark ('?'), exclamation mark ('!') etc. Finally, we obtained about 30 h of speech data, which were sampled at 44.1kHz with a sampling accuracy of 16bit.

**Corpus Structure and Statistics.** The file structure of the MnTTS2 corpus is shown in Fig. 1. Each speaker's recording file and the corresponding text collection are saved in a folder named after the speaker. All audios are stored in WAV format files , sampled at 44.10kHz, and coded in 16 bits. All text is saved

---

[2] https://www.adobe.com/cn/products/audition.html.

**Table 1.** The statistics results of MnTTS2 dataset.

| Statistical Unit | | Speaker ID | | |
| --- | --- | --- | --- | --- |
| | | F1 | F2 | F3 |
| Character | Total | 572016 | 459213 | 601366 |
| | Mean | 79 | 61 | 67 |
| | Min | 12 | 2 | 2 |
| | Max | 189 | 188 | 190 |
| Word | Total | 88209 | 71245 | 92719 |
| | Mean | 12 | 9 | 10 |
| | Min | 3 | 1 | 1 |
| | Max | 29 | 30 | 29 |

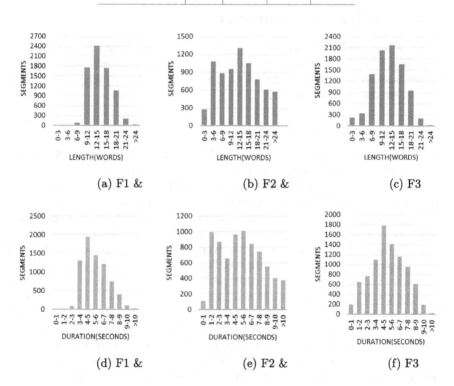

(a) F1               (b) F2               (c) F3

(d) F1               (e) F2               (f) F3

**Fig. 2.** Word number distributions (a, b, c) and sentence duration distributions (d, e, f) for all speakers of MnTTS2.

in a TXT file encoded in UTF-8. The file name of the audio is the same as the corresponding text file name, and the name of each file consists of the speaker, document ID, and corpus ID.

The statistical results of the MnTTS2 data are shown in Table 1 and Fig. 2. As shown in Table 1, the entire corpus has a total of 23, 801 sentences. Take the speaker F1 for example, F1 with a total of 572,016 Mongolian characters, and the average number of characters per sentence is 79, with the shortest sentence having 12 characters and the longest sentence having 189 characters. If words are used as the statistical unit, the total number of words in this dataset for F1 is 88,209, the mean value of words in each sentence is 12, the minimum value is 3, and the maximum value is 29. As shown in Fig. 2, we also counted the sentence duration to draw a histogram. Take speaker F1 for example, the word numbers of the sentences are concentrated in 12–15, and duration are concentrated in 4–5 seconds. In comparison, we found that the word numbers of sentences for F2 was not particularly concentrated, and the duration were relatively scattered. F3, on the other hand, is more similar to F1, with a more obvious concentration. The statistics of all three speakers are in line with the normal distribution.

## 4 Speech Synthesis Experiments

To verify the validity of our MnTTS2, we conducted Mongolian TTS experiments based on the FastSpeech2 model and HiFi-GAN vocoder and evaluated the synthesized speech using Mean Opinion Score (MOS) metric in terms of naturalness and speaker similarity.

### 4.1 Experimental Setup

We use the TensorFlowTTS toolkit[3] to build an end-to-end TTS model based on the FastSpeech2 model. The FastSpeech2 model converts the input Mongolian text into Mel-spectrogram features, and then the HiFi-GAN vocoder reconstructs the waveform by the Mel-spectrogram features. FastSpeech2 is a state-of-the-art non-autoregressive [28] speech synthesis model that extracts duration, pitch, and energy directly from the speech waveform and uses these features as input conditions in training. This model can effectively solve errors such as repetition and word skipping, and has the advantage of fast training speed. FastSpeech2 introduces more variance information to alleviate the one-to-many mapping problem. Also, the pitch prediction is improved by wavelet transform. Most of all, FastSpeech2 has the characteristics of fast, robust, and controllable speech synthesis. This is the main reason why we choose FastSpeech2. As shown in Fig. 3, based on FastSpeech2, we implement the multi-speaker FastSpeech2 by adding the speaker encoder module. The speaker encoder includes speaker embedding, dense, and softplus layers [3]. In the network architecture setting, the number of speakers is 3. The dimension of speaker embedding is 384. The hidden side of the text encoder is 384 and the number of hidden layers is 4, the hidden layer size of the decoder is 384 and the number of hidden layers is 4. The number of Conv layers of the variance predictors is 2 and the dropout rate is 0.5. The initial learning rate is 0.001 and the dropout rate is 0.2.

The HiFi-GAN vocoder builds the network through a generative adversarial network to converts Mel-spectrogram into high-quality audio. The generator of HiFi-GAN consists of an upsampling structure, which consists of a one-dimensional transposed convolution, and a multi-receptive filed fusion module,

---

[3] https://github.com/TensorSpeech/TensorFlowTTS.

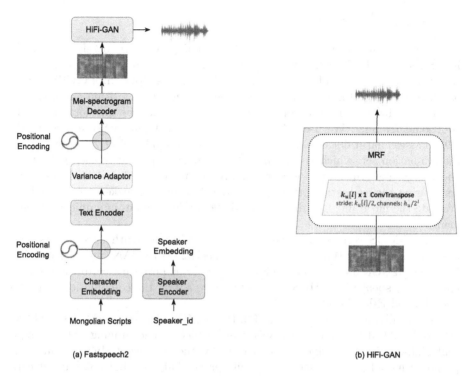

(a) Fastspeech2                                      (b) HiFi-GAN

**Fig. 3.** The structure of the FastSpeech2+HiFi-GAN model. We implement the multi-speaker FastSpeech2 by adding the speaker encoder module.

which is responsible for optimizing the upsampling points. HiFi-GAN, as a generative adversarial network, has two kinds of discriminators, including multi-scale and multi-period discriminators. The generagor kernel size of HiFi-GAN is 7 and the upsampling ratio is (8,8,2,2). The list of discriminators for the cycle scale is (2,3,5,7,11). The Conv filters of each periodic discriminator are 8. The pooling type of output downsampling in the melgan discriminator is AveragePooling1D, the kernel size is (5,3), and the activation function is LeakyReLU. HiFi-GAN is trained independently of FastSpeech2. For each speaker, the generator with only stft loss is first trained for 100 k steps, and then the generator and discriminator are trained for 100k steps. This gives us the corresponding vocoder for each of the three speakers.

Note that a teacher Tacotron2 model trained for each speaker was used to extract duration from the attention contrast for subsequent FastSpeech2 model training. For each speaker, the Tacotron2 model trained with 100 k steps of MnTTS was used to extract the duration. After that, the multi-speaker Fast-Speech2 model was trained with 200 k steps to do the final speech generation. 100 k steps were trained for HiFi-GAN's generator and 100k steps for jointly training the generator and discriminator. All the above models were trained on 2 T V100 GPUs.

## 4.2    Naturalness Evaluation

For a full comparison of naturalness, we compared our baseline system, **FastSpeech2+HiFi-GAN** with the **Ground Truth** speech. In addition, to verify the performance of HiFi-GAN, we added a **FastSpeech2+Griffin-Lim** baseline model for further comparison. The Griffin-Lim algorithm can directly obtain the phase information of the audio to reconstruct the waveform without additional training. We used the Naturalness Mean Opinion Score (SS-MOS) [29] to assess naturalness. For each speaker, we randomly select 20 sentences as the evaluation set, which are not used for training. The model-generated audio and the ground truth audio were randomly disrupted and distributed to listeners. During the evaluation process, 10 native Mongolian speakers were asked to evaluate the naturalness of the generated 400 audio speeches in a quiet environment.

The N-MOS results are given in Table 2. Ground truth speech gets the best performance without a doubt. FastSpeech2+HiFi-GAN outperforms the FastSpeech2+Griffin-Lim and achieves much closer performance to the ground truth. Each speaker's N-MOS score was above 4.0 on the combination of FastSpeech2 and HiFi-GAN.

Specifically, for the F1, F2, and F3, the N-MOS of FastSpeech2+HiFi-GAN achieved 4.02, 4.15, and 4.29 respectively, which is encouraging. This proves that high-quality Mongolian speech can be synthesized using MnTTS2 and the proposed model. In a nutshell, all results prove that our MnTTS2 dataset can be used to build a robust TTS system for high-quality speech generation.

**Table 2.** Naturalness mean opinion score (N-MOS) results for all systems with 95% Confidence intervals.

| System | Speaker ID | | |
|---|---|---|---|
| | F1 | F2 | F3 |
| FastSpeech2+Griffin-Lim | $3.56 \pm 0.18$ | $3.59 \pm 0.04$ | $3.86 \pm 0.12$ |
| **FastSpeech2+HiFi-GAN** | $\mathbf{4.02 \pm 0.18}$ | $\mathbf{4.15 \pm 0.06}$ | $\mathbf{4.29 \pm 0.11}$ |
| Ground Truth | $\mathbf{4.73 \pm 0.08}$ | $\mathbf{4.70 \pm 0.14}$ | $\mathbf{4.68 \pm 0.09}$ |

## 4.3    Speaker Similarity Evaluation

We further conduct listening experiments to evaluate the speaker similarity performance for the FastSpeech+HiFi-GAN baseline system. The Speaker Similarity Mean Opinion Score (SS-MOS) results are reported in Table 3.

We synthesized 20 audios for each speaker by FastSpeech2+HiFi-GAN baseline system. Ten native Mongolian-speaking volunteers were also invited to participate in the scoring. Each volunteer needs to evaluate whether the speaker is the same person or not in the synthesized and the ground truth audio.

The SS-MOS scores for F1, F2, and F3 are 4.58, 4.04, and 4.12 respectively, which is encouraging. The results show that the audio synthesized by the FastSpeech2+HiFi-GAN system performs good performance in terms of speaker similarity. The highest SS-MOS score was obtained for speaker F1. Auditioning the audio, we can find that F1's timbre has significant characteristics and the synthesized audio represents the speaker's voice information well. In a nutshell, this experiment shows that the MnTTS2 dataset can be used for speech synthesis work in multi-speaker scenarios.

**Table 3.** Speaker Similarity Mean opinion score (SS-MOS) results for FastSpeech2+HiFi-GAN system with 95% Confidence intervals.

| System | Speaker ID | | |
|---|---|---|---|
| | F1 | F2 | F3 |
| FastSpeech2+HiFi-GAN | $4.58 \pm 0.21$ | $4.04 \pm 0.16$ | $4.12 \pm 0.10$ |

## 5 Challenges and Future Work

With the development of "Empathy AI", the research of emotional TTS has attracted more and more attention [30]. Speech synthesis in conversational scenarios and emotional speech synthesis are hot research topics nowadays [31]. Furthermore, how to control the emotion category and the emotional intensity during speech generation is an interesting direction [32]. However, our MnTTS2 does not involve information related to emotion category and emotion intensity. In future work, we will carry out a more comprehensive and in-depth expansion of the data to serve the development of emotional Mongolian TTS.

## 6 Conclusion

We presented a large-scale, open-source Mongolian text-to-speech corpus, MnTTS2, which enriches MnTTS with more durations, topics, and speakers. Releasing our corpus under the Knowledge Attribution 4.0 International License, the corpus allows both academic and commercial use. We describe in detail the process of building the corpus, while we validate the usability of the corpus by synthesizing sounds with the FastSpeech2 model and the HiFi-GAN vocoder. The experimental results show that our system can achieve satisfactory performance on the MnTTS2, which indicates that the MnTTS2 corpus is practically usable and can be used to build robust multi-speaker TTS system. In future work, we will introduce emotional TTS dataset to further enrich our corpus. We also plan to compare the effects of different TTS architectures and model hyperparameters on the results and conduct subsequent analyses.

# References

1. Shen, J., et al.: Natural TTS synthesis by conditioning waveNet on mel spectrogram predictions. In: 2018 IEEE International Conference on Acoustics, Speech and Signal Processing (ICASSP), pp. 4779–4783. IEEE (2018)
2. Charpentier, F., Stella, M.: Diphone synthesis using an overlap-add technique for speech waveforms concatenation. In: ICASSP 1986. IEEE International Conference on Acoustics, Speech, and Signal Processing. vol. 11, pp. 2015–2018 (1986)
3. Rabiner, L.R.: A tutorial on hidden markov models and selected applications in speech recognition. Proc. IEEE **77**(2), 257–286 (1989)
4. Cho, K., van Merriënboer, B., Bahdanau, D., Bengio, Y.: On the properties of neural machine translation: encoder-decoder approaches. In: Syntax, Semantics and Structure in Statistical Translation, p. 103 (2014)
5. Wang, Y., et al.: Tacotron: towards end-to-end speech synthesis. Proc. Interspeech **2017**, 4006–4010 (2017)
6. Li, N., Liu, S., Liu, Y., Zhao, S., Liu, M.: Neural speech synthesis with transformer network. In: Proceedings of the Thirty-Third AAAI Conference on Artificial Intelligence and Thirty-First Innovative Applications of Artificial Intelligence Conference and Ninth AAAI Symposium on Educational Advances in Artificial Intelligence, pp. 6706–6713 (2019)
7. Arik, S., et al.: Deep voice: real-time neural text-to-speech. In: International Conference on Machine Learning (ICML) (2017)
8. Ren, Y., et al.: FastSpeech: fast, robust and controllable text to speech. In: Wallach, H.M., Larochelle, H., Beygelzimer, A., d'Alché-Buc, F., Fox, E.B., Garnett, R. (eds.) Advances in Neural Information Processing Systems 32: Annual Conference on Neural Information Processing Systems 2019, NeurIPS 2019, 8–14 Dec 2019, Vancouver, BC, Canada, pp. 3165–3174 (2019)
9. Ren, Y., et al.: FastSpeech 2: fast and high-quality end-to-end text to speech. In: 9th International Conference on Learning Representations, ICLR 2021, Virtual Event, Austria, 3–7 May 2021. OpenReview.net (2021)
10. van den Oord, A., et al.: WaveNet: a generative model for raw audio. In: The 9th ISCA Speech Synthesis Workshop, Sunnyvale, CA, USA, 13–15 Sep 2016, p. 125. ISCA (2016)
11. Kalchbrenner, N., et al.: Efficient neural audio synthesis. In: Dy, J.G., Krause, A., (eds.) Proceedings of the 35th International Conference on Machine Learning, ICML 2018, Stockholmsmässan, Stockholm, Sweden, 10–15 Jul 2018, vol. 80 of Proceedings of Machine Learning Research, pp. 2415–2424. PMLR (2018)
12. Kumar, K., et al. MelGAN: generative adversarial networks for conditional waveform synthesis. arXiv preprint arXiv:1910.06711 (2019)
13. Kong, J., Kim, J., Bae, J.: HiFi-GAN: generative adversarial networks for efficient and high fidelity speech synthesis. Adv. Neural Inf. Proc. Syst. **33**, 17022–17033 (2020)
14. Bulag, U.E.: Mongolian ethnicity and linguistic anxiety in china. Am. Anthropologist **105**(4), 753–763 (2003)
15. Hu, Y., Yin, P., Liu, R., Bao, F., Gao, G.: MnTTS: an open-source mongolian text-to-speech synthesis dataset and accompanied baseline. arXiv preprint arXiv:2209.10848 (2022)
16. Ito, K., Johnson, L.: The LJ speech dataset (2017)
17. Zen, H., et al.: LibriTTS: a corpus derived from libriSpeech for text-to-speech. In: Proceedings of Interspeech 2019, pp. 1526–1530 (2019)

18. Shi, Y., Bu, H., Xu, X., Zhang, S., Li, M.: AISHELL-3: a multi-speaker mandarin TTS corpus and the baselines (2020)
19. Barlow, H.B.: Unsupervised learning. Neural Comput. **1**(3), 295–311 (1989)
20. Zhu, X.J.: Semi-supervised learning literature survey (2005)
21. Weiss, K., Khoshgoftaar, T.M., Wang, D.D.: A survey of transfer learning. J. Big Data **3**(1), 1–40 (2016). https://doi.org/10.1186/s40537-016-0043-6
22. Huang, A., Bao, F., Gao, G., Shan, Y., Liu, R.: Mongolian emotional speech synthesis based on transfer learning and emotional embedding. In: 2021 International Conference on Asian Language Processing (IALP), pp. 78–83. IEEE (2021)
23. Liu, R., Bao, F., Gao, G., Wang, W.: Mongolian prosodic phrase prediction using suffix segmentation. In: 2016 International Conference on Asian Language Processing (IALP), pp. 250–253. IEEE (2016)
24. Liu, R., Bao, F., Gao, G., Wang, Y.: Mongolian text-to-speech system based on deep neural network. In: Tao, J., Zheng, T.F., Bao, C., Wang, D., Li, Y. (eds.) NCMMSC 2017. CCIS, vol. 807, pp. 99–108. Springer, Singapore (2018). https://doi.org/10.1007/978-981-10-8111-8_10
25. Liu, R., Bao, F., Gao, G., Zhang, H., Wang, Y.: Improving mongolian phrase break prediction by using syllable and morphological embeddings with BiLSTM model. In: Interspeech, pp. 57–61 (2018)
26. Zhi, T., Shi, Y., Du, W., Li, G., Wang, D.: M2ASR-MONGO: a free mongolian speech database and accompanied baselines. In: 24th Conference of the Oriental COCOSDA International Committee for the Co-ordination and Standardisation of Speech Databases and Assessment Techniques, O-COCOSDA 2021, Singapore, 18–20 Nov 2021, pp. 140–145. IEEE (2021)
27. Bao, F., Gao, G., Yan, X., Wang, W.: Segmentation-based mongolian LVCSR approach. In: 2013 IEEE International Conference on Acoustics, Speech and Signal Processing, pp. 8136–8139 (2013)
28. Gu, J., Bradbury, J., Xiong, C., Li, V.O.K., Socher, R.: Non-autoregressive neural machine translation. arXiv preprint arXiv:1711.02281 (2017)
29. Streijl, R.C., Winkler, S., Hands, D.S.: Mean opinion score (MOS) revisited: methods and applications, limitations and alternatives. Multimedia Syst. **22**(2), 213–227 (2016)
30. Liu, R., Sisman, B., Gao, G., Li, H.: Decoding knowledge transfer for neural text-to-speech training. IEEE/ACM Trans. Audio Speech Lang. Process. **30**, 1789–1802 (2022)
31. Liu, R., Sisman, B., Li, H.: Reinforcement learning for emotional text-to-speech synthesis with improved emotion discriminability. In: Hermansky, H., Cernocký, H., Burget, L., Lamel, L., Scharenborg, O., Motlícek, P. (eds.) Interspeech 2021, 22nd Annual Conference of the International Speech Communication Association, Brno, Czechia, 30 Aug - 3 Sept 2021, pp. 4648–4652. ISCA (2021)
32. Liu, R., Zuo, H., Hu, D., Gao, G., Li, H.: Explicit intensity control for accented text-to-speech (2022)

# Author Index

Printed in the United States
by Baker & Taylor Publisher Services